Multiple-Use Management: The Economics of Public Forestlands

Multiple-Use Management: The Economics of Public Forestlands

by

Michael D. Bowes and John V. Krutilla

Resources for the Future
Washington, D.C.

Printed in the United States of America

Published by Resources for the Future
1616 P Street, N.W., Washington, D.C. 20036

Books from Resources for the Future are distributed worldwide by The Johns Hopkins University Press

Library of Congress Cataloging-in-Publication Data

Bowes, Michael D.
    Multiple-use management: the economics of public forestlands / by Michael D. Bowes and John V. Krutilla.
      p. cm.
    "This study was undertaken as part of Resources for the Future's Forest Economics and Policy Program."
    Bibliography; p.
    Includes index.
    ISBN 0-915707-41-1
    1. Forest reserves—United States—Multiple use. 2. Forest reserves—Multiple use—Economic aspects—United States. 3. United States—Public lands.
    I. Krutilla, John V. II. Resources for the Future. Forest Economics and Policy Program. III. Title.
SD427.M8B68   1989   89-4018
333.75'0973—dc20   CIP

This book is the product of the Forest Economics and Policy Program, now in RFF's Energy and Natural Resources Division.

This book was edited by Samuel Allen, designed by Joan Engelhardt, and indexed by Margaret A. Lynch.

♾The paper in this book meets the guidelines for permanence and durability of the Committee on Production Guidelines for Book Longevity of the Council on Library Resources.

Contents

List of Tables   xi

List of Figures   xiii

Foreword   xv

Preface   xix

Part 1   The Historical, Institutional, and Theoretical Background

1   Development of the Concept of Multiple-Use Forestland Management   1

     1.   Timber Management or Wildland Management   2
     2.   Conservation or Economic Efficiency   3
     3.   Entering the Modern Era   5
        Notes   9

2   The Multiple Uses of the Public Lands   13

     1.   Introduction   13
     2.   The Provision of Market Commodities   14
     3.   The Provision of Nonmarket Services   20
     4.   Should Nonmarketed Services Be Priced?   24
     5.   Summary and Conclusions   27
        Notes   28

3 The Economics of Multiple-Use Forestry   31

1. Introduction 31
2. Multiple-Use Forest Management and Planning 32
3. The Benefits from Multiple-Use Production 37
4. Describing Multiple-Use Production:
   The Production Function 48
5. Describing Multiple-Use Production:
   The Cost Function 51
6. Special Features of Costs in the
   Multiple-Use Forest 71
7. The Choice of Product Mix 78
8. Summary and Conclusions 87
   Notes 89

4 Dynamic Models of Multiple-Use Management   91

1. Introduction 91
2. Single-Stand Models of Harvest Timing:
   Faustmann-Type Models 95
3. Multiple-Use Management of Related
   Forest Stands 120
4. Illustrations of Multiple-Stand
   Harvesting Solutions 132
5. Summary and Conclusions 140
   Notes 140

**Part 2 Applications of Multiple-Use Management in Forestry
Settings**

Introduction to Part 2   147

**5 Forest Management for Increased Timber
and Water Yields   149**

1. Introduction 149
2. Timber Considerations 150
3. Water-Yield Augmentation 155
4. The Economic Dimensions of Water 158
5. Testing Initial Conditions 167
6. Summary and Conclusions 172
   Notes 173

6 Valuing Recreational Quality: Hedonic Pricing 177

1. Introduction 177
2. Measuring Benefits of Changes in Site
   Attributes 179
3. Valuing Changes in Hunting Quality:
   The Black Hills National Forest 199
4. Summary and Conclusions 209
   Notes 210

7 Recreation Valuation for Forest Planning 213

1. Introduction 213
2. An Approach to Estimating the Demand for
   Forest Recreation 215
3. Sources of Forest Data 220
4. Demand Estimation 225
5. Application of the Analysis to Selected Issues 229
6. White Mountain Timber Resources and
   Management 233
7. Conclusions on Timber Management 242
8. The Relation between Recreation and Timber:
   Summary and Conclusions 244
   Notes 244
   Appendix 7-A. Prices and Per-Acre Timber Yields
   for Paper Birch and Northern Hardwoods 248

8 The Flip Side of Joint Production 251

1. Introduction 251
2. Conflicts in Land Use in the White Clouds
   Peaks 252
3. An Estimate of the Wilderness Recreation
   Value 255
4. The Place of Grazing Stock in the White
   Clouds 260
5. The Prospect of Mining 265
6. Summary and Conclusions 270
   Notes 271

9 Funding Nonpriced Resource Services 273

1. Introduction 273
2. The Budget and Appropriations 274

3. Some Critical Observations   280
4. A Digression on Capital and Cost Allocation   284
5. Summary and Conclusions   288
   Notes   290

10 Below-Cost Timber Sales and Forest Planning   293

1. Introduction   293
2. Below-Cost Sales in a Multiple-Use
   Framework   297
3. The Costs of Forest Outputs: A Capital
   Accounting Approach   302
4. Applying FORPLAN: The Shoshone and
   Mt. Baker–Snoqualmie National Forests   314
5. The Use of FORPLAN in Program
   Evaluation   325
6. Summary and Conclusions   328
   Notes   329
   Appendix 10-A. The Structure of FORPLAN
   Linear Programs   332

11 Conclusions: An Interpretive Synthesis   337

1. Introduction   337
2. Concise Review of Differences in Model
   Results   341
3. Further Observations on Realistic Multiple-Use
   Management   344
   Notes   349

Index   351

List of Tables

4-1 Definitions of Terms in Equation 4-7   125

4-2 Forest States, Per-Acre Amenity Values, and Decision Sets   133

4-3 Yields per Acre, Even-Aged Stands of Lodgepole Pine   135

4-4 Harvest Scheduling Solutions for State 20   137

4-5 Price Levels and the Corresponding Optimal Harvest Solution   138

4-6 Present Value Possibilities   138

5-1 A Timber Management Schedule   151

5-2 Yields per Acre of Even-Aged Stands of Lodgepole Pine, Site Index 60   152

5-3 Cost of Road per Acre Harvested by Type and Length of Road, Type of Terrain (discounted at 4 percent)   154

5-4 Projected Change in Yearly Water Yield from an Average Acre under the Water Augmentation Program   164

5-5 Gross Present Value per Acre of Augmented Water Flow, by Geographic and Functional Disposition   165

5-6 Present Value per Acre, with Timber and Water Values Discounted at 4 Percent   168

5-7 Present Value per Acre, with Timber and Water Values Discounted at 7 Percent   169

5-8 Cost of Road per Acre Harvested by Type and Length of Road, Type of Terrain (discounted at 7 percent)   169

5-9 Present Values per Acre: Effect of Rotation Age   171

5-10 Present Values per Acre by Rotation Ages, under Partial Area Management   171

6-1 Estimates of the Total Cost Relationship   204

6-2 Marginal Cost Estimates for MGDEAD   206

6-3 Demand Curves for MGDEAD   207

6-4 Comparison of Vegetation Characteristics by Subcompartment   208

6-5 Individual Consumer's Surplus Changes   209

7-1 Sample Comparison: Percentage of Visits by State Origin   223

7-2 Sample Comparison: Regression Results   224

7-3 Definitions of Variables Considered for Estimating Demand   226

7-4 Regression Results for LEMARS Subsamples   228

7-5 Demand and Benefit Elasticities for LEMARS Subsamples   230

7-6 Estimated Willingness to Pay for a Visit to White Mountain National Forest   231

7-7 Road Costs per Acre, White Mountain National Forest   236

7-8 Present Value of Paper Birch per Acre, Initial Harvest in Four Entries   239

7-9 Present Value of Paper Birch per Acre, Initial Harvest in One Entry   240

7-10 Present Value of Northern Hardwoods per Acre, Initial Harvest in Four Entries   241

7-11 Present Value of Northern Hardwoods per Acre, Initital Harvest in One Entry   242

Appendix 7-A. Prices and Per-Acre Timber Yields for Paper Birch and Northern Hardwoods   248

8-1 Cost of Trail Extensions for White Clouds Threshold Wilderness Recreation   258

8-2 Cost of Sanitary Facilities for White Clouds Threshold Wilderness Recreation   259

8-3 Estimated Costs of Management Programs for Big Boulder Creek Allotment   264

8-4 Estimated Forage Increase and Value on Big Boulder Creek Allotment   265

9-1 Relationship between Forest Service Functions, Budget Line Items, and Program Components   279

9-2 Program Planning, Region X   282

9-3 Allocation for Region X   282

10-1 Shoshone Results: Objective Function, Costs, and Separable Costs   318

10-2 Mt. Baker–Snoqualmie Results: Objective Function, Costs, and Separable Costs   323

10-A1 Example of a FORPLAN Linear Program for Multiple-Use Forest Planning   333

List of Figures

3-1 Isovalue curves: negative slope   39

3-2 Isovalue curves: positive slope   41

3-3 Measure of benefits from stumpage offering T 43

3-4 Benefits of recreation visits   46

3-5 Benefits of changed forest conditions   47

3-6 Concave isocost curves   62

3-7 Convex isocost curves   63

3-8 Site productivity: site A more productive in q_1 and q_2 64

3-9 Site productivity: site A more productive in q_2 only   65

3-10 Cost-minimizing product choice: both products on both sites   66

3-11 Cost-minimizing product choice: no product q_2 on site B 67

3-12 Cost-minimizing product choice with convex isocost curves   68

3-13 Isocost curves for timber harvest and wildlife services   72

3-14 Marginal costs of timber production   75

3-15 Separable costs of timber and wildlife services   77

3-16 Cost-constrained product choice   79

3-17 Cost-constrained product choice: negative marginal cost   80

3-18 Cost-constrained product choice: corner solution   81

3-19 Cost-constrained product choice: convex isocost curve   82

3-20 Locus of cost-constrained product choice solutions   83

4-1 The Faustmann solution: optimal rotation age and land value   98

4-2 The Faustmann rotation age   100

4-3 Faustmann rotation versus maximum sustained yield rotations   103

4-4 Value flow: water flow   110

4-5 Present value: water flow   110

4-6 Value flow: aesthetic   110

4-7 Present value: aesthetic   110

4-8 Value flow: all amenities   110

4-9 Present value: all amenities   110

4-10 Multiple-use rotation: timber and water flow   111

4-11 Multiple-use rotation: timber and aesthetic value   111

4-12 Multiple-use rotation: combined values   111

4-13 Multiple-use rotation: modified aesthetic value   112

4-14 Multiple-use rotation: modified timber value   112

4-15 Multiple use and the decision to manage   112

4-16 The tradeoff between timber and water-flow values   114

4-17 Present values   116

4-18 Convex tradeoff curve   116

4-19 Present values   116

4-20 Concave tradeoff curve   116

6-1 Per-capita demand and the choice of site   185

6-2 Valuing changes in site quality: the travel-cost method   186

6-3 The net benefit function   189

6-4 The marginal benefit of attribute quality   189

6-5 Travel-cost demand curves and site quality   194

6-6 The gross benefit function   195

6-7 The marginal value function   197

8-1 Location of Sawtooth and Challis national forests   253

8-2 Trail system of the White Clouds Peaks   256–257

8-3 Big Boulder C&H Grazing Allotment   261

8-4 Sites of proposed molybdenum mining and milling operations in the White Clouds area   268–269

Foreword

The phrase *multiple use* has been a powerful symbol in the lexicon of American forestry for many decades. Like all effective symbols, the phrase has many meanings. We encounter it in statutes, in economics, in political dialogues, in public policy and in resource management, to name a few of its many contexts. The meaning of multiple use depends in part on which of these contexts we are discussing. Moreover, in land-use controversies proponents of wilderness, on the one hand, and of free farms, on the other, have both called on the symbol of multiple use in support of their advocacy. Why such an apparent conflict in meanings? Because the symbol has at least some validity in describing aspects of these disparate forms of forest management and because its use has demonstrable power to convey political legitimacy.

Numerous attempts to define exactly what multiple use is have been made over the last half century, without producing general agreement. Nevertheless, as William Howard Taft said about conservation, "people are for it, no matter what it means." So it is with multiple use today. The public, for many of whom the phrase has obvious appeal, continues to use it in all of its grand ambiguity and symbolism to suggest something generally held to be central to wise use of forest resources.

In this book, *Multiple-Use Management: The Economics of Public Forest-lands*, Michael Bowes and John Krutilla present a masterful analysis and evaluation of multiple use as an economic concept, applied to the public forests of the United States. They bring eminent scholarly qualifications and much relevant experience to the task. For more than two decades Krutilla has been in the forefront of resource economists addressing the complex problems surrounding competition for the use of water, timber, wildlife, amenities, and other benefits derived from wildlands. His contributions to both the theory and the practice of analyzing such problems to throw light on policy decisions have been numerous and penetrating. Bowes has for several years directed superior mathematical and programming skills to careful study of the planning process as it is applied to the federal public lands. Now they have joined forces in the Forest Economics and Policy Program at Resources for the Future to bring these highly knowledgeable backgrounds to the writing of this book.

Even in an economic context, multiple use may be interpreted in more than one way. Bowes and Krutilla base their interpretation on the neoclassical economic theory of joint production. Forest economists began to develop this approach in the early 1950s; G. R. Gregory pioneered publication on the topic in 1955. Since then the joint-product approach has dominated thinking about the economics of multiple use. The present book integrates these earlier contributions in a single unified framework, and thus provides the definitive treatment of the theory of multiple use, viewed as a case of joint production.

But rigorous exposition of this theoretical material is only the foundation. The authors probe well beyond previous studies of the topic, particularly in two respects. First, they show that to restrict models and analysis to individual stands in the forest, without addressing multiple-use interdependencies among the several stands, may be misleading. So they extend their theory by including in it the current mix of stand ages in the existing forest. Stand age may then be used to take account of major variations in water flow, dispersed recreation, and wildlife services, as they are influenced forestwide by timber harvesting in particular stands. Although the authors find that a harvest solution for this more general model may be difficult to achieve, they demonstrate both the practicality and the policy value of solutions based on simplification of their general model.

This part of their analysis is quite complex, but it yields potentially rich results. For example, they suggest that in significant cases a policy of long, even-flow rotations is not a desirable one from the standpoint of maximizing aggregate multiple-use value. In these cases they conclude that such a policy may result in uneconomic timber and a balance of age classes which is inferior in the production of nontimber services. A distinctly shorter rotation, combined with permanent reservation from harvesting of selected portions of the forest, results in substantial increases in both timber and nontimber values over those attainable from long, even-flow rotations. Such conclusions suggest both the importance of the issues that their enlarged theoretical approach addresses and the diversification of policy options to which it points.

The importance of thus expanding the theoretical framework to permit analysis of the structure of the forest can hardly be overemphasized. Not only do the results challenge the conventional wisdom about multiple-use policies, but, as the authors indicate, optimal solutions may be sensitive to relationships about which there is little empirical knowledge. As a consequence, results of their analysis may be of great significance for planning research, not just in forest management, but also in the applied sciences which must provide the production functions on which multiple-use evaluations depend.

By showing how to include forest structure in the economic model, Bowes and Krutilla have taken a long methodological step toward providing economic analysis that can incorporate behavior of the entire forest as a biological system. For example, the Krutilla-Bowes approach, or an analogue of it, appears to provide a key to solving at least a number of aspects of the current puzzle surrounding the cumulative effects of private timber harvesting on watershed values. One may hope that the present authors, or other researchers following their lead, investigate such possibilities soon.

In the history of forest economics there are a number of instances where conceptually powerful theories have been developed which then lay fallow because the technique and information needed to apply the theories to practical problems were not available. A second major achievement of this book is to provide many examples of how to apply the theory to significant current policy problems. Indeed, a full half of the text is devoted to such examples. The results demonstrate the authors' great skill and imagination in developing practical methodologies. They also provide in their solutions much new and stimulating fodder for consideration by decision makers.

Anyone who tries to apply orthodox economic theory as a guide to decision making about issues of public forest policy and management faces a major difficulty. The physical and institutional realities of forestry practice on public lands usually differ so significantly from the assumptions underlying the theories of market economics that conventional market analysis and conclusions often have little or no relevance for forestry problems. Some policy analysts choose to assume away this difficulty either by ignoring it or by implying that the real world "ought" to perform in the way that market economics suggests. One of the great merits of this book is that it deals much more forthrightly with the matter.

The authors understand the nature of the physical processes and the institutional realities of public forest management today. They build their theory and methods of application in light of, rather than despite, these realities. Their discussion of the timber harvest/water flow issue in the subalpine forests of the Central Rocky Mountain region exemplifies the point. For example, they recognize the important differences for valuation purposes between changes in peak flows and off-peak changes, and they marshall substantial hydrologic data to evaluate the distinction. In valuing water yield their analyses reflect not only differences in value prevailing in different geographic areas, but also differences in appropriate valuation methods, depending on the economic characteristics of each of these areas. The meticulousness and the detail with which they treat the problems of application to this (and other) illustrative cases are impressive.

Some may disagree with the specific ways they have chosen to make such adaptations. But the efforts they have made in this direction will greatly enrich the field. They present many effective adaptations of theory to the particular circumstances of forestry. In other cases, the very challenge they offer to others to do better will continue to stimulate improvements.

Over the past several decades Resources for the Future has published a number of books which have become recognized as landmarks in the growth of the field of forest economics and have contributed significantly to improvement in the methods of forest resource management. Michael Bowes and John Krutilla, with their magnificent explication of theory and artful solutions for the problems of application, have added a new volume to that shelf of landmarks. Henceforth, their *Multiple-Use Management: The Economics of Public Forestlands* will be regarded as a classic in its field. Every researcher in forest economics and every forest planner will in the future owe them an enormous debt for effective exposition of multiple-use theory and for truly admirable guidance on how to use it effectively.

Henry J. Vaux
Berkeley, California

Preface

There are two basic reasons for our writing a book on the economics of multiple-use management of public forestlands. One is that public forestland is likely to be a feature of the American scene for the indefinite future. The Public Land Law Review Commission, a bipartisan body that considered the possible disposition of the public domain lands, elected in the late 1960s to recommend instead the retention of these and other public lands in federal ownership and multiple-use management. The recommendations of the commission were reinforced by passage of the National Forest Management Act (NFMA) of 1976 which, among other things, conferred on the national forests a legislative status they did not previously have. National forests historically were created by executive order and could be disposed of in the same way. After NFMA, the national forests could no longer be disposed of except by an act of Congress. The Bureau of Land Management (BLM) was created in 1946, also by an executive order, by combining the General Land Office and the Grazing Service. The Federal Land Policy and Management Act (FLPMA), also passed by Congress in 1976, conferred an even more significant authority for multiple-use management of the land under BLM jurisdiction because, unlike the National Forest System, until FLPMA's passage BLM lacked extensive multiple-use management authority.

The test of popular support for these public land systems came with the advent of the Reagan administration. Acting on a Heritage Foundation transition team recommendation, a public lands disposal policy was established in the hope that proceeds from the sale of these lands would offset, in part, the rapidly growing national debt. It is noteworthy that there was no enthusiasm for the policy even on the part of the commodity interests, such as the forest products industry. Encountering widespread and vigorous opposition, the policy of *privatization* (as it was dubbed) was rechristened *asset management* to assist its passing incognito to its quiet interment. Thus, having survived readily the most serious challenge that public ownership of the lands is likely to be exposed to for the foreseeable future, public land management will in all probability continue to be a matter of active concern to all of the multiple-use beneficiary interests. It is important, then, to address the problem of multiple-use management of the public forestlands with the hope that this effort may contribute to more efficient management. Needless to say, it is also our hope that this study will be used to complement the instructional material used in forestry curricula for upper-division undergraduates and graduate students of forestry economics.

The second reason we have written this book is that the new forest management legislation includes many references to economic considerations—substantially more than in any previous legislation affecting the public lands. This legislation permits the Forest Service to apply economic principles to its management of the national forests largely at its own discretion, and even may require it to do so in regard to certain forest planning and management activities. The staple economics of forest management at the time we began our study might best be characterized as predominantly the economics of managing timber stands rather than public forestlands having a number of priced and nonpriced forest outputs.

The need for an update and elaboration of the economic theory applicable to problems of multiple-use management of public forestland became apparent to us. How to accomplish this was not immediately obvious. It became clear in time, however, that such a theory would have to accommodate the consideration of multiple interdependent sites or stands in order to reflect the biological diversity on which multiple forest outputs depend.

Part 1 of this study (chapters 1 to 4) provides a brief historical and institutional background to set the stage. It also presents a rigorous theoretical treatment of multi-output production in a forest environment. Part 2 consists of a set of applied studies intended to illustrate multiple use and joint production in concrete forest settings. A number of lively issues are addressed in chapters 5 through 10. The final chapter attempts an interpretive synthesis of what we have learned, and we trust that it will be equally instructive to our readers.

The authors are grateful for the patience and forbearance of many persons over the long period required for the completion of this study. Chief among these were representatives of Resources for the Future (RFF), the USDA Forest Service, and the Weyerhaeuser Company Foundation, whose institutions funded the study, and members of the Advisory Committee of RFF's Forest Economics and Policy Program, who graciously shared their wisdom whenever required. A special debt of gratitude is owed to G. Robinson Gregory of the committee for a review of and constructive comments on an early draft of the study. Our debt to the academic fraternity is likewise notable. In addition to Bob Gregory, Peter Berck of the University of California-Berkeley has been a frequent reviewer of individual chapters in preparation, and William McKillop and his seminar students at the University of California-Berkeley provided a most useful set of constructive comments. And we have been encouraged throughout by the support that we enjoyed from Henry Vaux. The continuous support of every sort from colleague and friend William F. Hyde of Duke University is also warmly acknowledged.

Substantial aid has been provided on the various chapters by many persons. We are indebted to Charles F. Leaf and Charles A. Troendle, forest hydrologists, and Robert R. Alexander, silviculturist, of the USDA Forest Service's Rocky Mountain Forest and Range Experiment Station for assistance in connection with chapter 5. Paul B. Sherman also provided valuable research assistance with the water-routing task and market analysis in that chapter. Chapter 6 borrows heavily from Elizabeth A. Wilman's monograph, "Valuation of Public Forest and Rangeland Resources"; of great assistance to her were Robert Lynn, chief planning officer, Leon Fager, wildlife biologist, and other staff of the Black Hills National Forest. Invaluable assistance was also given to us by Kay Cool and Edward Neilson of the South Dakota Department of Game, Fish, and Parks, who provided access to data on hunters and to hunter report cards defining the populations from which samples were drawn. The study of the White Mountain National Forest treated in chapter 7 was greatly facilitated by Edgar Brannon, now forest supervisor, Flathead National Forest, and Gary Elsner, now director of the Recreation Management staff of the Forest Service. We are further indebted to the recreation and timber staffs of the White Mountain National Forest for generous aid and cooperation in providing forest data for the study. Acknowledgment is also due to Reed Johnson as principal author of "A Methodology for Estimating the Consequences of Forest Management on Recreation Benefits," the 1981 report to the Forest Service on which this chapter is based. The study also benefited greatly from review and constructive comments by Clark Binkley of Yale University; Ernest Gould, Harvard Forest; and Linda Hagen, Timber Management staff, USDA Forest Service.

Chapter 9 on funding the management activities that provide nonpriced or unmarketed forest resource services was difficult to handle satisfactorily because (among other reasons) the subject matter it addresses was undergoing substantive changes during the time of our study. The analysis has been done in the context of the state of budget preparation in late 1984 for fiscal year 1986. Although since then there has been some revision of the coding system used in the budget and appropriations processes, it has not been feasible to reopen the study to consider these recent changes. Accordingly, while the analysis in chapter 9 requires updating, we believe that the funding process and the nature of its independent impact on forest management, if not entirely consistent with the Forest and Rangeland Renewable Resources Planning Act of 1974 as amended by the NFMA, is likely to survive the revision in procedures and should be recognized as a complicating factor in multiple-use forest management.

In our preparation of chapter 9 we incurred many debts to many people. Chief among these are to the following individuals of the Forest Service: Alon Carter, Information Systems staff; Robert E. Gordon, Program Development and Budget staff; James Jordan, budget coordinator, National Forest System; Floyd J. Marita, regional forester, region 9, and former budget coordinator, NFS; James H. McDivitt, former budget coordinator, Wildlife and Fisheries staff, now with the Policy Analysis staff; and Fred Norbury of the Land Management Planning staff. Although great assistance was provided to us without which this chapter could not have been written, the responsibility for any errors of fact or interpretation rests with the authors.

Chapter 10, which presents a FORPLAN application in an analysis of the below-cost timber sales issue, has benefited greatly from assistance provided by the Planning and Operations Research staff of the Forest Service. We wish to thank particularly Chris Hanson of the Mt. Baker-Snoqualmie National Forest, Thomas R. Mitchell of the Shoshone National Forest, and Kim Cimmery of the Apache-Sitgraves National Forest for providing access to their FORPLAN data and giving us the benefit of their insights into FORPLAN modeling. The Land Management Planning staff at the Fort Collins Computer Center was most cooperative in enabling us to run the models. James McCallum provided assistance in getting us started, and Brad Gilbert was ever helpful in explaining to us the variations in particular model applications and the reasons for what otherwise appeared to be idiosyncratic results. And we must acknowledge that without the aid of Thomas B. Stockton, an extraordinarily capable research assistant, we would have been overwhelmed by the task of adapting and running these FORPLAN models. Finally, we need to acknowledge the contribution of Bjorn Dahl of the Policy Analysis staff for both his assistance in arranging funding and in providing the essential liaison with those national forests that served as subjects for analysis in chapter 10.

A volume such as this never gets carried to completion without its dedicated patrons. Ross Whaley, former director of the Forest Inventory and Economics Research staff of the Forest Service and now president of the State of New York College of Environmental Sciences and Forestry in Syracuse, is most deserving of recognition in this regard. His concept of a creative research relationship, incorporated in our first cooperative research agreement with the Forest Service, provided complete freedom for us to select the research topics for study under the multiple-use research effort, and a most hospitable environment in which to conduct this work during his tenure as director of the staff. Our debt to Ross is incalculable.

The second protector of our enterprise until it saw completion was Max Peterson, then chief of the Forest Service. Most generally, it seems, top administrators and research persons consider each other to be natural enemies. The long time needed to complete this undertaking would normally have overtaxed the patient good will of even the most charitable administrator. Yet in the circumstances of our study this was not to be the case, and it can be certified that without the support of Max Peterson this undertaking would not have survived to completion.

To acknowledge various persons in connection with their assistance in this project is necessary, for this study could not have reached a successful conclusion without that assistance. Yet in doing so we risk inadvertently omitting others whose assistance may have been no less essential. We urge those who have suffered our neglect to understand that our appreciation is no less great despite our lapses of memory. We reserve for them a special sense of gratitude.

Part 1

The Historical, Institutional, and Theoretical Background

1

Development of the Concept of Multiple-Use Forestland Management

Aside from the land in the Pacific Coast forests, in the forests of the panhandle of Idaho and western Montana and in some southern forests, the bulk of the land in the National Forest System is of indifferent quality for growing timber. In the higher elevations in the central Rockies, in much of the intermountain region and in the Southwest, and even in the mountainous areas of the Ozarks and Appalachians, with few exceptions strictly commercial timber operations are improbable economic undertakings. At the same time, many of these public wildlands are in great demand for forest-related outdoor recreation activities. Indeed, it is becoming obvious that the value of these lands depends more on their recreational potential than it does on the production of natural resource commodities. In spite of this, it may be that artful multiple-use management of these wildlands to meet demands for both timber and recreation, or perhaps other combinations of resource services, would represent the preferred management regime. This is the issue that we intend to explore.

In the United States there has persisted in public forest management a heavy reliance on the philosophy and norms of biological management. Recent legislation that acknowledges multiple uses and environmental constraints also shows an emphasis on the need for comparison of benefits and costs.[1] The emergence of economic concepts is a departure not yet fully integrated into the management philosophy of the U.S. Forest Service.

Historically, the public wildlands have not consciously been treated as economic assets. Although resource services other than timber harvests were recognized, they were not explicitly incorporated into the management regime. The method of management could perhaps best be described as a health care delivery system for the forest and range lands, with sensitive antennae tuned to the divergent interests of clients competing for use of the resource services of the land. The improvement and protection of the productive capability of the land were pursued with little attention to the economic worth or the costs of such undertakings.

Our study is intended to provide a background of economic concepts specifically focused on the management of the nation's public forestlands for the combined production of timber and those services such as recreation and water flow which are linked to the character of the land and its stock of vegetation. But first, in the following pages we will briefly trace the development of the current approach to multiple-use management.

1. Timber Management or Wildland Management

Forestry and the nature of the activities professional foresters are expected to engage in are more often than not rather ambiguously described. Webster defines forestry both as "the science and art of developing, caring for, or cultivating forests" and as "the management of growing timber." It is therefore not surprising that different views are held as to the true content of forestry. It is probable that the primary concerns of foresters differ depending on the type of forestland on which they work. Where productive forestlands are held principally by private individuals or industry, one may very well anticipate the emphasis tilting toward commercial management of growing timber. But, while management of the growing and harvesting of timber remains a significant function of the practice of forestry, there has always been a strong tendency on the part of some members of the profession in the United States to interpret the responsibility of foresters to be the management of wildland resources, and not management of timber alone or even of timber as a dominant objective. The historical antecedents of the formation and growth of professional forestry may suggest an explanation.

Forestry has been closely connected with the Forest Service in the United States. Indeed, the profession had its American origin through a cadre of federal *land* managers broadly conceived rather than timber managers per se. That forestry in the United States was intimately tied to land management is not difficult to understand in the light of the kind of lands that were set aside through "forest" reservations. A provision of the Creative Act of 1891 that established forest reserves authorized the president to withdraw from entry "public domain lands wholly or in part covered with timber or undergrowth, whether of commercial value or not, as public reservations."[2] This legislation, which did not provide for active management of the forest reserves, was superseded in 1897 by legislation commonly referred to as the Organic Act.[3] The Organic Act provided for active management to improve and protect the forest reserves, with explicit reference to watershed management as well as to timber. As further withdrawals occurred for what was to become in 1905 the National Forest System,[4] many withdrawals were motivated by concern for soil and water conservation. In time, under the Weeks Act of 1911,[5] such withdrawals were extended to eastern forestlands as well, predominantly in the mountainous areas of the Appalachians and the Ozarks. Much land was thus brought into the National Forest System for reasons other than, or in addition to, timber management.

The obligation of foresters in the Forest Service to manage nontimber land, and even land that was timbered, for more than a timber management objective posed problems for defining the qualifications for membership in the profession of forestry. We see appearing in the *Journal of Forestry* in the early 1940s a debate regarding criteria for membership in the Society of American Foresters, and discussion of a functional role for forestry—namely, Is forestry wildland resource management or is forestry confined to trees?[6] In his paper addressing this issue, H. T. Gisborne, a senior silviculturist employed by the Forest Service, pointed out that the professional forester had to devote time, energy, and thought to such other nontimber wildland resources as wilderness, water, forage, wildlife, and recreation. Accordingly, he argued for recognizing this fact in eligibility criteria for membership in the society, and, implicitly, in the society's responsibility for accrediting schools of forestry education.

Continuing differences regarding essential qualifications for a forester remain to this day. The broader view of forestry was nonetheless further advanced with the passage of the Multiple-Use and Sustained-Yield Act of 1960 (MUSY),[7] which explicitly listed objectives of multiple-use management consistent with the position that Gisborne advocated in the *Journal of Forestry* years earlier. Moreover, passage of the multiple-use legislation appeared to spur the Council of Forestry School Executives to recognize the need for curriculum revision in the training of professional foresters. Between 1966 and 1973 this interest succeeded in broadening the curricula of professional schools of forestry to meet the requirements of public wildland management.

2. Conservation or Economic Efficiency

While the Forest Service's decision to formalize its authorization to manage national forests for multiple uses was in general favorably received by the public, there has been another element of controversy regarding the management philosophy of the Forest Service.

Had the Forest Service and the Society of American Foresters, of which Forest Service personnel made up the preponderant contingent, opted to define forestry as the management of timber, it is possible that the management model would have reflected the contribution of the nineteenth-century German forester Martin Faustmann, who in 1849 perceived the management of a forest for timber production as essentially a matter of managing an economic asset.[8] Since a forest is a form of capital, producing its annual biophysical yield (and, usually, an annual appreciation of its monetary worth), it also represents an inventory of stock capable of conversion into saleable wood products. The financially optimal time to harvest was seen to involve the criterion of maximizing the present worth of the timber on a specific site, which given a positive rate of interest and the typically low growth-rate of mature trees would tend to reduce the rotation age, whereas the biological norm would be to maximize the volume of production from a given site. In the United States, Martin Faustmann's work was distinguished by its obscurity, having come to the attention of most American professionals only after publication of Mason Gaffney's work (1957),[9] in turn reformulated in a somewhat more accessible form by Peter Pearse (1967).[10] The genesis of the Forest Service, however, coincided in time with, or was the product of, the heyday of the early conservation movement in the United States. The initial support for establishing forest reserves and a cadre of conservation-oriented professional resource managers arose from the growing belief that the forests of the nation were being harvested too rapidly and carelessly, so that not only was a future "timber famine" in prospect, but also the productivity of the land was being destroyed by clear-cutting, improper roading and ditching, and other destructive logging practices that created excessive soil erosion. One estimate suggests that 150 million acres of forest were cleared, principally for agricultural purposes, by the end of the nineteenth century, and that as many as 600 billion board-feet of timber (a conservative estimate) were destroyed for lack of any constructive use for the timber.[11] Moreover, there was much illegal reckless logging on public domain lands and much unregulated grazing for want of any effective means of enforcing trespass laws.[12]

Early conservationists tended to place unwarranted faith in the doctrine of technical efficiency, which represented the application of principles of the natural sciences to matters encountered in resource management. As leading exponents of the conservation doctrine, early American foresters tended themselves to be products of the study of earth and life sciences. Being natural scientists by educational background and oriented toward natural resource management by professional interest, they were inclined to find technical, if not actually technocratic, models congenial. Among them Faustmann's work, had they been familiar with it, would not have been well received. Economic efficiency was not distinguished from market-driven commercial exploitation of natural resources, to which they perceived themselves as an appropriate long-overdue antidote.

Quite apart from ideology or management philosophy, the Faustmann criterion of financial maturity was concerned only with timber management, while as we have noted much of the land held by the Forest Service was not necessarily intended for, or capable of, economic timber production. The national forests of the Rocky Mountains were brought into the system, as were the forests of the Ozarks and the Appalachians, primarily for soil conservation and watershed management, although in some instances acquisition of private land suitable for timber production occurred under authority provided by the Clark-McNary Act of 1924.[13] Nevertheless, a management approach based on commercial timber management objectives was not congenial to many who had responsibility for the public lands, as noted in the discussions on the qualifications and activities of professional foresters in the *Journal of Forestry* in the 1940s. The national forests had been tended by personnel performing duties that were appropriate to Webster's definition of forestry as "the science and art of developing, caring for, or cultivating forests," and, we might add, wild forest (and even nonforest) lands. For this purpose they had developed norms and standards for a stewardship role, to a large extent using a biological mode of management to respond to the mandate in the Organic Act to improve and protect the forest and to secure "favorable conditions of water flow." The authority to manage national forests was interpreted to a significant extent as a directive to rehabilitate previously abused and other, often indifferently timbered or high-graded forest, as well as degraded rangelands.

3. Entering the Modern Era

Throughout much of its first fifty years the Forest Service performed primarily a stewardship role, with virtually all of the supplies of commercial timber coming off private lands. The use of the national forests for hunting-and-gathering recreational activities and by settlers for building materials was so limited relative to the National Forest System's capacity that a tradition of free usage was established for not-for-profit access to the national forests.

This type of use and management regime persisted into the period of World War II, after which, during a relatively brief time of transition, growth in demand for both the recreational services on the one hand and commodities—principally for timber and forage—on the other resulted in vigorous competition for the resources of the national forests or the uses to which the resources could be allocated. These had become scarce relative to the demand for them given the terms under which they were made available.

For a number of reasons the national forests became the primary purveyors of forest-related outdoor recreation. First, the size of the National Forest System is very large relative to the size of the National Park System, and thus the facilities for forest-related recreation are more abundant and often more readily accessible.[14] Second, the restrictions on recreational use in the national forests are less stringent; for example, hunting and gathering activities are permitted. This tends to foster more all-season recreation. Moreover, not only does the National Forest System satisfy the bulk of the total demand for forest-related outdoor recreation: that demand has grown phenomenally in the postwar period, multiplying about elevenfold since 1940. Recreational use of the national forests, then, is a truly significant economic use, and for many forests doubtless the most important.

The demand for wildland recreation has coincided with increased demands on national forests as a source of timber. There has been, roughly, a five- to sixfold increase in the volume of timber harvested from national forestlands since 1940.[15] Further, the public lands are likely to remain an important source of softwood supply for the foreseeable future. Although the productivity of the private lands is on average intrinsically higher,[16] inventories of old growth on private lands have been drawn down more than proportionately.[17] The result has been that the stock of standing softwood sawtimber remaining at the present time in the Pacific Coast and northern Rocky Mountain national forests represents almost half of the nation's total.[18] Accordingly, despite the more favorable distribution of high-productivity sites among private holdings, the large inventories of old-growth softwood sawtimber on some two or three dozen coastal and northern Rocky Mountain forests establishes these national forests as significant for the nation's softwood sawtimber supply in the intermediate term, until the growing younger stock on private lands becomes large enough to harvest.

As increased demand began to press the capacity of the national forests, conflicts arose between those who would manage the stands for commercial timber purposes and those who would preserve forest environments in an aesthetically pleasing way for outdoor recreation enthusiasts. So intense was the conflict between those with an interest in having the national forests produce a larger volume of natural resource commodities and the growing body of militant outdoor recreationists that the issues which divided them were increasingly taken to court for adjudication. The solution advanced politically was to attempt to wrest a larger volume of production from the fixed land base by means of more intensive management and enlarged manpower and budget resources. With the assistance of the late Senator Hubert H. Humphrey and others, the Forest Service, through passage of the Forest and Rangeland Renewable Resources Planning Act of 1974 (RPA),[19] was provided with authority to prepare a program to achieve these goals.

The RPA legislation directed the Forest Service to undertake a broad assessment of supply and demand relationships, to be followed by the preparation of a program for setting goals and production targets and achieving them. The mechanisms established by the legislation were rather technocratic in conception, and not uncongenial to the spirit of the early (Theodore Roosevelt) and latter-day (Franklin Roosevelt) conservation philosophy. They were technocratic in the sense that the demand postulated was interpreted as a technical requirement which the supply provided by the program would bring "into balance" without explicit price- or cost-equilibrating mechanisms. Further, the goals to which Congress and the Forest Service were presumably committed were arrived at without knowledge of whether those goals were consistent among themselves—that is, whether the land and the biophysical production relationships would permit achievement of the goals.

It follows that without any means for testing the goals and related program elements for compatibility, the result was essentially a technical exercise relying more heavily on arithmetic than on an adequate analysis of the underlying production relations. This deficiency is particularly acute in circumstances where jointness in supply is encountered; that is, where the cost of providing a given level of some forest output is not independent of the production level of one or more of the interdependent (multiple-use) outputs. The level of understanding of National Forest System production capabilities that is required before the RPA goals and programs can command much conviction, however, requires careful examination of the production capabilities "on the ground" in the various forests comprising the National Forest System. Authorization for a mechanism to achieve this was not long in coming.

The RPA apparatus was scarcely in place before the National Forest Management Act of 1976 (NFMA) was passed.[20] The NFMA was motivated by a need to respond to the decision of the courts in the Monongahela timber harvesting case, which demonstrated that the 1897 Organic Act did not provide an adequate basis for successfully managing the national forests under present conditions.[21] Among other provisions, NFMA addresses the need to state production targets that recognize the interrelationships of outputs in production. Moreover, while there are many references in the act to biological or physical (noneconomic) criteria for achievement of objectives, there is also a considerable emphasis on economic criteria and the comparison of benefits and costs interpretable as economic concepts.[22]

In response to the NFMA legislation, the Forest Service has begun an ambitious program of planning for some 190 million acres of national forestlands by using linear programming methods. By 1983 large linear programming models were in use or under development on many forests. They are employed to aid in the selection among numerous options for scheduling harvests and land use over time. The linear programming approach would seem well-suited to developing an overall management program that is coherently related to the production possibilities of individual forest areas.

Still, the planning effort under NFMA is not free of some of the more restrictive elements of the biological-technocratic management philosophy.

The treatment options considered reflect not particularly well-defined prior restrictions on allowable harvest ages, as well as mandatory standards imposed on the various management activities. Such restrictions limit the possibility for advantageous specialization of uses on different areas of the forest and tend to raise the cost of management. The management standards and the production targets themselves seem to be set with insufficient knowledge of the relative economic values or costs of outputs and services. The planning effort has served to highlight the need for adding an economic element to complement the biophysical considerations employed in public land management.

This is a large problem, and we do not wish to suggest that all the answers to the many issues raised in the management of public lands are lurking in this volume, only to be discovered by the perceptive reader. On the contrary, we know such an objective is unrealistic, but there is some contribution to be made through an economic treatment of the problem. This is the task we set for ourselves.

In part 1 we have sketched out the historical and institutional background and the economic theory appropriate to the problem of public forestland management; chapters 2, 3, and 4 address this task. In part 2 we look at a number of resource management problems to see the various guises in which multiple use, or joint production, can arise. In chapter 5 we examine the possibility that combined management for timber and augmentation of water yields may be economically justified where the management for timber alone is not. In chapter 6 we view another joint production problem in a different context—the way in which forest practices may affect the value of recreational hunting, through vegetation manipulation favorable to wildlife habitat. This chapter offers a discussion of theoretical and empirical issues in valuing the recreational resources of multiple-use forests.

In chapter 7 we look at joint recreation and timber management in the context of a planning/management exercise. National forests do not have resident researchers and thus cannot be expected to undertake the kind of data assembly and analysis that is possible in one of the forest-science research labs or by cooperative research units. We address the issue, then, by specifying a model that can use data available to the forest supervisor, whether on the forest or at the regional or Washington offices. That is, the analysis makes do with data that are available within the National Forest System, but the approach is one in which better data can be assimilated. Indeed, there are prospects of better data accruing through the routine administrative functions of a national forest.

Discussions in chapters 5 through 7 illustrate the prevalence of joint production in one of its aspects. We deal there with cases in which the interdependence in production was such that when a given forest output was produced at a higher level, the jointly produced forest outputs either increased or were able to be increased at lesser unit costs. There is, of course, the possibility that the interrelations between one forest output and another have a less happy outcome. It is possible that the production of one output either increases the cost of the other or serves to preclude it altogether. This case occurs frequently where mining and wilderness uses compete for the land. We address that problem in chapter 8.

In treating the management of public land for multiple uses we encounter the question of the costs of providing nonmarketable resource services (public goods). Now if the nonmarketed resource services have a large component of public goods characteristics, they cannot be allocated efficiently by pricing. We must therefore address the issue of how they are funded—an issue that requires attention quite apart from the fact that not all nonmarketed services are inherently nonmarketable (for example, developed-site recreation). We take up this issue, in part, in chapter 9.

In chapter 10 we look at the companion issue when programs are undertaken with public funds—the issue of public accounting—and we do so in the context of the below-cost sales issue. This chapter also provides a discussion of the linear programs used in national forest planning, illustrating how these models might be used to address questions of multiple-use policy. In chapter 11 we conclude with an interpretation and synthesis of the implications of the multiple-use management model developed and applied in this volume.

Notes

1. J. A. Haigh and J. V. Krutilla, "Clarifying Policy Directives: The Case of National Forests Management," *Policy Analysis* vol. 6, no. 4 (Autumn 1980).

2. 16 U.S.C. sec. 471 (1891). See U.S. Department of Agriculture, Forest Service, *The Principal Laws Relating to Forest Service Activities,* Agricultural Handbook no. 453 (rev. ed., Washington, D.C., September 1983) p. 5 (hereafter cited as USDA Forest Service, *The Principal Laws*).

3. 16 U.S.C. secs. 473–475, 477–482, 551 (1897). See USDA Forest Service, *The Principal Laws,* pp. 5–8.

4. Under the Transfer Act of 1905 (16 U.S.C. secs. 472, 524, 554 [1905]). See USDA Forest Service, *The Principal Laws,* pp. 11–12.

5. 16 U.S.C. secs. 480, 485, 500, 515, 517, 517A, 518, 519, 521, 552, 563 (1911). See USDA Forest Service, *The Principal Laws,* pp. 19–23.

6. H. T. Gisborne, "The Challenge to the Society of American Foresters," *Journal of Forestry* vol. 41 (November 1943). This historical exchange was brought to our attention by Donald Duncan.

7. 16 U.S.C. secs. 528(note), 528–531 (1960). See USDA Forest Service, *The Principal Laws,* pp. 156–157.

8. M. Faustmann, "On the Determination of the Value which Forest Land and Immature Stands Possess for Forestry," in M. Gane, ed., *Martin Faustmann and the Evolution of Discounted Cash Flow*, Institute Paper no. 42, Commonwealth Forestry Institute (University of Oxford, 1968) pp. 27–55; originally published in German in *Allegemeine Forest und Jagd Zeitung* vol. 25 (1849).

9. M. Gaffney, "Concepts of Financial Maturity of Timber and Other Assets," *Agricultural Economics Information*, series 62 (Raleigh, North Carolina State College, September 1957). William McKillop of the School of Natural Resources at the University of California–Berkeley takes exception to this account, writing (in correspondence with the authors, September 10, 1985) that Faustmann's work was known to professional foresters and taught in schools of forestry as early as the 1920s. Perhaps the observation of Peter Berck (also of the School of Natural Resources) on the matter will help to reconcile the two perceptions: he asserts that extensive survey research has established that the typical forestry student forgot Faustmann's lesson in 2.3 beers (private communication from Professor Berck, n.d.), suggesting that Faustmann's influence has been small.

10. P. H. Pearse, "The Optimum Forest Rotation," *Forestry Chronicle* vol. 43, no. 2 (June 1967).

11. P. W. Gates, *History of Public Land Law Development* (Washington, D.C., GPO, 1968) p. 531. This work was written for the Public Land Law Review Commission.

12. See H. Clepper, *Professional Forestry in the United States* (Baltimore, The Johns Hopkins University Press for Resources for the Future, 1971) chaps. 1, 2, and 10.

13. 16 U.S.C. secs. 499, 505, 568, 568a, 569, 570 (1924). See USDA Forest Service, *The Principal Laws*, pp. 48–50.

14. For example, the National Forest System has a trail network of close to 94,000 miles, which compares to the fewer than 9,000 miles of trails in the national parks. See USDA Forest Service, *An Assessment of the Forest and Rangeland Situation in the United States*, Forest Resource Report no. 22 (Washington, D.C., 1981) pp. 82–83.

15. See USDA Forest Service, *An Analysis of the Timber Situation in the United States, 1952–2030*, Forest Report no. 23 (Washington, D.C., 1982). Data on harvests for 1940 were obtained in correspondence with Rena Gary, Timber Management Staff, USDA Forest Service, Washington, D.C.

16. Roughly 45 percent of the private industrial forestlands are of high productivity, capable of producing more than 85 cubic feet of industrial wood per acre per year. Of the national forestlands considered potentially commercial (capable of producing 20 cubic feet per acre per year), only 30 percent are of high productivity. See USDA Forest Service, *Analysis of the Timber Situation*, pp. 350–359.

17. Of the Forest Service's commercial timberlands, 64 percent are stocked with older sawtimber stands, while only 45 percent of the private industrial forestlands are stocked with sawtimber. See ibid., pp. 360–362.

18. Ibid., p. 371.

19. 16 U.S.C. secs. 1601(note), 1600–1614 (1974). See USDA Forest Service, *The Principal Laws,* pp. 353–369.

20. 16 U.S.C. secs. 472a, 476, 500, 513–516, 518, 521b, 528(note), 576b, 594-2(note), 1600(note), 1601(note), 1600–1602, 1604, 1606, 1608–1614 (1976). See USDA Forest Service, *The Principal Laws,* pp. 441–460.

21. Stipulations in the 1897 act were to the effect that only individually marked or designated trees that were fully grown, or mature, could be harvested. See *West Virginia Division of the Izaak Walton League v. Butz,* 522 F.2d 945, 955 (4th Cir. 1975).

22. See Haigh and Krutilla, "Clarifying Policy Directives," pp. 409–439.

2

The Multiple Uses of the Public Lands

1. Introduction

In this chapter we will provide brief descriptions of the conditions of sale or provision for the various outputs of the public forests. It is quite clear that were the public lands to be judged by criteria appropriate to lands under private ownership—namely, the earning of maximum financial returns, or maximization of the positive difference between discounted market receipts and expenditures—the bulk of the national forests would fail such a test. This can be explained in part, but only in part, in terms of the intent of the various legislative acts to have public land managers provide a mix of unpriced resource services in addition to timber. Perhaps largely because of the complexity of the production process and limited knowledge as to the value of many of the resulting services, the Forest Service has remained reluctant to emphasize economic criteria alone in making its supply decisions.

Certain complicating features of the public land management problem stand out. Most obviously, there is multiple output production, with the forest providing a wide range of services even when unmanaged. But such joint production of several outputs is not in itself unusual, and neither is it necessarily a source of great difficulty. Many firms make choices about production levels and quality for a wide mix of goods and services. For the public forest manager, complications arise from a number of other related sources.

First, there is a strong interdependence between current output choice and the level of capital stock. On the one hand there is little flexibility in the output mix except through alteration of the stock of timber. The nontimber outputs such as dispersed recreation, wildlife, and water flow are largely dependent upon the characteristics of the land and its stock of standing

13

timber. On the other hand there is very limited opportunity to adjust stock except through timber harvests. The choice of harvest timing and location, coupled with the slow process of biological growth, provides the means for adjustment in the age and diversity of timber stock. It may take a considerable time to attain any particular desired stock condition. The stock of timber cannot always be immediately adjusted to best meet current output needs. Further, the effect of any treatment of the land may be long-lived, affecting outputs in a complex manner for many years. The manager must anticipate the effect of his current decision on the potential future flow of products and services. Under these conditions there can be no meaningful separation of the decision on current product mix from the long-run decision on the holding of capital stock in timber.

Second, there is a traditional expectation of the right to enjoy many services of the forest without charge, or at nominal cost. This expectation extends to certain commercial users, such as those seeking access to mineral deposits, as well as to users of the recreational or amenity services of the forest. The lack of competitively determined market prices is a source of great difficulty in the land manager's choice of an appropriate mix of products and quality of services. Without prices to make users aware of the social costs of their activity, there is a tendency to overuse and an accompanying pressure to increase the level and quality of service. On the other hand, those activities which do not produce revenue seem likely to be less than preferentially treated in the political budget allocation process. There also seems to have been limited incentive to explore the potential for managing the forest so as to promote those services which provide no revenue. The manager, without prices as a clear signal of relative value, and without profitability as a goal, faces a complex planning process, balancing demand pressures against the realities of a limited budget and the professional paradigms of good forest management.

In the following brief description of how the various outputs of the public forest are provided,[1] we also try to provide a sense of the relative economic importance of each output.

2. The Provision of Market Commodities

The policy of the Forest Service has been to charge for those products and services that are provided to commercial enterprises. Timber is the primary commercial product, yielding roughly $770.5 million in receipts in 1985 sales.[2] The range program is much smaller in scale, with approximately $9.0 million in yearly receipts from grazing fees. Both timber and range receipts have dropped sharply from the level obtained in 1980. Energy and minerals permits or leases provided about $159.4 million in 1985.

In addition, fees may be charged to users of certain specialized services. Fees from users of developed recreation sites and receipts from permits for the operation of commercial recreation facilities on Forest Service lands yielded about $31 million in 1985.

The circumstances of the sales of these commodities and services vary widely. Although the prices at which some of these products are provided are determined competitively, the level of supply is strongly influenced by criteria other than economic efficiency, such as traditional rights of access, concern for local economic stability, and various conservationist principles of land management.

Timber

About 90 million acres of the total 190 million acres in the National Forest System meet the administratively determined minimum productivity requirements for timber management and have not been withdrawn from such use. These lands for the most part tend to be of lower quality and in poorer locations relative to private commercial timberlands.[3] Nevertheless, the large inventories of older-growth softwood establish the national forests as important to the nation's sawtimber supply. Because of past conservative harvest policies, the national forests—primarily those of the Pacific slope coniferous region—now contain about 41 percent of the standing softwood sawtimber in the nation, and currently provide more than 20 percent of the timber harvested in the United States.[4]

Timber sales from Forest Service lands are in general made by public, competitive offerings. Once it has been determined that a market for a potential sale exists, the Forest Service will undertake considerable planning. The sale area is inventoried and marked, an appropriate method of roading and harvesting is determined, and the harvest volume and the potential sale value are estimated. This information is made known to potential purchasers. Once publicly offered, sales may be made either by sealed bid or by open auction. The purchaser has the option of harvesting at any time within a specified number of years following the sale date. Payment, at prices determined by the auction, is for the actual harvest volume removed.[5]

There is occasional criticism of the timber sale procedure suggesting that alternative mechanisms might yield either higher receipts or lower administrative expenses.[6] Certainly for many sales there are few bidders, but there seems to be little evidence that reasonable receipts are not obtained. In 1985 the national forests in total had sales resulting in $558.2 million in cash receipts and $192.3 million "in kind"—the value of roads constructed at timber purchasers' expense.[7]

Whether or not the auctions are as competitive as might be desired, it should be understood that the Forest Service is not offering timber supplies in the manner of a profit-maximizing firm selling in a competitive market. Indeed, at the time of this writing a controversy has developed over the apparently extensive practice of offering for sale timber that is not expected to yield prices sufficient to cover even the costs of sales preparation and harvest administration. The offer of timber for sale is not motivated simply by the potential for maximizing monetary receipts. This is not surprising given the charge to provide various nonreimbursable amenity services, but there are other reasons worth noting.

The Forest Service gives as other reasons for selling timber at less than the direct management costs of the sale: (1) meeting the silvicultural needs of the individual timber stands, (2) the salvaging of damaged timber, and (3) satisfaction of the needs of communities and timber purchasers who are dependent on national forest timber sales. Also, the Forest Service states that there have been high costs associated with bringing extensive areas of land into a program of management to ensure a continuous, relatively even flow of future harvests.[8] The institutional focus on the stability of harvest levels and on biological criteria for timber treatment is apparent. It should be noted that in many of the geographical markets in which national forest timber is sold the volumes supplied represent the dominant market force, and stability of the local economy is real cause for concern.

The costs of timber harvesting, as presented in Forest Service records, are not appropriately compared to the revenues received in any given year. This is so because current and capital expenditures are not separated, and because some expenditures, as for roads, arguably also serve other multiple-use ends. Considering the capital category, funds expended in previous years that may have improved the quality of the timber or provided more ready access to the harvest sites affect the current year's expenditures and receipts, but they will not show up in the current year's account categories. Similarly, some expenditures made in connection with the current year's harvest will have utility in future years. Without a proper capital account, the mixing of current and capital expenditures, as done in cash-flow analysis, vitiates the cost/returns analysis. As to common costs for joint outputs, some roads will serve public recreation (see chapter 7), some harvests will alter the condition of stream-flows in a favorable way (see chapter 5), and some sanitation harvests will improve the conditions for wildlife (see chapter 6). Thus the problem of common costs of joint outputs adds a dimension of complexity to accounting; this will be treated in chapters 9 and 10. It should be said here that Congress has requested the Forest Service to develop a more adequate method of accounting, and work on such an accounting system is currently well advanced.

Other aspects of the timber harvest policy have raised concern among forest economists. First, there is the slow removal of inherited stocks of older timber. Under the policy referred to as nondeclining even flow, these stocks are liquidated slowly so as to sustain an approximately constant level of harvest over time. Second, the age at which trees are harvested is considered excessively old. As a general rule timber is to be cut at no less than the age of culmination of mean annual increment. Considered the age of biological maturity, this is the age at which the average annual harvest volume (harvest volume divided by age of the stand at harvest) is at a maximum. The concern in both cases is that the Forest Service is accepting too low a rate of return on the asset value of the forest. These criticisms of Forest Service timber supply policy implicitly focus on the failure to appropriately consider interest rates. There is an opportunity cost to holding inventories of standing timber. With the poor growth rate of older stocks, the potential increase in value of such stocks when held over time may be insufficient to compensate for passing up the prompt receipt of revenues from an earlier harvest. A greater net value to the public might be achieved by an earlier removal of timber and the subsequent regrowth of more vigorous stock.

The nondeclining even-flow policy, and similar policies which limit the rate of harvesting from existing stocks of older timber, can be costly in other ways. An attempt to stabilize Forest Service supply in the face of demand fluctuations may simply add to price instability.[9] The policy may also shift the costs of demand instability onto smaller private suppliers of timber. It can also be demonstrated that even-flow policies lead to timber management activity on lands where this activity is not economically justified. This last argument, illustrating what is referred to as an allowable cut effect, can be easily understood.

Consider a single forest, one half of which is covered with old growth and the other half with younger timber stands,[10] and suppose that activity to promote timber growth on the younger stands would increase their future harvest yield, but that the discounted present value of this increased future yield would not justify the necessary current expenditure. Now consider the effect of an even-flow constraint. Implementing growth-improving activity on younger stands would increase the average sustainable yield available from the area as a whole. Under the even-flow constraint, this higher potential sustainable yield would allow an immediate increase in the yearly harvest, with the harvest increases initially coming from the existing old growth. Although not economically sensible in itself, intensive management and the planned later harvest of the presently younger stands might now be justified, since they allow an immediate increase in revenue from the more rapid harvest of old growth. Far from implying that the intensive management is desirable, though, the result is best viewed as an indication of the cost of an even-flow policy. It would be better to relax the even-flow constraint, avoiding the unprofitable activity on the younger stands and still gaining the immediate benefit from early harvesting of the older stock.

Range

In contrast to other uses of the forest, there has been a gradual reduction in the intensity of grazing since about 1920. This was, at least initially, a response to the deterioration of conditions caused by heavy grazing before that time.[11] About 100 million acres of national forest land is now in grazing allotments, with about half of that suitable for grazing. The roughly 14,000 paid permits for commercial grazing result in about 10 million animal unit months (AUMs) of livestock use each year.[12] Although this represents a small fraction of the total forage requirements for livestock, the Forest Service lands can be locally important as a complement to private lands, particularly in certain seasons of the year.

Grazing permits for commercial use are not allocated by competitive bidding. Typically, they are held by historical privilege. Permits, which run for ten-year periods (for shorter periods in the eastern forests), are reissued to the existing permit holder except under unusual circumstances. Few new permits become available. Existing permits are in theory not transferable, but the purchaser of a ranch is given first priority for receipt of a new permit to replace that relinquished by the seller.

The Forest Service has the authority to control the intensity of grazing in accordance with multiple-use goals and in order to maintain land productivity. In cooperation with the permittee, the Forest Service may also actively manage to improve range quality. This might include structural improvements such as water development or fencing, and also nonstructural improvements such as brush or weed control and seeding with better varieties of range vegetation. The authority to regulate grazing, and to charge fees, was first confirmed by a Supreme Court decision in 1911.[13] Funding for range management and improvement in 1985 was $32.1 million, while receipts from permittees were about $9 million.[14]

The fee for commercial grazing on Forest Service lands is intended to reflect a fair market value comparable to the charge for use of private lands. Because of the very limited market in federal grazing privileges, poor information on rental of private grazing land, and differences in the nature and use of private versus public lands, it has proven difficult to determine such a fair market value. The current fee, found by a formula authorized under the Public Rangelands Improvement Act of 1978,[15] is considered to be below a fair market value. The Forest Service in its 1980 planning process estimated the value of an AUM at two to three times higher than the grazing fee.[16] In 1983 grazing fees on private lands averaged $8.85 per AUM, which compared to about $1.40 for the public grazing fee. There have been reports of leaseholders of Bureau of Land Management (BLM) grazing lands subletting those lands for fees considerably higher than the federal grazing fee.[17] On the other hand, there is some evidence that the fees for some grazing are at least as high as the value of the forage. If the permittee elects to, he can declare "non-use" of an AUM for a limited period of time. He would do so if its value were below the fee and associated costs of grazing. On some public lands, nonuse is reported for a very large proportion of the total allocated AUMs.[18]

Energy and Minerals

The role of the Forest Service in the development of energy and minerals is limited. The Department of the Interior is responsible for subsurface resources, while the Forest Service manages only the surface resources. The role of the Forest Service is largely limited to the review of applications for exploration, development, or production activity to ensure that the surface resources are adequately considered. Ongoing mineral activity further requires that the Forest Service incur expenses for the control or mitigation of damage to surface resources.

There are three categories of minerals. Under the Mining Law of 1872,[19] mining for hardrock minerals (gold, silver, zinc, and others) enjoys a preferential status on public lands, except those lands explicitly withdrawn from such use. The filing of claims for the development of hardrock minerals on public land requires little expense and no royalty payment. Certain materials are not covered by the Mining Law. Construction materials such as sand, gravel, and building rock may be directly sold by the Forest Service. Energy minerals such as oil, gas, and coal, as well as certain other minerals including phosphates, sodium, and potassium, are leasable. Competitive leasing is required for access to these minerals on lands within known producing geologic structures. In 1985 receipts from sales, rents, and royalties totaled about $160 million. The funding for expenses incurred in the processing of applications and in administering ongoing activity was about $27 million in that same year.[20]

A variety of minerals, including some of strategic importance such as chromium, tungsten, and molybdenum, are found under national forestlands. There is currently relatively little oil and gas produced on Forest Service lands. It is anticipated that western national forests will become significant sources for coal and phosphate rock production.

Fee Recreation and Occupancy Permits

The authority to charge fees for recreational use of Forest Service land is greatly restricted by law. Under the Land and Water Conservation Fund Act of 1965 there are to be no admission fees simply for access to the land (except in national recreation areas).[21] Recreation fees may be charged, however, for the use of certain specialized facilities and services; these fees are intended to reflect the cost of providing facilities. Actually, little of the recreational use of the forest is at fee sites: a recent report indicates that less than 15 percent of the use was at fee management sites.[22] While the proportion of campgrounds at which fees are charged is increasing and is now close to one-half, the use of all such developed campground facilities accounts for only one-third of the recreational use of the forest.

Special use permits provide another source of receipts largely related to recreational use. Under a 1915 statute,[23] permits may be issued for a period of up to thirty years allowing use of Forest Service lands for commercial operations or vacation homes. Such uses must not interfere with the general public enjoyment of the national forest. Today, permits are primarily issued for stores, gas stations, and ski areas where these will improve public enjoyment of the forest. Where possible, competitive bids to provide services and facilities are considered. The greatest revenue from special occupancy permits is provided by ski areas. Such land use is of importance to local economies. A large proportion of the ski areas in the western states use at least some Forest Service lands.

3. The Provision of Nonmarket Services

In the early years of custodial management, many services of the forest were abundantly provided as secondary by-products of the natural forest condition or as a result of the limited management for timber. Demands for the amenity services such as recreation, wildlife habitat, and preserved natural areas were met with little cost and without need to forgo production of the timber or forage which were of primary interest. In more recent years, with rising demands, such services began to compete directly for use of the land and have required costly modification in timber harvesting practices. The increasing scarcity of amenity services relative to rising demands has not been matched with higher prices. Many of the resource services which the public looks to the national forest to provide are still available without charge, or at nominal fees.

Water

One of the early concerns that led to the establishment of the Forest Service was that unregulated timber harvesting and grazing were causing the siltation of streams, ponds, and reservoirs, and in some areas increasing the potential for flooding. The Organic Act of 1897 had, as one of its two main criteria for the establishment of managed national forests, the goal of "securing favorable conditions of waterflow." The Weeks Act of 1911 further emphasized this management goal, authorizing land acquisition expressly for the purpose of regulating water flow.

The early concerns over the effects of timber harvests on water quality and flow are still reflected in Forest Service activity. In 1983 almost $30 million was budgeted for expenditure on soil and water management. The primary activity was the provision of advice on the management of timber and minerals. As preventative measures, hydrologists or soil scientists assist in the planning of timber harvests and road networks. A secondary responsibility, at least in terms of funded activity, is the restoration or improvement of the quality and quantity of water flow. To restore more favorable water-flow conditions, flood prevention measures and activities to stabilize stream banks, gullies, and old roads are undertaken.[24] In arid areas of the country there has long been an interest in methods to increase the quantity of water flow, which may be achieved through removal of vegetation to reduce transpiration and improve the timing of snowmelt (see chapter 5).

Fish and Wildlife

On the national forestlands, as on all federal lands open to fishing and hunting, responsibility for the management of wildlife resources is shared with the states. The management of wildlife stocks, particularly through the licensing and regulation of hunting and fishing activities, has always been the prerogative of the states. The Forest Service manages habitats. This role of the Forest Service was formally recognized in legislation with the passage of the Multiple-Use and Sustained-Yield Act of 1960. The national forests, of course, have always provided significant areas for fish and wildlife populations; active management for the improvement of habitat for fish and game is a more recently accepted role.

The improvement of habitat and the protection of water quality through suitable timber management are the primary tools of wildlife management. The forest wildlife specialist's function is largely to advise in the planning of timber harvests. An estimated 40 percent of the wildlife habitat improvement reported in 1985 was a result of timber management.[25] In certain locations a more direct role in promoting wildlife may be undertaken. Prescribed burning to promote forage, wetland improvement for waterfowl, control of competition from livestock, and improvement of access to water may be important activities. Further effort is directed toward the mitigation of existing damage, an effort that includes advising on the regeneration of poorly stocked lands.

Active habitat improvement is largely directed to species of high interest to hunters and fishermen. The wildlife program, prepared under requirements of the Resources Planning Act of 1974 (RPA), anticipates eventually providing increases of 20 percent or more in the populations of many game species.[26] In addition, the Endangered Species Act of 1973 requires special attention to the protection and promotion of the habitat needs of threatened or endangered species. Management to improve or retain suitable habitat is reported to have stabilized or increased the populations of some species. Special effort has been directed toward reducing unnecessary mortality of grizzly bear populations.

Non-priced Recreation

The authority to manage the forests for recreation was also formalized with passage of the Multiple-Use and Sustained-Yield Act of 1960.[27] Even before passage of that act, the provision of access and the management of recreational facilities were recognized as significant functions of the Forest Service. With the value of recreation taken by the Forest Service to range between $6.50 and $12.50 per visitor day, depending on location and other related conditions, it can be seen that recreation is the largest single source of value in the National Forest System, amounting to an annual value of around $1.5 to $2.0 billion, compared to less than half that amount for timber.[28]

The earliest role of the Forest Service was simply the provision of access. Even today, two-thirds of the some 230 million annual recreation visitor days (RVDs) of use is dispersed use, not associated with developed sites. Forest Service policy in providing recreation has always been to emphasize opportunities consistent with the natural setting. The practice of actively providing public facilities and service for recreational users developed slowly, initially more for the protection of the forest environment than for the convenience of users.[29] The major effort by the Forest Service to provide public recreation facilities began in the late 1950s. This was a period of growing interest in public outdoor recreation, as reflected in the formation of the Outdoor Recreation Resources Review Commission and the rapid increase in demand for outdoor recreation.[30] Through this period, and until recently, Forest Service policy had been to attempt to provide facilities as needed to meet peak demand. Now, however, budget limitations have resulted in reductions in the maintenance and services at many existing facilities and little construction of new facilities.

Recreation facilities now provided by the Forest Service include some 90,000 miles of trails and more than 4,000 campgrounds, along with trailhead facilities, picnic areas, boat ramps, and visitor information centers. More than half the campgrounds are available at no fee. Most other facilities are also provided at no charge. Funding for recreation management in 1985, including management of wilderness and fee recreation sites, was about $102 million. Trail construction and maintenance funding was an additional $7 million. About $12 million was allocated for construction of recreation facilities. These amounts do not reflect any allocations for such items as roads and fire protection, expenditures for which are at least partly attributable to recreational use.

Wilderness

In 1924 the nation's first wilderness area was set aside in the Gila National Forest of New Mexico. This was primarily the result of efforts by Aldo Leopold, then an assistant district forester, who was later to develop the first curriculum in wildlife biology, at the University of Wisconsin. Other lands were soon added to the Forest Service's system of "primitive" areas. At first, the protection accorded these lands varied widely.

A process of review of the primitive areas was begun by the Forest Service in 1938 because of concern over standards for protection. Some lands of commercial value were removed from the system. Other lands were administratively classified as either wild or wilderness lands and given a high degree of protection from road construction and timber harvesting. The process of classification proved remarkably slow, at least partly because of some reluctance within the Forest Service to forfeit its flexibility in management.

Concern by various groups that there be some permanent statutory wilderness designation led eventually to passage of the Wilderness Act of 1964,[31] which established the National Wilderness Preservation System (NWPS) made up of all federally administered wilderness lands. Forest Service lands which were at the time classified as wilderness or wild were now designated as part of the NWPS. The remaining primitive areas were promptly reviewed and, for the most part, designated as wilderness areas. Independently, the Forest Service began a study of an additional 62 million acres of roadless lands to determine their suitability for inclusion in the wilderness system.[32]

As of 1985, the Forest Service administered 32.1 million acres of wilderness. This is about 17 percent of the total National Forest System, and 84 percent of the NWPS lands within the forty-eight contiguous states. Legislative action on the disposition of the roadless areas may eventually result in from 34 to 44 million acres of wilderness within the national forests.[33]

Management of the wilderness emphasizes the protection of the resources and the reduction of conflicts between use of the wilderness and the wilderness character of the lands. Wilderness, as defined by the Wilderness Act, should generally appear to be natural, not substantially altered by man, and should provide outstanding opportunities for solitude and primitive recreation. The Wilderness Act strictly limits most uses of the wilderness which are not consistent with the wilderness character. Roads, timber harvesting, construction of facilities, and motorized vehicles are excluded with few exceptions. Prospecting for minerals was allowed until the end of 1983, and mining is allowed on valid claims. Existing grazing privileges were continued. Such uses may be regulated, within reason, to ensure protection of the wilderness.

In 1985 recreational use of the wilderness areas was estimated to be about 12.7 million recreation visitor days.[34] There is heavy use of certain areas,

particularly the John Muir Wilderness in California and the Boundary Waters
Canoe Area in Minnesota, each having about 1 million RVDs of use a year.
Such high-density use makes the preservation of wilderness conditions troublesome. Management to maintain the quality of the wilderness recreation
experience includes both rationing of access and measures to ensure greater
dispersion of use over trails.[35]

The statutory provision for wilderness areas on the national forests distinguishes this land use from the other multiple uses of forest. Some initial
reluctance on the part of the Forest Service to endorse a wilderness act was
largely based on the concern that its lands would be further divided into such
dominant-use areas. Indeed, the Multiple-Use and Sustained-Yield Act of
1960, which endorses a very flexible view of administrative multiple-use
management, was passed with the support of the Forest Service to provide
some assurance that, apart from wilderness, further statutory allocation of
land use would be avoided. A good deal of the discussion as to what
constitutes multiple-use management can be traced to concerns raised by the
debate over the wilderness legislation.

4. Should Nonmarketed Services Be Priced?

Economists would generally recommend that publicly provided goods and
services be priced so that demand equals supply, with the received payment
per unit of output equal to the marginal cost of supply. This represents an
attempt to approximate ideal competitive market pricing so as to provide
maximum net public benefits. For a variety of reasons, services of the forest
are neither priced in such a manner nor actually supplied in functioning
markets. Typically, some combination of high transaction costs, institutional
limitations, the "public good" nature of certain amenity services, and a
reluctance to interfere with what is by now viewed as a vested right to free
access perhaps have made market-type pricing impractical. The absence of
market pricing is not altogether surprising in these instances. It is of course
also possible that zero prices are occasionally justified because of the absence
of any social marginal costs for a service. This might still occur in connection
with remote-area primitive camping and wilderness recreation on some forests, provided there is excess capacity. A zero price would permit some
individuals to enjoy beneficial use of those amenities who would not have it
were a positive price in effect. Their incremental use would entail no corresponding increase in costs.

There do arise conditions in which the marketing of certain services of the
forest is technically possible but the transaction cost of ensuring collections
exceeds the norm for a reasonable ratio of collection costs to receipts. With
recreational use, the great number of access points to a forest means that

collection of fees will be impractically expensive and incomplete. The voluntary payment of fees for recreational use is possible, but fees to be left in collection boxes under the honor principle may be either unpaid or misappropriated. With increments in water flow resulting from forest management, the nature of water rights and uncertainty as to the source of increments in water supplied to a point of storage or consumption make it unlikely that a pricing agreement could be arranged. With other services, such as the retention of an old-growth area to permit survival of an endangered species, the value of the service may not be strongly associated with visitor use. The identification, for pricing purposes, of those who value habitat protection would be difficult and costly.

In other cases, institutional arrangements limit the possibility for market pricing. There are no means for the national forests to appropriate the value of services supplied through the management of forestland for desirable conditions of wildlife habitat. Under the institutional arrangements whereby wildlife is perceived to be the collective property of citizens of the states, with the Forest Service having jurisdiction only over the land and vegetation, the value of the increased browse, forage, and improved cover provided by wildlife management programs could not be captured by the Forest Service were it motivated to do so. Also to be considered are the benefits received by landholders adjacent to the national forests. The value of these lands, often attractive for vacation homes or recreation-related business, results in large measure from their proximity to the protected national forestlands. The Forest Service has no means of receiving payment for the benefits received by these landowners.

The public-good nature of many of the amenity services of the forest presents another problem. Recreational use illustrates some aspects of the public land pricing problem. There can be an important distinction between setting up a functioning market (or pricing in a manner analogous to an idealized market) and determining those prices which are consistent with maximizing the economic value of the public lands. In the case of recreation, the forest manager is directly concerned with selecting a management program that will determine the quality of the recreation site. He must choose whether to alter the stock of standing vegetation, to construct facilities, or to modify the network of roads and trails. These physical attributes of a recreation site have a public-good nature. That is, once the site quality is selected, it is available largely undiminished to all and may provide value to many. Now the manager, even though "producing" site quality, will not find it easy to directly market such a service. Instead he will find it most practical to market "site entry." A market price on entry would in general be an imperfect device for aiding the manager in the selection of both the level of visitation and the quality of service that maximize the economic value of the land. To illustrate the problem we need only present a simple example.

Consider the case of a site in which there is excess capacity, and an additional user imposes no additional management or congestion costs. In such an instance, any market-determined entry price that is above zero would lead to needlessly forgone benefits by reducing use, with no corresponding savings in cost. On the other hand, the market price of zero which is consistent with optimal use provides no information as to the appropriate level of site quality. In fact, such a price would give the manager no incentive to provide any improvements in site quality, even if the overall consumer benefits from such improvements would exceed the cost. In general, a market price on site entry alone is not going to be sufficient to guarantee that consumers and the producer choose both the economically efficient level of site use and the efficient level of site quality.

As is well known, the pricing of public goods in a manner consistent with economic efficiency would require separate prices for each consumer, with the sum of these prices equaling the marginal cost of production and each individual's price equal to his marginal valuation of the public good.[36] Individuals are almost certain to differ in their views on the marginal value of the quality or character of a forest site. In general, no single price that is uniform across individuals, whether an entry price or a one-time access rights fee, can lead the producer and all individuals to agree on the appropriate level of service to be supplied.

It is not suggested that all amenity services must inherently be nonrevenue-producing. In many cases one could find some arrangement for collecting fees earmarked for the provision of a particular service. Such arrangements might include excise taxes on related equipment, fees for club membership giving access rights to particular areas, and market-determined or administratively set prices. Whatever the limitations of various pricing strategies, it could well be that practical concerns for reducing the costs of planning might favor the pricing of many of the presently nonmarketed services of the public forests.

A primary reason for instituting a market pricing system is that it can provide signals to consumers, encouraging the selection of an appropriate level of goods and services while at the same time reducing some of the conflict currently associated with forest management decisions. The provision of free amenity services promotes an excessively high level of demand for those services. Any pricing of these services would provide a means for making the consumer aware of the costs of provision, which include both the direct management expense and the forgone uses of the land, and would serve to reduce the demand pressures for uneconomic increases in the level and quality of service. Pricing can also promote equity and reduce conflicts among users. By reducing the perception of subsidy, any fee receipts from the users of non-priced services of the forest are likely to reduce the opposition by commodity users to the provision of amenity services. Ideally, a pricing system also aids the manager in choosing the appropriate levels and quality of commodities and services, thus reducing the costly role of planning. Further, under ideal circumstances market pricing might provide the appropriate signals to both producers and consumers to ensure that land is managed to provide maximum economic value.

For a variety of reasons, fees for users of national forest resource services have not been introduced as a matter of policy. Traditional privileges may be internalized over time as individual rights and become so deeply or indelibly etched into a cultural norm that the political cost of abandoning a tradition established when it was economically justified is retained long after that justification has passed. Prices and user fees for services provided by agencies of the public tend to be interpreted by beneficiaries as an attack on their collective birthrights. They are thus often eschewed by managers of public facilities as a means of avoiding difficulties in their relations with the public. In some cases, however, the resistance to pricing for services provided results from concern about the public agency's image and public support for its programs; that is, the tendency may be to provide services free of charge even when it is feasible to do otherwise.

5. Summary and Conclusions

The picture we have presented is of a largely demand-driven supply decision, modified to reflect traditional policy concerns. The Forest Service has attempted to meet a high level of demand for all goods and services. In doing so, it has not always been mindful of the costs of activities relative to their benefits. Traditional policies with respect to timber harvesting and protection of the resource base have added to costs. Such policies may also have led to a more extensive area being under timber management than might be desirable.

Timber activity, although the major source of revenue, has been particularly costly. It has been costly both directly and in terms of the associated expenses incurred to avoid damage to other resource services. Offsetting this cost, to some unknown degree, are the indirect multiple-use benefits of the timber program. Certainly these must include some significant benefits from improved access for recreational use.

Mineral activities and livestock grazing provide lesser sources of revenue. These programs, as a result of traditional privilege or legal arrangement, are somewhat less under the control of the Forest Service than is timber. Although not especially costly relative to timber activity, the mineral and range programs do have the potential for significant impact on water quality and localized impact on the aesthetic condition of the forest.

Among the nonmarket services, recreational use and water flow are probably the highest-valued. It should be noted that some of the value from these services would be available even in the absence of further management of the forest. Indeed, some of the current value derives from the fact that some of the National Forest System receives only custodial management. More than one-third of the forest system is still either in wilderness or unroaded. The extent to which expenditures on behalf of the nonmarket services have resulted in corresponding increments in value is unclear. Much of the current expenditure is directed toward improvement or mitigation of the effects of timber, range, and mineral activities.

Notes

1. For a more detailed discussion, see G. O. Robinson, *The Forest Service: A Study in Public Land Management* (Baltimore, The Johns Hopkins University Press for Resources for the Future, 1975).

2. Including purchaser road credits. This, and other statistics on receipts, expenditures, and level of service used in this chapter can be found in USDA Forest Service, *Report of the Forest Service: Fiscal Year 1985* (Washington, D.C., February 1986).

3. USDA Forest Service, *An Analysis of the Timber Situation in the United States, 1952–2030,* Forest Report no. 23 (Washington, D.C., December 1982) pp. 119–120.

4. USDA Forest Service, *Report of the Forest Service: 1985,* p. 16.

5. Contracts may allow for price adjustments indexed to the current market at the time of harvest. Some sales—more so in the eastern than western forests—may be based on the initially estimated volume rather than the measured volume.

6. For a number of comments and references related to timber sales, see Robinson, *The Forest Service.*

7. See USDA Forest Service, *Report of the Forest Service: 1985,* pp. 25–27; and USDA Forest Service, "Cash Flow Analysis at National Forest Timber Sales in Response to HR-5973 and Associated Report Language," draft working paper, July 23, 1984.

8. USDA Forest Service, *Report of the Forest Service: Fiscal Year 1983* (Washington, D.C., February 1984) p. 24.

9. See T. R. Waggener, "Community Stability as a Forest Management Objective," *Journal of Forestry* vol. 75, no. 11 (November 1977) pp. 710–714.

10. For another, somewhat different illustration, see chapter 10.

11. For information on grazing use and receipts, see M. Clawson, *The Federal Lands Since 1956* (Baltimore, The Johns Hopkins University Press for Resources for the Future, 1967) p. 58. Surprisingly, before 1920 grazing provided greater revenue than timber harvests. It was not until the 1940s that timber revenue began to greatly exceed grazing revenues. See also P. W. Gates, *History of Land Law Development* (Washington, D.C., GPO, 1968) p. 607.

12. USDA Forest Service, *Report of the Forest Service: 1985,* p. 531. An animal unit month is the forage amount needed to support a 1,000-pound animal for one month.

13. *United States v. Grimaud,* 220 U.S. sec. 506 (1911).

14. USDA Forest Service, *Report of the Forest Service: 1985,* p. 27.

15. 43 U.S.C. secs. 1752–1753, 1901–1908 (1978); 16 U.S.C. sec. 1333(b) (1978). See also USDA Forest Service, *The Principal Laws Relating to Forest Service Activities,* Agricultural Handbook no. 453 (rev. ed., Washington D.C., 1983) pp. 527–534.

16. USDA Forest Service, *A Recommended Renewable Resource Program—1980 Update,* FS-346 (Washington, D.C., September 1980) pp. C-4 through C-8, and C-20.

17. See *Public Land News* vol. 9, no. 13 (June 21, 1984).

18. We owe this point to an anonymous reviewer.

19. 30 U.S.C. secs. 22, 28, 28b (1872). See also USDA Forest Service, *The Principal Laws,* pp. 1–3.

20. USDA Forest Service, *Report of the Forest Service: 1985,* pp. 4, 11.

21. 16 U.S.C. secs. 4601(note), 4601-4–4601-6, 4601-6a, 4607-7–4601-11 (1965); 23 U.S.C. sec. 120(note) (1965). See also USDA Forest Service, *The Principal Laws,* pp. 187–190.

22. USDA Forest Service, *An Assessment of the Forest and Rangeland Situation in the United States,* Forest Resources Report no. 22 (Washigton, D.C., 1981) p. 77.

23. 16 U.S.C. sec. 497 (1915). See also USDA Forest Service, *The Principal Laws,* p. 27.

24. For details on watershed protection and flood prevention, see USDA Forest Service, *Report of the Forest Service: 1983,* pp. 138–139.

25. USDA Forest Service, *Report of the Forest Service: 1985,* p. 28.

26. USDA Forest Service, *Recommended Renewable Resource Program—1980,* pp. 47–48.

27. Authority to acquire lands specifically for their recreational value was first provided by the Land and Water Conservation Fund Act of 1965, which also limits the pricing of recreational services on federal lands.

28. See USDA Forest Service, *Report of the Forest Service: 1985,* p. 64, table 2.

29. At first it had been left to the private sector to provide facilities to support recreation. Land could be leased for hotels, resorts, recreational homes, and other such facilities under the 1915 statute 16 U.S.C. sec. 497. See USDA Forest Service, *The Principal Laws,* p. 27.

30. Outdoor Recreation Resources Review Commission, *Outdoor Recreation for Americans* (Washington, D.C., GPO, 1962). For data on recreational use and expenditures, see Clawson, *The Federal Lands,* pp. 60, 66.

31. 16 U.S.C. secs. 1121(note), 1131–1136 (1964). See also USDA Forest Service, *The Principal Laws,* pp. 177–185.

32. See USDA Forest Service, *RARE II: Summary—Final Environmental Statement—Roadless Area Review and Evaluation,* FS-324 (January 1979). An earlier roadless area review (RARE) was begun in 1972.

33. See USDA Forest Service, *Recommended Renewable Resource Program—1980,* pp. 525–526.

34. USDA Forest Service, *Report of the Forest Service: 1985,* p. 35. A recreation visitor day is defined as twelve hours of use, whether by one individual for twelve hours, or twelve people for one hour, or any equivalent combination of individual or group use.

35. As of 1979, six national forest wilderness areas had some form of use limitation. See USDA Forest Service, *Assessment of the Forest and Rangeland Situation,* p. 103.

36. See P. A. Samuelson, "The Pure Theory of Public Expenditure," *Review of Economics and Statistics* vol. 36 (November 1954); P. A. Samuelson, "Aspects of Public Expenditure Theories," *Review of Economics and Statistics* vol. 40 (November 1958).

3

The Economics of Multiple-Use Forestry

1. Introduction

This chapter is intended to provide a background in economic concepts specifically addressing the multiple-use management of public forestlands.* We will look at economic approaches to describing multiple-use management that take into consideration the jointness in production of forest outputs, and that reflect the values of both the market and nonmarket services of the forest. Our goal is to provide a fairly elementary discussion that still captures many of the essential and interesting elements of the multiple-use problem. Among issues of concern are the compatibility of various land uses and the potential effect of the nontimber multiple-use values on economically desirable levels for timber harvest. For the present, we will largely ignore the issues raised by the dynamics of forest growth, withholding discussion of such concerns until chapter 4.

In the present chapter, section 2 gives introductory comments on multiple-use planning and the issues raised by past and present planning practices. Section 3 provides a general discussion of the means for representing benefits from forest outputs and services. More specific details on measuring such benefits can be found throughout the later chapters of the book. Section 4 is a brief comment on the production function approach to representing the multiple-use forest. An alternative approach, using the cost function, proves

*Portions of this chapter have been adapted, with permission, from M. D. Bowes and J. V. Krutilla, "Multiple-Use Forestry and the Economics of the Multiproduct Enterprise," in V. K. Smith, ed., *Advances in Applied Micro-Economics*, vol. 2 (Greenwich, Conn., JAI Press, 1982).

in many ways more convenient in representing the economically relevant features of the multiple-use forest. In section 5 this cost function approach is developed and various characteristics of multiple-use production are described. In section 6 we provide illustrations designed to highlight a few important features of production costs special to the multiple-use forest. In section 7 economically efficient land-use solutions are described, presented in terms of the benefit and cost characteristics developed in this chapter.

2. Multiple-Use Forest Management and Planning

The Multiple-Use and Sustained-Yield Act of 1960 provides a definition of multiple-use management as the judicious use of land so as to best meet the needs of the American people. Such management, it is stated, may result in the production of either some or all of the resources and services on each particular area of land. The chosen combination of uses need not provide the greatest possible financial return or provide the greatest unit output. However, such management should reflect consideration of the relative values of the multiple-use outputs. It would seem that this legislative definition of multiple-use management is sufficiently broad as to encompass any number of approaches to management.

In our discussion, multiple-use management is to be viewed as the purposeful choice among the set of possible land-use activities. Economic multiple-use management further requires that land-use activities be selected so that each area of land is maintained to the extent possible in its highest-valued use, whether that use is single-purpose or multiple-purpose. That is, under economic multiple-use management, land should be treated over time with the sequence of activities that is expected to provide the greatest discounted net present value from the resulting flow of goods and services. Not simply a financial criterion, economic management requires that the value of the nonmarketed amenity services be considered along with that value attributable to the marketed commodities and services.

Dominant-Use Management

In past practice, multiple-use management often meant dominant-use management. Under such management, land areas are allocated to a primary purpose and the overall pattern of land use across the forest provides for the multiple uses. In theory, the allocation might begin with the classification of lands according to their timber productivity, on the one hand, and their recreational suitability, on the other. Areas of adequate timber productivity and low recreation value might be assigned to active timber management. Areas of obvious scenic value and little commodity value might be set aside for recreational use. Within a dominant-use area, other joint services might be considered to the extent that they add to or could be accommodated without significant interference with the primary management function. In practice, land allocations seem to have been largely based on the good judgment of the forest manager.

While a dominant-use allocation is not necessarily inconsistent with economic multiple-use management, economics has not been the basis for the acceptance of such procedures. Rather, the dominant-use allocation of lands is accepted, to some degree even today, because it has the appeal of simplicity. For the forest managers, the approach may seem desirable because it relies to a great extent on their judgment and knowledge and seems to promise relief from the many complications of planning with full consideration for jointness in production. Further, with the Forest Service staff and budgets organized by functional skills, the dominant-use approach to management tends to be the path of least internal resistance. For the users of forest services and outputs, dominant-use allocations probably seem to offer protection from uncertainty—better perhaps to accept a rigid allocation of land to specific purposes than to face the outcome of a poorly understood and feared-to-be easily manipulated process of integrated multiple-use planning.

Not surprisingly though, the dominant-use approach to land allocation has also met with much skepticism. As rising demands increase the pressure on the supply potential of the public forestlands, more creative approaches to management seem likely to become necessary. While some outputs, such as wilderness recreation and timber production, might be mutually exclusive and necessarily provided separately, other products are joint and either cannot or need not be separated in production. The provision of watershed and wildlife services is not precluded by active timber management. With artful management of the timing and location of harvests and roading activity, it might even be true that considerable recreation could be provided in areas under timber management. The relegation of one such joint service to a secondary role in the choice of land management practices can result in a loss in the overall value from the forest outputs. Dominant-use management implies the systematic exclusion from consideration of many production possibilities.

Perhaps equally troubling, the basis for the allocation for lands under a dominant-use regime remains unclear. There is no easy means for ranking lands by best use, without undertaking the same kind of comprehensive economic analysis that most seek to avoid in calling for the dominant-use approach to management. Otherwise, conflicting rankings seem inevitable, whether for the lands with high potential value in both recreation and timber, or for the wide areas of land that are of low value in all uses. It is inevitable that there will be concern over accountability for the forest manager's judgment, if that is the primary basis for allocating lands. Indeed, in the past there was certainly great concern among environmental groups that dominant use would mean, increasingly, commodity use. Whether or not this concern was justified, it is clear that by the 1960s a simple dominant-use approach to land management planning was becoming untenable.

The Transition to Comprehensive Planning

The 1960s, on until as late as the mid 1970s, were transition years for forest planning. This was a period of piecemeal attempts to modernize forest planning.[1] Dominant use and "good judgment" slowly merged with computerized and comprehensive planning. In this period, despite often-stated Forest Service concerns for the coordination of resource management, the functionally separated planning and budgeting processes in the Forest Service tended to promote a continued dominant-use view of land management. There was further a lack of balance in the functional (timber, recreation, and wildlife) plans that gave at least the perception of bias toward timber management. Among the functional plans, typically only the timber plan would address the entire national forest. Neither timber plans nor other functional plans (if prepared) addressed in a comprehensive fashion the basis for land-use allocations across the forest. Even the so-called multiple-use plans of the 1960s, prepared by national forests as supplements to the functional plans, tended to focus on general policies for timber harvests.

The National Environmental Policy Act (NEPA) of 1969 formalized a requirement for accountability in the public forestland allocation decision. Following NEPA, the forests began to prepare "unit plans" for localized areas of the forest. The unit plans represented perhaps the first broad-scale attempt to systematically evaluate local options for multiple-use forest management. They amounted to local land-use zoning.

Under the unit plans, forest lands were categorized into a variety of types. Some of the land might be classified as general timber lands. Other areas might be classified as special zones, for example as recreation units, critical wildlife zones, riparian zones, or transportation corridors. For such special protection areas, standards would be developed to ensure that any timber activity and other management practices were in conformity with the goal of protecting or promoting the relevant multiple-use services. For example, a watershed that was determined to be of scenic interest might be protected against degradation by the imposition of restrictive standards on timber harvesting practices within the area.

Unit plans were developed in addition to separate functional plans. In practice they served an unclear role as supplements to the functional plans. The link to the forestwide timber plans in particular was unclear. Each unit plan was developed at the local level, with little immediate attention to consistency with goals for a forestwide pattern of land use or for the overall supply of timber and services.

Certainly by the 1970s the resulting inconsistencies in the planning process were a source of great concern to the Forest Service. The history of poor management decisions and the resulting rise in public concern are evidence of the overall failure of this period of piecemeal planning. The acrimonious debates in the late 1960s over the clear-cutting of timber on the Monongahela and Bitteroot national forests provide the most striking examples of the rising public concern with multiple-use forest management and planning. As detailed in the Forest Service's own task-force review of the Monongahela situation,[2] multiple-use management was proceeding with little sense of the available inventory of forest amenities, with almost no awareness of the demands for multiple-use services, and with no comprehensive approach to planning. To be fair, the focus on timber management in the 1960s was perhaps a reflection of the sophisticated (but badly flawed) computerized models that were then becoming available to aid in the scheduling of timber harvests. There was little knowledge then of the possibilities for managing the forest to promote the joint production of multiple-use purposes, nor were there practical models to support integrated multiple-use planning.

Comprehensive Planning Under NFMA

With the passage of the Forest and Rangeland Renewable Resources Planning Act of 1974 (RPA) and then, more significantly, the National Forest Management Act of 1976 (NFMA), multiple-use management entered its modern era. The RPA (also known as the Renewable Resources Planning Act) addressed the need for an assessment of forest inventories and the demands for services of forestlands. The NFMA required an economic evaluation of explicitly documented alternative management strategies for each land area, with a balanced consideration of the potential mix of all goods and services from these lands. Planning under NFMA differs from past practice in two primary ways. First, local unit planning is no longer distinct from forestwide planning. Unit multiple-use goals are now determined with consideration for best meeting the broader goals of the forest as a whole. Second, NFMA established FORPLAN, the linear programming package now used by the national forests to develop their multiple-use plans.[3] FORPLAN has offered the forest manager a tool for choosing among a wide range of multiple-use land-management options for each land area in order to best meet the production goals of the forest. The process for allocating lands does not inherently give priority or dominance to any one product. While NFMA and the pursuant regulations fall somewhat short of mandating economic multiple-use management, it seems fair to say that the required planning process must give considerable attention to economic efficiency in multiple-use management, if only as a benchmark against which to judge the selected management actions.

Planning Needs for Multiple-Use Management

Despite the development of practical tools for multiple-use management, or perhaps because of this development, one must have much sympathy for the forest planner, who faces a difficult task. The initial requirement of any multiple-use planning process is an assessment of the production possibilities. It is apparent, though, that there is still a great lack of knowledge of the production capabilities of the public lands. Not surprisingly, planning has had to proceed using piecemeal evidence combined with the best estimates of the forest planning staff as to the production possibilities. There is an inherent uncertainty attached to natural processes. The long time horizons between regeneration and harvesting of trees make for a particularly difficult problem in projecting the outcome of forest management actions. But apart from this natural uncertainty, it must be recognized that there can be very practical reasons for not attempting to fully investigate the multiple-use production possibilities of many forest and wild lands. There are large areas of land to be managed, and in many cases the potential gain in value from improved knowledge of the management possibilities will not be great. Indeed, simply taking inventory of the current timber stocks and the features of the land can be a costly venture. To further develop information on how timber stocks, the condition of the land, and the related multiple-use services might develop over time in response to a variety of management techniques would require systematic and perhaps unjustifiably expensive experimentation across forest sites over a number of years. Planning with limited knowledge of the production possibilities may be both inevitable and sensible.

The second essential element in a multiple-use planning process is an assessment of relative economic values of, and of the level of demand for, the various resource services of the forest. However, many of the services of the forest are not marketed commodities, and little is known about the economic value of providing these outputs. Although there have been advances in the methodologies for determining demand-based values for non-marketed recreational and amenity services, there remains a great deal of doubt as to our ability to estimate these with accuracy. The difficulties facing the forest planner are perhaps more pronounced than for most other resource managers. The slow growth rate of trees results in long-lived effects of management activity. Consequently, there is a need to determine the demands and relative values for resource services many years into the future—an almost insurmountable task.

A further complication is the scale of the planning problem. A national forest might encompass more than a thousand square miles. For such an area, one could not reasonably expect a detailed evaluation of even the known multiple-use management options. The scale of the planning problem raises other difficulties on the demand side. The demands for services from the many different areas of a forest are likely to be closely linked. Management actions on one unit will potentially alter the use and value of other areas of the forest in a manner that can be troublesome to estimate empirically and equally hard to account for in a linear programming model, or indeed in any practical planning model. Given the difficulties facing the forest planner in adequately representing either the supply potential of the forest or the demands for forest resource services, it is to be expected that any planning effort would have its critics.

In fact, the linear programs used for national forest planning have proved to be expensive both to formulate and to solve. In practice, they often tend to be heavily constrained both directly and implicitly through the type of activities considered. In this manner the solutions can be forced to make "sense," but at the cost of introducing some doubts about the unbiased nature of the planning exercise. In addition, it is as yet unclear whether the structure of the linear programs is adequate to reflect the nontimber services, or whether the solutions to these linear programs can be sensibly implemented. In early applications, the linear programs were often lacking in spatial definition and were too obviously descendant from earlier timber harvest scheduling models (see chapter 10 for a detailed discussion of FOR-PLAN). One thing is clear: the FORPLAN linear programs now used for making forestland allocation decisions are immense and complex. On the other hand, much of the available literature on the economics of multiple-use forestry is oversimplified, providing the student with little basis for evaluating methods used in actual planning.

In this chapter and the next we will try to find a middle ground. The goal is to allow the development of intuition and judgment rather than to provide practical models for multiple-use management. Nevertheless, it is hoped that the presentation is sufficiently rich as to be clearly related to, and useful in the evaluation of, working methods for multiple-use forestland allocation.

3. The Benefits from Multiple-Use Production

We begin with a general benefit-function approach to the value of the multiple-use outputs. Because this general description tends to obscure many underlying features of interest, a somewhat more specific discussion of the economic benefits from forest outputs will follow. Our focus in that later discussion will be on two sets of outputs, the timber harvests and the amenity services of the forest land and its standing stock of vegetation. These two sets of goods represent extremes among the multiple-use outputs. While timber is a marketed commodity, sold at a price determined by competitive bidding, the amenity services of the forest are for the most part nonmarketed and made available with no pricing. Further, the amenity services are in some ways public goods in that many people can benefit from the services without significantly depleting their availability for others.

The Benefit Function

Let $B(Q)$ represent the aggregate benefits that would result from the supply of outputs Q, where $Q = (q_1, \ldots, q_n)$ is a list of outputs and services offered from the forest lands. As is usual in studies of economic efficiency, demand-based measures of willingness to pay will be accepted as appropriate and meaningful indicators of economic welfare. In particular, the benefit function $B(Q)$ should be considered to include direct expenditures and consumer's surplus-type measures of value which sum areas under the demand curves for each output, up to the quantity supplied. Our benefit function $B(Q)$ will sometimes be used to represent the aggregate willingness to pay for the outputs of the forest as a whole, with Q then being a list of the outputs from every single subarea of the forest. For convenience, we will also use the expression $B(Q)$ to represent the benefits from smaller forest units, with Q then being the outputs of this area under consideration.

Marginal Benefits.   Marginal benefits are the per-unit value of a very small increase in the provision of an output. The marginal benefits for the ith output or service will be represented by B_i with

$$B_i(Q) \equiv \frac{\partial B}{\partial q_i}(Q)$$

For those outputs supplied to the highest bidders in a competitive marketplace, the marginal benefits correspond to the expected unit price that would be received by the forest for a small increment in supply. For the nonmarketed outputs, the marginal benefit may be thought of as the amount that individuals would be willing to pay for a marginal increase in the product or service supplied. In the case of services with a public-good nature, this should be taken to mean the sum of the amounts that each individual would be willing to pay.

Marginal benefits can often be assumed to be positive given a suitable definition of the outputs (for example, erosion reduction, not erosion), although a positive marginal value cannot be guaranteed. For instance, timber offered on inaccessible forest units might have negative marginal value, with hauling costs exceeding the value of the timber at the processing mill.[4] The forest would have to pay for the removal of such timber. For the amenity services, negative marginal values are also possible. There are two sources

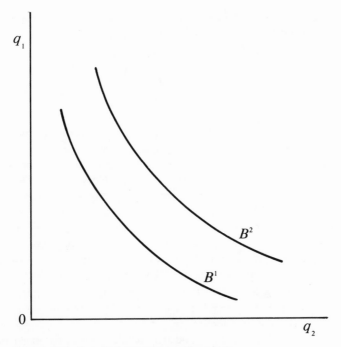

Figure 3-1. Isovalue curves: negative slope

of difficulty here. First, an amenity to some is a disamenity to others. Extensive stands of older trees may make attractive landscapes for general recreation, but may not best provide for grazing or water flow. Second, there are limits to a good thing. While initial reductions in timber stand density might favor game, further cuts could eliminate the habitat, destroying the balance between shelter and feeding.

Isovalue Curves. Typically there will be many ways in which a forest land unit can be managed that result in the same overall public benefits. Isovalue curves illustrate such constant-value output combinations. An isovalue curve can be represented by the expression $B(Q) = B^i$, where B^i is a given level of benefits. In figure 3-1, an isovalue curve $B(q_1,q_2) = B^1$ has been drawn, illustrating how society might trade product q_1 for q_2 while maintaining the constant level of benefits B^1. In a similar manner, a whole family of curves could be drawn to reflect the tradeoffs at other levels of benefits. The curve B^2 gives one such isovalue curve for a level of benefits greater than B^1. Here, as usually expected, increases in the availability of an output lead to greater benefits.

The slope of an isovalue curve indicates the rate at which one output may be traded for another in consumption while holding the overall level of benefits constant. The slope of isovalue curves can be expressed in terms of a ratio of marginal values. For any pair of outputs i and j, with the other outputs held fixed, the slope of an isocost curve is

$$\frac{dq_i}{dq_j} = \frac{-B_j(Q)}{B_i(Q)}$$

For example, the sacrifice of one unit of good q_1 (worth B_1) would require an extra B_1/B_2 units of q_2 in order to maintain current benefits.

Isovalue curves are usually anticipated to be shaped much like those illustrated in figure 3-1. There, the slope is seen to flatten (slope $-B_2/B_1$ gets less negative) as we move along a curve to higher levels of output q_2. Put another way, these isovalue curves are convex functions with respect to the axes (the curves bow toward the axes). This curvature is expected when the marginal values of output are positive but declining with increased consumption of the output. If the marginal values of both products were positive but constant, unaffected by consumption, then the isovalue curves would be straight lines sloping downward from upper left to lower right. Such linear isovalue curves might represent the benefits from a forest unit that is a small supplier relative to the total market.

In figure 3-2, a pair of more unusual isovalue curves has been drawn, having a positive slope at all output combinations. Such isovalue curves would result if q_1 was an output with a negative marginal value (timber too expensive to remove), while q_2 was any other output with positive marginal value. The isovalue curve labeled B^2 in figure 3-2 represents higher-valued product mixes than those on curve B^1. Here increases in the provision of product q_1 lead to a reduction in benefits. We might jump ahead of ourselves to point out that production of the product q_1 would be economically sensible only if there were also a negative marginal cost to providing q_1. Since a timber sale will always require some significant expenditures, such negative marginal costs could result only if timber harvesting greatly reduced the costs of providing other nontimber outputs.

Economic Efficiency and Income Distribution

In a general sense, the isovalue curves are analogous to indifference curves of individuals. There is, however, an important distinction. Switching between different alternatives on a single isovalue curve will have distributional consequences. That is, a shift in the mix of output provided will inevitably make some groups better off while leaving others worse off. The monetary measurement of economic benefits $B(Q)$ is silent as to the desirability of such distributional effects. More positively stated, the use of our benefit

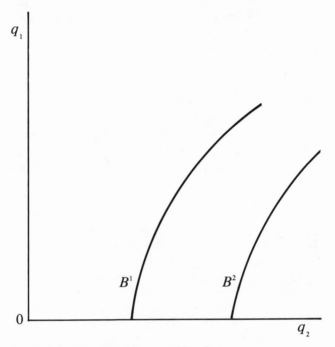

Figure 3-2. Isovalue curves: positive slope

function can be taken to imply an acceptance of the existing distribution of wealth (or of external mechanisms for equitably redistributing wealth), along with an implicit assumption that the forests will not be considering actions that lead to great changes in existing distribution.[5] We will not comment further on the distributional consequences of forest management choices, considering them better dealt with in the political arena within which forest planning proceeds. We focus instead on economic efficiency as a primary benchmark in the management decision.

With this general description of the means of representing the economic benefits from multiple-use supply, it will now be helpful to directly relate the expression $B(Q)$ to the demands for the timber and amenity services of the forestland.

The Benefits from Timber Harvests

The demand for timber from a forest is derived from the value of wood as an input in the production of final products. The mill price, the price per unit volume for wood delivered as logs to a lumber mill, will depend upon the overall volume and quality of timber stumpage offered for sale and on the price available for processed timber. The latter price in turn depends upon the strength of the demand for final wood products, and on the volume of stumpage sold from this forest and others in the market area.[6] The price that would be offered per unit of volume for harvested logs on site will equal the mill price, less an amount to reflect the costs to be incurred in hauling.

Timber Benefits. The benefits from an overall offering of harvested logs can be measured as the area under the mill-price demand function up to the harvest volume supplied, minus the costs that would be incurred in hauling the cut logs. It should be noted that the actual costs of harvesting, although in fact often borne by the processor, are better treated as if incurred by the forest.

Let $H = (h^1, h^2, \ldots)$ represent a list of the sales volumes from each subarea of the forest, let $T = \Sigma h^i$ be the total harvest volume from the forest, and let k^i represent the cost per unit volume of hauling logs from the ith forest subarea. If $P = P(T)$ is the mill-price demand function, giving the mill price P that would result from the harvest volume T, then the overall benefits from log sales on the forest are given by $\beta(H)$ with

$$\beta(H) \equiv \int_0^T P(x)\,dx - \Sigma(k^i h^i)$$

Function $\beta(H)$ provides us with the economic willingness-to-pay measure of the benefits that would result if harvested volume H were to be offered across the various units of the forest. This timber component of the overall benefits from forest production is illustrated in figure 3-3. The shaded area is the benefit from an offering of total amount T of harvested logs. The area below the stepped line is the purchasers' cost incurred in hauling these logs from the various forest units.

It is sometimes mistakenly felt that focusing on the demands of the immediate purchaser of logs results in ignoring some economic value generated in the final products market. In fact, if we accept the assumption of a competitive timber-products sector, the measure of benefits above can be shown to appropriately reflect all changes in net income and consumers' willingness to pay throughout the timber sector that result from the timber offerings.

Marginal Value of Timber. The marginal value of timber, the value of a small increment in timber harvest sold, will depend upon where the timber is taken within the forest. The marginal value of logs cut from one forest unit can be expressed as a slope of the timber benefit function $\beta(H)$:

$$p^i(H) \equiv \frac{\partial \beta}{\partial h^i}(H) = P(T) - k^i$$

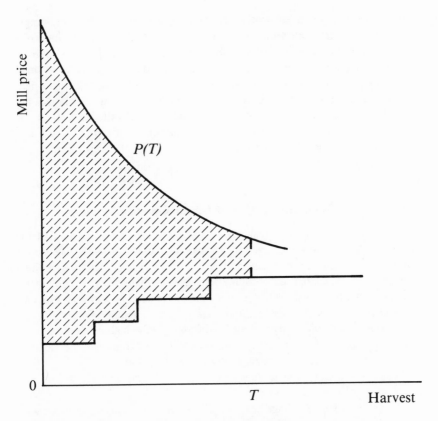

Figure 3-3. Measure of benefits from stumpage offering T

This defines the on-site offer price for cut logs from any given forest unit as equal to the mill price P less k^i, the local costs of hauling.

The marginal value functions $p^i(H)$ can also be thought of as giving the site demand functions, the demand functions for logs from the ith forest area. This is useful because our primary concern often will be with the benefits of timber sales from a particular forest area, rather than from the forest as a whole. To focus on the harvest from one forest unit, we will write a site demand function as $p(h)$, with h the stumpage sale from the unit. The other (now suppressed) elements of H should be considered to be fixed at anticipated sales levels. The benefits from a harvest offered on this one forest unit can be calculated as the area under the site demand function up to the volume offered. The benefits $\beta(h)$ are represented by

$$\beta(h) = \int_0^h p(x)dx$$

It is quite likely that the willingness-to-pay measure of value for a single unit $\beta(h)$ will differ little from its first order approximation, the revenues $p \cdot h$ received from the sale of harvested logs from the forest unit.[7] The approximation is appropriate because the volume of timber offered from a small forest area is unlikely to have any significant effect upon the log price. Indeed, for some national forests as a whole, the changes in supply that are under serious consideration can be expected to have little impact on the received prices. In such a situation, the timber benefits from the forest as a whole would be closely approximated by the sum of the log sale revenues from each forest unit.

Benefits from Recreational Use of Forest Amenity Services

It is the condition of the forest stocks and facilities from each unit of the forest that we consider to be the produced amenity outputs. The forest conditions, or amenity services of the forest, are valued for their indirect service as inputs into final consumption. It is perhaps obvious, but worth stressing, that the forest itself does not produce recreation; rather, it provides the scenic qualities and the facilities that attract recreation. Forest amenity conditions will be variously valued by different individuals. The forest conditions will play a role in determining both wildlife habitat and the value of the forestland for livestock grazing. The amenity conditions of the forest will influence the quality and quantity of water flow and also the nature and distribution of recreation use across the forest. They may have direct effects on the value of land holdings near the forest. Indeed, some forest lands with especially attractive or unique features may have value to individuals who neither use nor expect to visit the forest. It is the sum of these values, over all the relevant parties, that determines the benefit from the forest amenity services.

Typically, many of the demands for forest amenity conditions will not be revealed directly in market transactions. As a result, determining the values of services can be a formidable task. But even if hard to measure, there can be little doubt that there are economic benefits from the nonmarketed services. It seems reasonable to expect that people would consider themselves to be less well off were the amenity services of the forests reduced, and that they would be willing to pay to avoid such an occurrence. In the case of recreation, the measurable increases in amenity value resulting from management actions will be a combination of the very real cost savings that result from improved access to desirable forest lands plus the more elusive benefits associated with the improved quality of forest conditions.

Recreation Benefits.   The forest manager, through decisions as to the timing, location, and nature of timber harvests and the development of facilities, alters the condition of the forest lands and the resulting recreational services (a topic treated in more detail in chapters 6 and 7). Broad changes in the local forest condition, as from the building of a road into a presently un-roaded area, obviously can significantly alter the recreational opportunities provided. Less dramatic changes also have an effect, altering the quality of service and the level and distribution of recreational use across the forest.

It often proves possible to estimate the value of changes in amenity serv-ices through observations on closely related economic activity.[8] One conven-ient method for valuing changes in the recreational services is to observe how changes in the forest alter the demand for recreational travel. Increased willingness to pay for recreational travel to improved forest lands may provide an adequate proxy for the value of the underlying improvement in conditions. To illustrate, we will focus on the demand for the use of a single forest unit, neglecting the complexities that result when other sites must be considered.

Suppose that $V_j = V_j (k_j, z)$ expresses the jth individual's demand for visits to a forest site, a function of k_j the per-visit cost (travel cost), and z the conditions of the unit. For this individual, the benefits from visits to the forest site are found as the area under the trip demand curve up to V the visits demanded, minus the necessary cost of these visits (excepting any entry fees). This willingness-to-pay measure of value, for one individual, is illustrated by the shaded area in figure 3-4.

The total value of the visits to the forest site is given by the sum of each individual's benefits or willingness to pay for the site. Now, this total value might reasonably be attributed to the facilities and condition of the forest area. However, it will often be desirable to focus on the net increase in benefits of the managed forest as compared to the benefits from some base level of service z^0 that might have resulted even without management. To accommodate this focus, we can measure benefits by $A(z)$, where

$$A(z) = \sum_j \left[\int_{k_j}^\infty V(x,z)dx - \int_{k_j}^\infty V(x,z^0)dx \right]$$

The recreational benefits resulting from management are found by subtract-ing the gross recreational benefits of the forest in its base state z^0 from the benefits of the forest in the new condition z. In figure 3-5, the change in one individual's benefits from an improvement in the site amenity services from z^0 to z is illustrated by the shaded area.

Marginal Value of Recreational Amenities. Marginal values of changes in amenity services of the forest unit can be found by measuring the resulting change in individuals' willingness to pay for the recreation visits. The mar-ginal willingness to pay for a change in amenity service, represented by $\alpha(z)$,

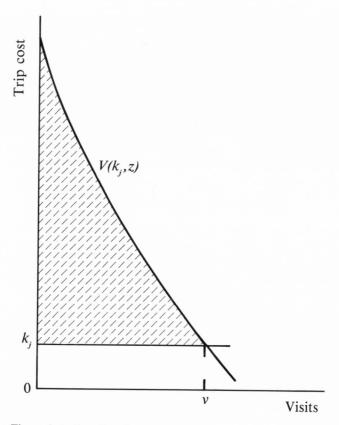

Figure 3-4. Benefits of recreation visits

is the derivative or slope of the recreation benefit function $A(z)$ with respect to an element of the forest condition:

$$\alpha(z) = \frac{\partial A}{\partial z}(z) \equiv \sum_j \int_{k_j}^{\infty} \frac{\partial V}{\partial z}(x,z)\ dx$$

This marginal value should be thought of as the sum of the prices each individual might be willing to pay for a small improvement in site conditions. For a single individual, this "price" is the area between two travel demand curves, much like that area in figure 3-5 which gives the value of a larger, incremental change in forest conditions.

One occasionally hears it stated that the marginal value of an additional visit to a forest recreation site is zero. This in some sense is true—although irrelevant! It is irrelevant because we are not trying to value additional visits

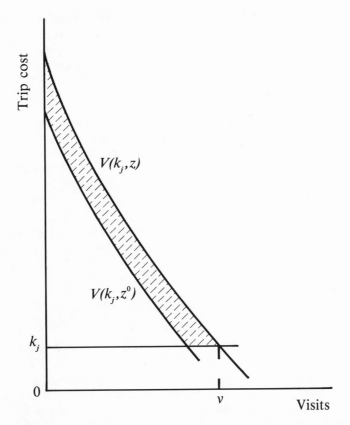

Figure 3-5. Benefits of changed forest conditions

to the forest in its current condition. It is true that, with no site-capacity constraints, individuals can be expected to use sites until the value of an additional visit just equals the cost of the visit. If no entry fee is received by the forest, and if the site is unchanged in quality, there would be no benefit gained from an individual taking an additional trip. Indeed, there is no reason for the individual to take an extra trip. However, marginal changes in the amenity services of a forest site (and the extra visits that result from such changes) will have value, as measured by $\alpha(z)$ above. As an example, consider what would happen if an area of the forest were to be developed as a new recreation facility considerably closer to major population centers than existing facilities. Not the least of the benefits would be the saving in travel costs by those who switch to use of the closer site. Further benefits would then be associated with the increased use that is likely to result from the more convenient location of the new facilities.

The easy approximation of value by revenues that applied in the case of the marketed commodity timber simply does not apply in the case of recreation. By no means should this be taken to suggest that incompatible measures of value are being used for these two services. In both cases, the consumer surplus measure of value is appropriate. For timber, this measure happens to be often closely approximated by market revenues.

Several Forest Sites. When there are several forest areas offering recreational opportunities, the demands for use of one forest area are likely to depend in a complex manner on the management and the access to other areas of the forest. In later chapters we will talk more about measuring the overall recreational benefits of forest management actions, accounting for this interrelation. For now, we will represent recreational value from the forest as a whole by $A(Z)$, where $Z = (z^1, z^2, \ldots, z^n)$, with z^i giving the condition of the ith area. This value will also depend on the trip costs $K = (k_1^1, k_2^1, \ldots; k_1^n, \ldots)$, with k_j^i being the trip cost for individual j to use the ith land area.

The Combined Benefits of Timber and Amenities

The benefit function $B(Q)$ can be taken to represent the sum of the benefit functions for all relevant outputs. For example, with timber and recreation from the whole forest considered, $B(Q)$ would then represent the sum of willingness-to-pay benefits $A(Z) + \beta(H)$, with harvests H and site conditions Z making up the outputs represented in general form by Q. Let us now turn to a description of multiple-use production and the costs of providing such goods and services.

4. Describing Multiple-Use Production: The Production Function

The U.S. Forest Service is by no means alone among government-owned or regulated industries in providing multiple services from single facilities. Water development projects provide irrigation, flood control, and recreation to various localities, with seasonal fluctuations in the quality of these services; the Postal Service provides parcel post, first-class delivery, and magazine delivery from its offices, and is faced with the further choice of the size and location of these offices; and telephone companies provide several communications and information services.

Recent literature in the econometrics and theory of the firm has recognized the importance of the multiple-output production problem.[9] Such processes have become of interest partly because of their importance for regulatory management and partly because of the fairly recent development of suitable frameworks for empirical analysis. These frameworks are based on the dual relationships between cost functions and production functions.[10] The focus in economic studies of multiple-output, joint production problems has fallen on some areas also of particular concern to the forestry student: the optimal scale and scope of production, and the desirability and sustainability of joint production by one enterprise rather than by a set of specialized enterprises. Essential to an understanding of the analysis and modeling of such problems is a background in the basic economics of the multiproduct firm. What follows is a fairly elementary discussion of the primary means for characterizing the features of multiple-output production processes through the production function.[11]

The Production Function Defined

As for any multiproduct enterprise, the relationship between inputs and outputs for a multiple-use forest can be summarized through a multiple-output production function. This production function can be written in general form as $T(Q,X) = 0$, where Q again represents the list of production levels for all outputs and services and X lists the amounts of each input used in production. The production function provides a convenient means for representing the set of efficient input-output combinations.

The expression $T(Q,X)$ can always be standardized so that infeasible production combinations have a value $T(Q,X)$ less than zero. The production possibility set, defined as the set of all feasible input-output mixes, may then be represented as those input-output combinations for which $T(Q,X) \geq 0$. We reserve the term "production function" to refer only to the combinations of inputs and outputs for which $T(Q,X)$ just equals zero. This is the technically efficient frontier of the production possibility set. Inefficient, but otherwise feasible, production mixes within the production possibility set would have a value $T(Q,X)$ greater than zero. Technical efficiency in production is defined by two conditions, both of which must be true. Given a set of inputs X and outputs Q, production is technically efficient if: (a) with the inputs X, the forest can increase production of any one output only by reducing the production of others; and (b) it is not possible to produce the output Q using less of any one input except with corresponding increases in the use of other inputs.

Joint and Nonjoint Production

In a special case of multiple-output production, each unit of input can be uniquely associated with a particular output. That is the case of nonjoint production. Nonjoint production implies that the production technology could be equivalently represented by independent production functions, one for each output. Joint production, on the other hand, means that technology cannot be represented by a set of independent production functions. In effect, joint production means that some inputs are shared, having an effect on the production of more than one output. For example, the feeding of cattle provides both hides and beef. These are joint products. On the other hand, inputs applied to a farmer's soybeans have no influence on the output of corn from his other fields. These are probably nonjoint products—even if both are produced on the same farm.

It might be noted that the general form of the production function $T(Q,X)$ is capable of representing most cases of either joint or nonjoint production. To be interesting, jointness in production should be taken to mean a less extreme product linkage than the classic examples of joint products that are necessarily provided in fixed proportions. Joint production is most usually consistent with the possibility of production in varying proportions.

Observed Production Data and the Production Function

Although there are many useful characterizations of the features of multiple-output production functions, it is easier to use the cost function to describe such features. There is however one aspect of the production function approach that warrants comment. Often some confusion arises in relating the idea of a production function to the description of multiple-use management practices found in the forestry literature.

In practice, we might never explicitly be presented with the overall production function. Instead, our knowledge of multiple-use production develops from on-the-ground experience with a limited number of specific management practices. For example, one researcher might investigate different spatial patterns of timber harvesting, identifying their effects on the production of timber and water flow. Another might consider the effect of timber management practices on the volume of herbage available for livestock or wildlife. Studies such as these provide us with partial production relations, or at best a few elements of the production possibility set. A production function should be thought of as a convenient way to describe the most effective combinations of all such management practices, whether for the forest as a whole or for particular units of the forest.

While the explicit production function might never be developed, the existence of such a set of production possibilities is certainly at the bottom of any multiple-use planning exercise. One increasingly common planning technique is to represent the production technology for a forestland unit by a discrete set of multiple-use production activities which have been developed on the basis of research studies, judgment, and experience. These activities then become elements of choice in a mathematical programming problem (as in the linear program FORPLAN). The planning problem is to choose the appropriate forest production activities for each of the several management units of land into which a forest has been divided. The ability to form combinations of activities or to switch among activities allows flexibility in the overall production mix from the land unit. A production function can in this case be thought of as representing all potential technically efficient solutions to the mathematical programming problem.

A production function may then be simply a shorthand representation of the huge amount of data embedded in a forest management linear program. In a study of the petroleum industry, Griffin demonstrates one method of deriving a multiple-output production function from a linear program.[12] Griffin takes a linear programming model to be a representation of the true multiproduct refining technology. His procedure for developing a production function requires solving the linear program repeatedly, systematically varying output and input prices, each time generating a different technically efficient solution. These solutions for inputs and outputs are then used as data in a statistical procedure to estimate an approximation to the underlying production function.

A similar approach might be appropriate for the national forests, given the widespread availability of linear programming process models for forest planning. An explicitly estimated production function can allow us to easily determine features of the multiple-use forest technology that cannot be apparent from even the most patient inspection of the individual activities.

5. Describing Multiple-Use Production: The Cost Function

Despite the generality of the production function, the cost function—an alternative description of technology—often proves much more useful. It is quite clear that, given input prices, the nature of the production relationship between inputs and outputs determines costs, as long as inputs are purchased according to some identifiable strategy such as cost minimization. Not as obvious is the dual relationship. Full knowledge of the costs of production implies knowledge of all economically relevant characteristics of the underlying production function. The detailed clarification of this and similar dual relationships has been an important recent contribution to microeconomics. The cost function proves more useful than the production function in part because it incorporates a great deal of information on optimizing behavior with respect to the purpose of inputs. The cost function allows us to avoid the usually somewhat arbitrary listing of inputs, focusing instead on the more clearly measurable costs. By doing so, it provides a concise tool to highlight issues in multiproduct production. Indeed, many features of the technology turn out to be most naturally described in terms of costs.

Many basic cost measures of interest in investigating production plans and pricing policy are readily computed from the cost function. These would include measures of marginal costs and incremental costs, well known from study of the single-product firm. Other features of multiproduct costs also have direct analogies in the single-product case, but now prove to have rather different implications. These would include measures related to economies of scale and average costs of production. Further features of cost are specific to the multiple-output case, including the measures of product jointness or complementarity. They are important elements in the decision as to whether specialized production or combined production on a single planning area is most desirable.

The Cost Function

The multiple-output cost function $C(Q;R,M)$ indicates the least cost of producing any mix of output Q. These costs will depend upon R, a list of prices for the variable inputs, and on M, a list of the fixed inputs. The cost function should be thought of as representing the sum of variable input costs $\Sigma_i[r_i x_i^*(Q; R,M)]$, where r_i is the price of the ith input and $x_i^*(Q; R,M)$ is the corresponding input demand function.

In order to focus attention on the multiple-use outputs, we will from now on represent the cost function by $C(Q)$, suppressing the notation for both input prices and fixed factors of production. It should always be understood that any changes in prices or fixed inputs would alter costs. For convenience, we will use the same notation $C(Q)$ when referring to either the cost function of an individual forest planning area or the cost function of the forest as a whole. Our primary focus will be on the cost function for the smaller forest unit.

A word of caution is due. The characteristics of the cost function depend critically upon the cost-minimization assumption. When observed costs do not reflect cost-efficient behavior, it would prove most difficult, unless the actual input purchasing strategy can be identified, either to estimate a true cost function or to interpret that function which might be estimated using the observable cost data. Nevertheless, an underlying true-cost function representing the least costs of production still exists and serves for our expository purposes.

The Isocost Curves.   Isocost curves represent those output mixes that, when produced by least-cost means, are attainable for the same level of expenditure. The isocost curves are contour curves of the cost function. They will be represented by $C(Q) = C^i$, where C^i is any given constant level of cost.[13] The isocost curves provide a convenient means of showing the set of cost-efficient production possibilities. They are of particular relevance if management is in fact constrained to production choices at a fixed budget level. Even without such a rigid budget constraint, the development of a display of several equal-cost management options at each of various candidate budget levels can be of great help to decision makers.

Marginal Costs.   Marginal cost is the per-unit cost of a very small increase in the production of one output, holding other outputs fixed. The marginal cost of the *j*th output, which will be represented by $C_j(Q)$, is found by taking the corresponding partial derivative of the cost function:

$$C_j(Q) \equiv \frac{\partial C}{\partial q_j}(Q)$$

We will also be interested in the effect of changes in the production mix on the level of marginal costs. The second-order partial derivatives of the cost function

$$C_{ij}(Q) \equiv \frac{\partial^2 C(Q)}{\partial q_i \, \partial q_j} \qquad \textit{for all } i, j$$

identify how the marginal cost of product *i* is altered by a small increase in an output *j*.

It would usually be anticipated, in the forestry case, that marginal costs will not decrease with increased production of the same good. That is, we expect the second-order direct-partial derivatives $C_{ij}(Q)$, for $i=j$, to have a value greater than or equal to zero. On the other hand, the second cross-partial derivatives $C_{ij}(Q)$, for $i \neq j$, may variously be positive or negative in value. The sign of the cross-partial derivatives is a measure of compatibility between products, as we shall describe.

It should be emphasized that in the multiproduct case, marginal costs depend upon the production levels of other outputs (including, in the more general case, future levels of output). This interdependence of costs is the basic element of joint production. It is also important to understand that marginal cost is defined for an adjustment in one output alone, with other outputs held at fixed production levels. Implicit in the definition of marginal cost is the adjustment of input use in a cost-minimizing manner as needed to increase the production of just the single output.

There is a temptation to take a single-product view of marginal cost. One might think that the marginal cost of timber could be found as the immediate expense of those activities directly required for the sale and harvest of a unit of timber. This would be correct under single-product timber management. However, under multiple-use management, the effect of an increase in timber sale activity on the production of other services must be accounted for. The marginal cost of timber must reflect the costs of a package of management practices best designed to provide the extra sale volume and to offset the effect of the harvest on other present and future outputs.

Often the true marginal cost of timber harvests is higher than might be anticipated under the single-product view. It is higher when it must include some expenditures on mitigatory activity. In some cases, though, if increased harvesting tends to promote other multiple-use services, the marginal cost of timber harvesting can actually be lower than anticipated under the single-product view. It is even possible for marginal costs to be negative, with the direct cost of an increase in the timber harvest offset by cost savings in the provision of other services. In general, we may conclude that marginal costs will not be immediately apparent from inspection of the individual production activities, if there is jointness in production.

Incremental and Separable Costs.   Incremental cost, as used here, is the least cost of making any nonmarginal adjustments in the set of outputs. The incremental cost of the change in output from a mix Q to some alternative output mix Q' is defined as the cost difference $C(Q') - C(Q)$. It should be emphasized again that in the multiproduct case these costs will be affected by the chosen level of production for other outputs.

One special case of incremental cost is the separable cost of an output. The separable cost of an output can be defined as the incremental cost associated with including that output in an overall production mix. Let $C(Q)$ be the least cost of providing a chosen mix of outputs Q, and let $C(Q^{\{j\}})$ be the least cost of providing the same levels of all outputs except q_j, with output q_j at a level, perhaps zero, which is incidental to least cost production of the other goods. Then

$$SC(q_j) = C(Q) - C(Q^{\{j\}})$$

defines the separable cost of output q_j.

Separable costs have taken on importance because of the role they play in many cost allocation schemes. As a convenience in the budget process, it may be that certain costs are allocated to specific outputs. A common error is the identification of these budget cost allocations with separable costs. In fact, budget allocations, no matter how logical, usually have little to do with separable costs.

Suppose that by modifying timber harvesting practices, a forest provides improved wildlife habitat but in doing so makes it more costly to harvest a given volume of timber. Under current budget procedures, the full cost of all timber harvesting activity is likely to be charged to timber, and none of these costs would be allocated to wildlife. However, the extra costs of modifying the harvest are separable costs of wildlife, and the decision to make the improvement in habitat should be judged by comparing its value against these separable costs.

Suppose further that with no timber harvesting on this land unit there would be no expenses, and the quality of the wildlife services would equal that provided by the modified harvest program. With the least cost of providing the planned wildlife services at zero, the separable costs of the timber harvesting are the full expenditure under the modified program. And so it should be noted that the separable cost calculation does not uniquely assign input activity cost to individual output. In fact, here separable costs sum to more than total costs, a result of diseconomies to joint production. In general, with jointness in production there can be no clear assignment of input costs. In a discussion of the below-cost timber sales issue in chapter 10, separable costs are treated in greater detail.

Characterizing Jointness in Production

The essential element that makes for true multiple-output production is jointness in production. As we have said, this term is best used to refer to something other than the usual example of products which are necessarily provided in fixed proportions. That case is the extreme limit of jointness in production. We might further note that the fact that production of several outputs occurs together does not necessarily indicate jointness. We have also said that a technology is joint when it cannot be described by a set of independent production functions. This proves not to be the most convenient definition for empirical purposes.[14] When the cost function is known, however, testing for jointness of outputs is easy.

A technology is nonjoint if and only if the overall cost function can be expressed as the sum of independent cost functions, one for each output.[15] That is, the cost function for a nonjoint (in all outputs) technology can be expressed as

$$C(Q) = C^1(q_1) + C^2(q_2) + \ldots + C^n(q_n),$$

where $C^i(q_i)$ is the cost function associated with the ith output. Put another way, if (and only if) there is nonjoint production, then there is an unambiguous assignment of overall costs to individual outputs. On the other hand, if production is joint, it is not possible to express the overall cost function as the sum of individual-product cost functions. So, when production is joint, there can be no unique measure of the cost of producing a single output. Rather, the cost of including one output in the overall product mix will depend upon the level at which the other goods are provided. This is the basis for the commonly stated conclusion that a unique cost allocation is not possible under joint production.

As a direct, and easily testable, implication of the definition of nonjointness, we may say that the ith output is nonjoint with the other outputs if at every output level (and input price level) its marginal cost is unaffected by changes in production levels of the other outputs, with

$$C_{ij}(Q) \equiv \frac{\partial^2 C(Q)}{\partial q_i \, \partial q_j} = 0 \qquad \text{for all } Q \text{ and all } j \neq i$$

On the other hand, a product does exhibit jointness with other outputs if its marginal costs are found to be influenced by changes in the production levels of those other outputs $C_{ij}(Q) \neq 0$, at some Q, for some $i \neq j$.

Is Forest Production Joint? There seems to be little question that there is jointness in production among the outputs of a single forest planning unit. The possibilities for providing both timber harvests and the various amenity services of the forest are related in common to the conditions of the standing stock of timber. Any change in the stock of timber through harvests will directly alter the forest conditions and so affect the opportunity for providing amenity services. Of course, the nature of this jointness among the forest outputs through stock conditions can be complex.

Up to a point, increased timber sale volume may improve habitat for some wildlife, improve the volume of water flow, and even improve scenic vistas. Low levels of harvesting, by improving visibility and habitat, may reduce the cost of providing increased amenities. Further harvesting of stocks might then begin to reduce wildlife habitat, add to stream siltation, and spoil scenic quality. At high levels of harvest, timber and amenity services will almost certainly be competitive. More careful planning of harvest location, rerouting of trails, extra cleanup of harvest debris, and any number of mitigation activities may be required to provide increments of harvest without impairing the amenity services. The possibility of keeping activities spatially separated on a given land area is perhaps present but is eventually limited, and the resulting independence of products cannot continue at very high levels of production. Only products which are independent at all levels of output can be considered nonjoint. It seems likely that there will be few significant forest management activities that do not eventually have some effect in common on several of the multiple uses of the forest.

For the multiple-use forest site, jointness in production can be shown if we accept a stylized representation of the above description of the forest production process. Let us say that the costs of providing each forest output depend both upon some output-specific factors and upon the condition of the forest unit. The ability to achieve a particular forest condition, defined somehow in terms of the mix and type of vegetation, we will suppose to depend primarily upon the current timber harvest policy. Let us express this view in terms of costs. We define output q_1 to be the timber harvest, q_2 through q_n to be the amenity outputs, and we use z to be the indicator of the forest condition. Our hypothesized cost function for a forest area can then be expressed as

$$C(Q) = \operatorname*{Min}_{z} \left[c^1(z,q_1) + \sum_{i=2}^{n} [c^i(q_i;z)] \right]$$

where the first term $c^1(z,q_1)$ represents the cost of achieving a forest condition z while also providing a harvest level q_1. The remaining $n-1$ terms $c^i(q_i;z)$ each indicate the costs of providing some nontimber output q_i, under given forest conditions z. With this cost function, we may show that the cross-partial derivatives $C_{ij}(Q)$ are generally not equal to zero, and therefore that production is joint. This is hardly surprising with the forest condition providing a common link among all outputs. For any pair of outputs i and j:[16]

$$C_{ij}(Q) = \frac{- c^i_{qz} c^j_{qz}}{\Sigma_k c^k_{zz}}$$

where, for convenience, we have treated z as a single-element descriptor of the forest condition. The numerator in the equation is made up of terms which measure the effect of changes in the forest conditions on the marginal cost of producing the individual products. For forest outputs, these terms are unlikely to always be zero, and therefore production is almost certain to be joint.

Complements, Substitutes, and Independent Goods. Closely related to the concept of jointness in production is the characterization of outputs as complements, substitutes, or independent goods. In standard economic usage, these expressions are defined in terms of the effect of an increase in the production of one good on the marginal costs of providing another. In contrast to the definition of jointness, the classification of products as complements or substitutes is a local measure of interaction, meaning that it is specific to the particular level of output. The relation between outputs may well differ with a change in the overall production mix. In contrast, jointness is a global measure. Products are nonjoint only if their costs are unrelated at every possible output level.

Complements (or complementary products) are those output pairs for which an increase in the production of one leads to a reduction in the marginal cost of providing the other. That is, products i and j are complements at a given output mix Q if $C_{ij}(Q) < 0$. Substitutes (or competitive products) are those output pairs for which an increase in the production of one results in higher marginal costs for the other. That is, products i and j are substitutes at output

Q if $C_{ij}(Q) > 0$. Independent outputs are those for which marginal costs are unrelated, with $C_{ij}(Q) = 0$.

A more intuitive understanding of the nature of complementarity can perhaps be gained from the following equivalent definition. Two products are complements if the cost of adding small increments of both outputs together is less than the sum of the costs of adding those increments separately. That may result when adding to the production of one output requires the use of a commonly beneficial input, so that the subsequent cost of adding the other output is reduced.

For example, wildlife habitat and water flow are both likely to be improved by the provision of numerous relatively small openings in the forest cover. A timber sale and a developed recreation site may be able to use in common a newly constructed road spur. For these complementary outputs, because of the commonly beneficial input requirements, the actual cost of a combined increase in all outputs may be much lower than the sum of the costs of separately increasing each output alone.

Now, generally in the case of the forest production system there will be both commonly beneficial practices and other practices which tend to promote one output at the expense of others. The combined effect of practices chosen in the least-cost manner will determine whether goods are complements or substitutes at a particular output mix. It is to be stressed that implicit in the definitions of complements and substitutes is the understanding that changes in production will be made in the least-cost manner. There is a natural tendency to look at the more apparent links between products such as might be present in particular practices or in a single activity of a linear programming model. It is, however, the adjustment in output mixes by efficiently shifting among such practices which is of concern, not the response of outputs to an increase in one particular practice.

Economies or Diseconomies of Jointness.   The classification of products as complements or substitutes is local and may differ with changes in the overall production mix. In many cases, though, the relationship among products will hold over a broad range of possible production mixes. If independence among outputs is global, occurring at every output mix, then production is nonjoint by the earlier definition. When a set of products are global complements—complements over the full range of feasible output mixes—we will find there are economies of jointness (also referred to as economies of scope). That is, combined production of outputs will be cheaper than production of each individual output on a separate (but otherwise identical) land area. Suppose, for example, there were just a pair of outputs q_1 and q_2, and that these were complements over the full range of production; then we would find economies of jointness with

$$C(q_1, q_2) < C(q_1, 0) + C(0, q_2)$$

Conversely, if two outputs are everywhere substitutes, then there will be diseconomies of jointness with

$$C(q_1, q_2) > C(q_1, 0) + C(0, q_2)$$

That is, the cost of producing the two substitute outputs together on one land unit would be greater than the cost of separately producing the outputs on two different land areas. In an extreme case, two products might be called incompatible or mutually exclusive if the cost of combined production is infinite or impractically high in comparison to the costs of separate production.

Do diseconomies of jointness suggest that we should spatially separate the production of outputs which are generally substitutes? Perhaps surprisingly, the answer is not necessarily. It is after all hardly surprising that we could provide the same total product more cheaply on twice as much land area. We will return to this matter shortly.

Average Costs and Scale Economies

Average cost in the multiproduct case turns out to be an ambiguous concept. Certainly we cannot find average costs of production for specific outputs when total costs, which may be joint, cannot be allocated unambiguously. On the other hand, any measure of overall average costs (dividing total cost by total output) is also ambiguous, since any measure of total output must rely upon arbitrary weights to combine the individual outputs. We can, however, sensibly ask the following: By what proportion would costs increase if we proportionately expanded the levels of all outputs? That is, we can usefully talk about relative changes in average costs when outputs are maintained in fixed proportion.

Ray Average Cost. Following Baumol, ray average costs at a given output mix Q can be defined by $RAC(\lambda Q) = C(\lambda Q)/\lambda$, where λ is a factor of proportionality taking a value $\lambda = 1$ at the arbitrary output mix Q.[17] Ray average costs provide a natural generalization of the single-product average cost. The cost measure is related to the profitability of marginal cost pricing and also to economies of scale in the same manner as are average costs under single-output production.

Ray average costs are said to be globally increasing if for all Q and all expansion factors, $\lambda > 1$:

$$C(\lambda Q)/\lambda > C(Q)$$

That is, relative average costs (along a "ray" of expansion) are increasing everywhere if expanding output by a common factor of proportionality always leads to a more than proportionate increase in costs.

Ray average costs are increasing locally at the particular output Q (where $\lambda = 1$) if:

$$\frac{dRAC(\lambda Q)}{d\lambda}\bigg|_{\lambda=1} = \Sigma_i[q_iC_i(Q)] - C(Q) > 0$$

The term on the right side of the equality gives the sum of marginal costs times their respective outputs, minus the overall costs of production. The term can be viewed as measuring the potential profitability from marginal cost pricing. With locally increasing ray average costs, marginal cost pricing of outputs would be profitable. Conversely, with decreasing ray average costs, marginal cost pricing could not be profitable.

Returns to Scale.   The extension of returns to scale from the single-product to the multiproduct case is straightforward. There are different implications to the measure of scale economies, however. In the multiproduct case, increasing returns to scale does not guarantee that production by a single unit is necessarily the least-cost arrangement.

The most useful measure of returns to scale is the proportionate increase in costs arising from a marginally small equiproportionate increase in all outputs. This measure will be represented by S, with

$$S \equiv \frac{d[ln\ C(\lambda Q)]}{d(ln\ \lambda)}\bigg|_{\lambda=1} = \frac{\Sigma_i[q_iC_i(Q)]}{C(Q)}$$

For $S < 1$ there are increasing returns to scale at Q, or economies of scale; for $S > 1$ there are decreasing returns to scale; and for $S = 1$ there are constant returns to scale. Decreasing returns to scale are seen to be associated with the possibility of profitable marginal cost pricing. Note that the degree of scale economies is a local measure (specific to a particular output mix) and may change with adjustments in the output mix. The link between the measure of scale economies and ray average cost should be apparent. Locally decreasing returns to scale are equivalent to locally increasing ray average costs.

It is well known that for the single-product firm economies of scale at every level of output would constitute conditions for a natural monopoly. Under such cost conditions, a single firm can produce more cheaply than can any group of firms attempting to provide the same overall level of output. That is not necessarily true for the multiproduct firm, because of the opportunity for firms to separately specialize in different output mixes. The presence of scale economies for the multiproduct firm guarantees only that any proportionate division of output among separate firms (with each firm providing the same proportionate mix of outputs) would be more costly than production by the single firm.[18]

With regard to the multiple-use forest, a few points should be noted. First, it should be realized that we cannot unambiguously talk about scale economies of timber production alone without consideration of other multiple-use services. Second, there should be a strong presumption of eventual decreasing returns to scale for the multiple-use forest. The reason is that there must be an ultimate limit to the potential for increasing outputs from one given area of land. The assumption that marginal costs are increasing ($C_{ii}(Q) > 0$) is consistent with the belief that there are decreasing returns to scale at all output combinations.

The Slope and Curvature of Isocost Curves

The slope of an isocost curve indicates the technical rate at which we are able to substitute one output for another in production, while holding cost constant (it should be recalled that earlier the slope of isovalue curves was similarly described as indicating the rate at which consumers were willing to substitute outputs). The slope of isocost curves can be defined in terms of a ratio of marginal costs. For any pair of outputs i and j, with the other outputs held fixed, the slope of the isocost curve is

$$\frac{dq_i}{dq_j} = \frac{-C_j(Q)}{C_i(Q)}$$

That is, if we gave up one unit of good j, reducing costs by $C_j(Q)$, we could then increase production of good i by $C_j(Q)/C_i(Q)$ units at a unit cost of $C_i(Q)$ in order to maintain the original expenditure level.

Isocost curves may be variously shaped. They may look like those illustrated in figure 3-6, for example. There the slope of the curve is seen to decrease (steepen, as $-C_2/C_1$ gets more negative) as we move down the curve to higher levels of output q_2. These isovalue curves are concave functions with respect to the axes (the curve bows away from the axes). In figure 3-7, isocost curves convex to the axes have been drawn; the slope is increasing (flattening) as output q_2 is increased.

An isocost curve $C(q_1, q_2) = C^i$ can be shown to be convex (concave) to the axes if $[(C_2/C_1)C_{11} + (C_1/C_2)C_{22} - 2C_{12}]$ is less than (greater than) zero. The concavity or convexity of isocost curves is related to the complementarity or competitiveness of outputs and to the economies of scale, but not simply. Under an assumption of positive and nondecreasing marginal costs ($C_{ii} \geq 0$), the convexity of isocost curves (as in figure 3-7) must reflect an underlying competitiveness of the two outputs. The two outputs must be substitutes, with $C_{ij} > 0$. On the other hand, under the same assumptions, concavity of the isocost curves does not rule out the possibility that the

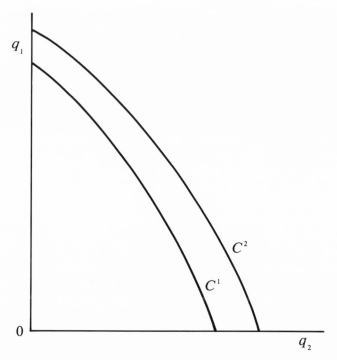

Figure 3-6. Concave isocost curves

products are substitutes. However, if two products are complementary with $C_{ij} < 0$, then, under our assumptions about marginal costs ($C_i > 0$, $C_{ii} \geq 0$), isocost curves will certainly be concave to the axes. Curvature of the isocost curve is a major determinant of whether specialized or integrated production is least costly.

Site Productivity

The best measure of productivity is a comparison of the costs of production. A site is more productive than another if it is capable of producing the same outputs at less cost. Take $C^A(Q)$ to give the costs of producing Q on site A, and $C^B(Q)$ to give the cost of producing Q on site B. The relative productivity of two forest sites A and B in producing Q can then be measured by the ratio of respective production costs, $C^A(Q)/C^B(Q)$. If this cost ratio has a value less than one, then site A is more productive in providing output Q. It should be noted that the productivity of sites may well depend upon the mix and the level of outputs provided.

In figures 3-8 and 3-9, isocost curves for two forest units A and B have been drawn. The curves labeled C^{A1} and C^{B1} should each be taken to

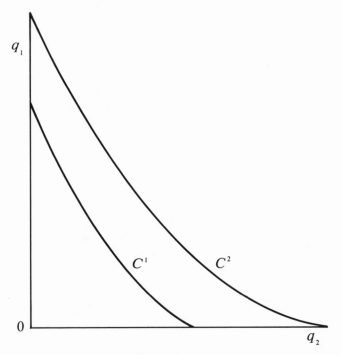

Figure 3-7. Convex isocost curves

represent the same level of costs. Figure 3-8 illustrates a case in which unit A is absolutely more productive, capable of providing any mix of outputs at less cost (or more outputs for the same cost). Such an across-the-board productive advantage is not always to be expected. Figure 3-9 illustrates an alternative case in which site B is more productive than A for q_1 (timber), while A is more productive in q_2 (amenities).

Not surprisingly, differences in productivity must play a primary role in the choice of product mix for each forest site. More precisely, differences in the choice of product mix for two sites will depend on the relative slopes of the isocost curves, and these slopes in turn depend in a simple manner upon the relative productivity of the sites. Notice that in figures 3-8 and 3-9 the relative productivity of site A increases (the cost ratio $C^A(Q)/C^B(Q)$ decreases) as we move along the isocost curve C^{A1} to become more specialized in q_2. If this same observation were to hold for every isocost curve, we could be sure that the isocosts curves for site A have a flatter slope (dq_1/dq_2) at any output mix than do the site B isocost curves. Under such cost conditions, it proves better to have site A relatively more specialized in the production of q_2, while site B is relatively specialized in the production of q_1. Let us look more closely now at the cost-minimizing allocation of production.

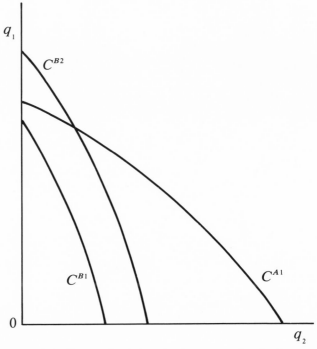

Figure 3-8. Site productivity: site A more productive in q_1 and q_2

Specialization or Integration in Production

Specialization in product mix across sites may be desirable for a variety of reasons. The decision to specialize is favored by diseconomies of jointness, by economies of scale, and by differences in site productivity. Beyond these cost-based factors, variations in the marginal value of outputs across sites will also influence production choices. Variation in product value across sites can often be expected, especially when there are accessibility differences. In the present discussion we will focus on the purely cost-based reasons for specialization or integration in production.

Let us consider how to produce a given overall mix of two outputs $(\bar{q}_1, \bar{q}_2)$ at least cost from the two forest sites A and B. We must find output allocations $Q^A = (q_1^A, q_2^A)$ and $Q^B = (q_1^B, q_2^B)$, which solve the problem[19]

$$\underset{q}{\text{Min}} \ \{C^A(q_1^A, q_2^A) + C^B(q_1^B, q_2^B)\}$$

$$\text{subject to: } q_i^A + q_i^B = \bar{q}_i \qquad \text{for } i = 1 \text{ and } 2$$

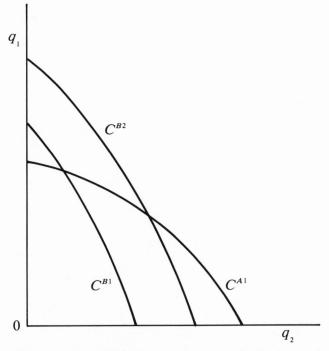

Figure 3-9. Site productivity: site A more productive in q_2 only

The cost-minimizing solution to this problem must meet the following first-order conditions:

1. If the ith output is to be produced on both sites, the marginal cost of production should be the same on each forest site, with

$$C_i^A(Q^A) = C_i^B(Q^B) \qquad \text{if } q_i^A > 0 \text{ and } q_i^B > 0$$

2. If, on the other hand, it is less costly to provide the full amount of the ith output $\bar{q}^i$ on one site, then either

$$C_i^A(Q^A) > C_i^B(Q^B) \qquad \text{if } q_i^A = 0 \quad \text{and } q_i^B = \bar{q}_i; \qquad \text{or}$$
$$C_i^B(Q^B) > C_i^A(Q^A) \qquad \text{if } q_i^B = 0 \quad \text{and } q_i^A = \bar{q}_i$$

The cost-minimization conditions are illustrated using isocost curves. Figure 3-10 illustrates the case where overall output $\bar{q}_1$ and $\bar{q}_2$ is best provided

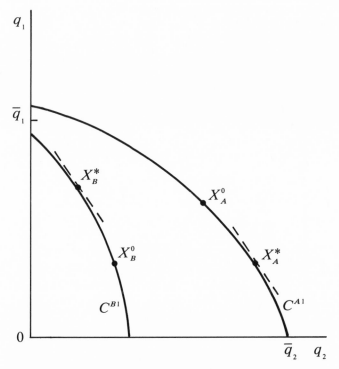

Figure 3-10. Cost-minimizing product choice: both products on both sites

with some production of both goods on each site. We must produce at positions such as X_A^* and X_B^* on the isocost curves where the slopes are identical, with $C_2^A/C_1^A = C_2^B/C_1^B$. Figure 3-11 illustrates another case, now with no output q_2 produced on site B. Here the slope of the isocost curves at the optimal solution satisfies the condition $C_2^A/C_1^A < C_2^B/C_1^B$.

These slope conditions for optimality are fairly easy to understand. Suppose that we had initially chosen to produce the same proportionate mix of outputs on each site, as at X_A^0 and X_B^0 in figure 3-10. This could not be a least-cost means of production. Overall output can be increased at no extra cost by a shift toward more specialized production. On site B we could give up one unit of q_2 and add C_2^B/C_1^B (the slope of the isocost curve) units of q_1, with no change in cost. On site A, we could add the one unit of q_2 while giving up C_2^A/C_1^A units of good q_1, again with no increase in cost. Since the isocost curve for site A is relatively flatter than that for B, the sacrifice C_2^A/C_1^A is less than the gain C_2^B/C_1^B, and overall output of q_1 has been increased with no cost. Further moves toward specialization would prove to be beneficial, up to points X_A^* and X_B^* at which the slopes of the two curves are equal, or (as in figure 3-9) no further specialization is possible.

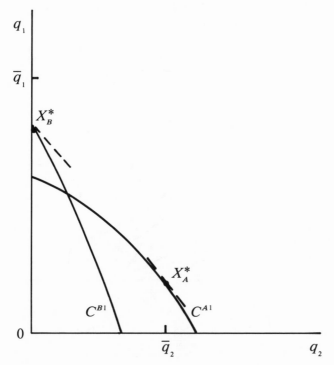

Figure 3-11. Cost-minimizing product choice: no product q_2 on site B

Site Productivity and Specialization. In the two cases illustrated in figures 3-10 and 3-11, the move toward specialization was a result of inherent differences in site productivity. Site B, which has the steeper isocost curves (more negative slope dq_1/dq_2) at any given mix of outputs, is seen to specialize in product q_1, while site A is relatively more specialized in output q_2.

At the bottom of arguments in favor of dominant use is a presumption that it is easy to allocate lands to different purposes on the basis of productivity. For example, it is suggested that we should array lands in order of the costs of producing timber, and allocate the lands high in timber productivity to timber and those low in timber productivity to other uses. Unfortunately, it is not that simple.

Let us suppose that in figure 3-10 the isocost curves C^{A1} and C^{B1} duplicated those from figure 3-8, so that the curves represent equal levels of expenditure. Site A is then more productive than site B in providing timber (q_1), and, for that matter, more productive in providing any output mix. However, the best allocation of land requires that site A, the productive timber site, be relatively less specialized in timber than site B. The point is that a site capable of producing timber at lower cost should not necessarily be more

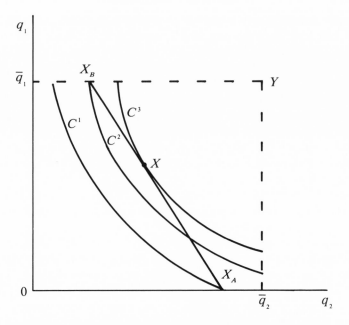

Figure 3-12. Cost-minimizing product choice with convex isocost curves

specialized in timber than other sites. Rather, differences in cost-efficient production emphasis between sites should depend upon differences in the relative slopes of the isocost curves. In both figures 3-8 and 3-9 the isocost curves of site A are seen to be less steeply sloped than are those of site B at any given proportionate output mix. In both cases, site B should provide a product mix proportionately higher in q_1 than should site A.

Convex Isocost Curves and Specialization. Convexity in the isocost curves of individual sites favors specialization across a forest, even in the absence of differences in site productivity. Suppose a forest were made up of two identical land areas, each with the same convex isocost curves. A balanced overall mix of output is then best provided by having each unit of the forest specialize, rather than by producing the same mix on each unit. The convexity in site isocost curves thus proves to be perhaps the most relevant indicator of incompatibility between products.

In figure 3-12 we have labeled an overall output mix $\bar{q}_1$ and $\bar{q}_2$ as Y. The production technologies of two identical sites are represented by isocost curves. Three convex isocost curves C^1, C^2, and C^3 represent the outputs producible at increasingly higher levels of cost. A balanced production mix for the two sites, each site providing $\bar{q}_1/2$ and $\bar{q}_2/2$, is labeled X. It is seen that X is on the highest isocost curve C^3. With this balanced mix X chosen for both sites, overall costs of production would be $2 \cdot C^3$.

In figure 3-12 a line is drawn tangent to isocost curve C^3 through point X. Because of the convexity of the isocost curves, any output mix on this line other than X itself must cost less than C^3 to produce. In fact, the further we move along this line away from X, the lower the costs of production will be. Looking at the points X_A and X_B on this line, it can be seen that together they provide the desired total output $\bar{q}_1$ and $\bar{q}_2$, but they do so at lower overall cost. Overall production costs are reduced to $C^1 + C^2$, which is less than two times C^3. Some such specialized allocation of production targets across sites will always prove cheaper than balanced production on both sites, if isocost curves are convex.

It would be appropriate to add a few comments here, the proofs of which can be easily verified by example, using isocost curves. First, even with convex isocost curves, complete specialization by output might not be the least-cost arrangement. That is, with the overall output $\bar{q}_1$ and $\bar{q}_2$ to be produced from two identical land areas, we should not expect that it would be least costly to provide $\bar{q}_1$ on one unit and $\bar{q}_2$ on the other. Second, concavity of the isocost curves does not guarantee that identical balanced production on each land unit is cheaper than complete specialization.

Substitute Products and Specialization. It is occasionally suggested that if two products are always substitutes, they are best produced on separate land areas. As noted earlier, the fact that two products are substitutes may tend to favor, although not guarantee, the convexity of isocost curves. Some separation of substitute products is appropriate if the extent of the products' competitiveness is sufficient to result in convex isocost curves. More generally, though, the determination of whether combined production or spatial specialization of substitute products is least costly will depend in a complex manner on both the diseconomies of jointness and the diseconomies of scale associated with specialized production.

The reason that there is some confusion is probably attributable to the following result: if we have two products which are substitutes at any output mix, then there are diseconomies of jointness with

$$C(q_1, q_2) > C(q_1, 0) + C(0, q_2)$$

This seems to suggest that costs can be reduced by spatially separating the production of outputs which are generally substitutes. That conclusion is not always correct. The confusion comes from not paying close attention to the land areas under consideration. The inequality really says it is cheaper to provide the overall output separately on *two* identical land units than it is to provide the same total production from just *one* of these land units. It is not surprising that we can produce more cheaply with twice the land area, since no rental charge for land is included in the costs.

Let us consider a more relevant question. Suppose that we have two identical land units and want to produce q_1 and q_2 in total from these two areas. With the two goods substitutes, is it ever cheaper to provide an identical balanced product mix on each unit, rather than to fully specialize in production across sites? That is, will we ever find

$$2 \cdot C(q_1/2, q_2/2) < C(q_1, 0) + C(0, q_2)$$

We want to show that this second cost condition can be consistent with having outputs that are substitutes at all production levels.

Combining the two cost conditions, it can be seen that in order for integrated production of substitute products to be cheaper than fully specialized production, costs must satisfy

$$2 \cdot C(q_1/2, q_2/2) < C(q_1, 0) + C(0, q_2) < C(q_1, q_2)$$

For this to be possible, we must have $2 \cdot C(q_1/2, q_2/2) < C(q_1, q_2)$. In other words, if the two substitute products are to be produced together, there must be increasing ray average costs through any particular output mix q_1 and q_2—there must be decreasing returns to scale. Combined production of substitute products is desirable when diseconomies of scale outweigh the diseconomies of joint production. To understand this conclusion, it must be realized that specialization requires an increase in the scale of production for each individual output. Specialization may lead to increased costs as a result of diseconomies of scale which are sufficient to offset any diseconomies of jointness from combined production.

In many cases there will be opportunities to reduce overall costs by spatially separating the production of substitute products. This would be particularly appropriate for those extreme cases of substitutes, the incompatible products.

Site Costs versus Forestwide Costs. The unit of land for which the cost function is considered affects the classification of products as complements or substitutes. The discussion to this point has largely focused on the more natural description of management costs for individual planning units of the forest. Once the management of larger land areas is considered, there are greater opportunities for spatially (and temporally) organizing production so as to avoid significant diseconomies of jointness. A cost function for the aggregated outputs of the forest would reflect a cost-minimizing arrangement of productive activity across forest units. The separation of incompatible products would be built into such a forestwide cost function. As a result, we should expect to observe few incompatible or strong substitute products in the cost function for the aggregated outputs of the forest. For the same reason, we would not expect to find convex isocost curves for the aggregated outputs of the forest.

One of the difficulties in talking about a forestwide cost function is that the aggregation of outputs across planning units is not always appropriate. At least, aggregation cannot usually be based on knowledge of costs alone. The amenity services of different land units must be to some extent viewed as distinct products. Further, the outputs from different planning units will differ in value, with the value depending upon the accessibility of the site. For these reasons, we will not explicitly consider a forestwide cost function, except as the summation of the cost functions from individual sites. This means that planning units must be formed so that production is reasonably assumed to be nonjoint across units. The output of each site is treated as distinct. Any aggregation of outputs that is appropriate will be reflected in the structure of the forestwide benefit function $B(Q)$.

6. Special Features of Costs in the Multiple-Use Forest

Three characteristics of forest production can give a somewhat unusual structure to the forest-site cost function. First, there is the potential to produce some of the services at no variable cost, the result of natural forest growth and conditions. Second, the provision of one primary output by least-cost means often results in incidental production of other outputs and services. Third, there is a dominant influence on overall costs from timber harvesting practices. The cost associated with harvesting will be a major component of overall costs, and harvest practices have a strong influence on the potential for providing other services. With these features in mind, let us turn to a graphic illustration of the costs of providing timber and wildlife services from a single forest area. It should be noted that although we have tried to include some realistic elements in this example, it remains simplistic. In particular, production is treated as if it takes place within a single time period.

The Forest Site Isocost Curves Illustrated

In figure 3-13, isocost curves have been drawn to represent the costs of providing two outputs q_1 and q_2 from a single forest planning unit. Output q_1 represents the volume of harvested timber offered; output q_2 represents the wildlife services of the land. Isocost curve C^1 shows the various combinations of outputs that are attainable for an expenditure of C^1 dollars. The curves labeled C^2, C^3 and C^4 are similarly defined for increasingly higher

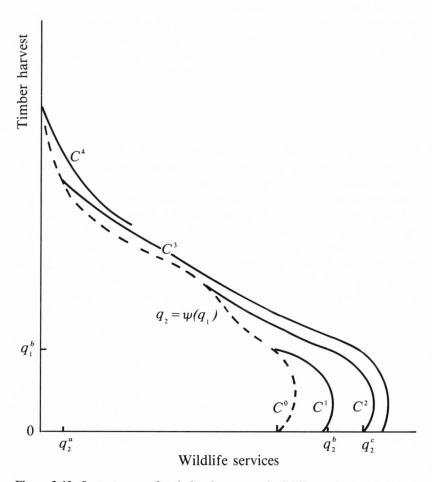

Figure 3-13. Isocost curves for timber harvest and wildlife services

levels of expenditure. The single point marked C^0 represents the wildlife services of the land in its current condition. These services are attained at no cost, with the cost C^0 equal to zero. All cost increments $(C^i - C^{i-1})$ are to be considered equal. Measured costs include logging expenses and other variable costs of on-site management, but not the cost of hauling logs from the site.

The curve $q_2 = \psi(q_1)$, marked by a broken line in figure 3-13, plays a major role in determining the unusual arrangement of the isocost curves. This curve indicates the wildlife service q_2 that would result when this land area is managed to provide timber q_1 by least-cost means. That is, it gives the wildlife services that would be incidental to single-purpose timber management.

Incidental Production of Wildlife. Looking more closely at the curve q_2 = $\psi(q_1)$ in figure 3-13, it can be seen that it shows the incidental wildlife services q_2 that result from the least-cost, single-purpose production of timber output q_1. The curve begins at C^0; here there are no timber harvests and no costs, but there are wildlife services from the land in its unharvested condition. As we move up curve $\psi(q_1)$, the volume of harvest increases, and so do the management costs. In fact, we have assumed that under single-purpose management for timber, costs would increase in proportion to the harvest volume. Initially increased timber harvesting also leads to higher incidental production of wildlife services. In actuality, low levels of harvest do often increase wildlife services, improving the balance of forage relative to protective cover. As we move further up the curve to higher harvest levels, it is seen that incidental production of wildlife services begins to decline rapidly. Finally, at the top of the curve, with 100 percent of the timber stands on the unit harvested, there are seen to be no remaining wildlife services.

The shape of curve q_2 = $\psi(q_1)$ has a dominant influence on the position and shape of the isocost curves. In particular, the economically relevant portion of each isocost curve begins on the curve $\psi(q_1)$. Production combinations to the left of $\psi(q_1)$ are not relevant, since there can be no cost savings from reducing wildlife services below a level incidentally provided under least-cost timber management. From the base level $\psi(q_1)$ improvements in wildlife services (with timber held constant) can be achieved at some cost, either through shifting to alternative, more costly harvest methods or by increasing nontimber activity intended to directly promote wildlife services. Movement along an isocost curve, from curve $\psi(q_1)$ toward the lower right, might reflect a reduction in volume harvested, coupled with a shift to more costly wildlife-promoting activity and harvest practices.

Features of the Isocost Curves. On land areas that have been heavily cut over there is likely to be little wildlife and little impact from expenditures to improve wildlife services above base levels. The isocost curve C^4 is drawn to reflect these limited, high-cost options for improving wildlife services at high harvest levels. The steepness of the curve, almost following the underlying q_2 = $\psi(q_1)$ curve, indicates that at these levels of production we would have to sharply reduce timber harvests in order to allow any increase in wildlife services, if costs are to be held constant.

In contrast, wildlife management might be very effective when timber harvests are moderate. Increasing wildlife services, while holding timber harvest constant, might then require only careful attention to the dispersion and shaping of openings. We see on the lower isocost curve C^1 both a high base level of wildlife service $\psi(q_1)$ and a high impact per dollar from attempts to improve these services. On curve C^1, little timber harvest need initially be sacrificed as we move from single-purpose timber management on $\psi(q_1)$ to add improvements in wildlife service.

What is most unusual in figure 3-13 is the slope of the isocost curves C^1, C^2, and C^3 at the lowest levels of harvesting. The lower segments of these curves slope upward to the right. This slope matches, although to a lesser degree, the curvature in the base-level curve $q_2 = \psi(q_1)$. These isocost curves show that, at this level of production, we can actually increase both timber harvest and wildlife services with no increase in costs. For this to occur, harvesting must be not just a wildlife-promoting activity, as indicated by the slope of $\psi(q_1)$, but must also be cheaper than some other nonharvest activities that could be used to promote wildlife.

Production in the positive-sloping segments of the isocost curves is not economically sensible if timber and wildlife are positively valued. If, in that case, more of both timber and wildlife could be provided at no extra cost, then more should be provided. But it should be remembered that there is a possibility that timber may have a negative marginal value. It can be too expensive to take timber from some remote and poorly stocked areas. On such lands it might make sense to produce in an upward-sloping region of the isocost curves. That is, we might choose to pay to have timber removed, in order to promote other amenity services. The timber harvest might then be better thought of as an input to production, rather than as an output.

Marginal Costs Illustrated

In figure 3-13 the costs of increments to the timber harvest are constant when harvests are increased by least-cost means along the timber management curve $q_2 = \psi(q_1)$. If this were a private timber forest, the land area would be considered to exhibit constant marginal costs of timber production. However, with multiple-use outputs considered, the true marginal costs of timber harvests are not constant, but rather are rising rapidly. The computation of marginal cost in the multiple-use context requires that other outputs be held constant. In figure 3-13, when wildlife services are held constant, the cost of adding increments to the timber harvest can be seen to rise with increased harvesting.

The three marginal cost curves for timber drawn in figure 3-14 have been roughly derived from figure 3-13. Each curve associates a marginal cost on the vertical axis to a level of harvest on the horizontal axis. The curves, labeled $C_1(q_1 | q_2)$, show the marginal cost of timber when the harvest is q_1 and the wildlife service is constant at a level q_2. The three curves are based on the successively higher levels of wildlife q_2^a, q_2^b, and q_2^c marked in figure 3-13. Our marginal costs were approximated by evaluating

$$C_1(q_1 | q_2) \simeq [C(q_1', q_2) - C(q_1, q_2)] / [q_1' - q_1]$$

where $q_1' - q_1$ is a small change in timber harvest level. For comparison,

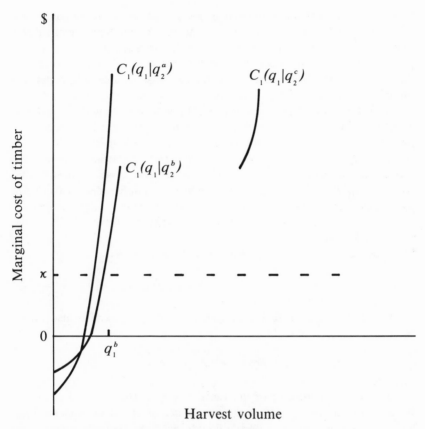

Figure 3-14. Marginal costs of timber production

the unit cost of harvesting under single-purpose timber management (along curve $\psi(q_1)$) is marked by the broken line in figure 3-14 at cost level κ.

In figure 3-14 the marginal cost of timber is seen to be increasing with the harvest level. This increase in marginal cost reflects the shift to intensive wildlife management activity and more costly harvest practices required to maintain the level of hunting quality while increasing timber harvest. The marginal costs of providing timber can also be seen to be generally higher as the level of wildlife service is increased. That is, the two products are generally substitutes. However, it should be noticed that at low levels of harvesting the marginal cost of timber harvesting decreases with higher wildlife service. Here, the two products are complements. In either case, there is evidence of jointness in production.

At the lowest levels of timber production, the marginal cost of harvest is negative. This negative marginal cost corresponds to the sections of the isocost curves in figure 3-13 with a positive slope. There adding to the timber harvest lowers overall costs by reducing the need for direct wildlife improvement activity. On the other hand, at the highest levels of timber production, the marginal cost of harvest is positive and greatly exceeds κ, the marginal cost of harvesting as it would be measured under single-purpose timber management.

It is useful to think of the cost of timber harvests in this multiple-use forest as made up from two components. First, there is κq_1 the cost of harvesting timber by least-cost means (on curve $q_2 = \psi(q_1)$) without regard to the impact on wildlife. Second, there is the cost of making improvements to the wildlife services above the base level $\psi(q_1)$ that results from single-purpose timber management. Providing increments $q_2 - \psi(q_1)$ of wildlife service adds to cost an amount we will represent by $c(q_2 - \psi(q_1))$.[20] The greater the increment in wildlife provided, the higher this component of cost will be. That is, c' the slope of function c is positive.

The overall costs of production are now decomposed as

$$C(q_1, q_2) = \kappa q_1 + c(q_2 - \psi(q_1))$$

and the marginal cost of timber harvests can itself be viewed as made up of two components, with

$$\partial C / \partial q_1 = \kappa - c' \psi'$$

where ψ' is the derivative of the function $\psi(q_1)$. The marginal cost of a multiple-use harvest includes κ the marginal cost of harvest under single-purpose timber management. In addition, it includes a term $c'\psi'$ that reflects the change in the cost of maintaining wildlife services at the chosen level. This second component of marginal cost measures the expense of compensating for changes in the base level of wildlife services caused by increased harvests.

The true marginal cost of harvest may be less than κ. This is the case at lower harvest levels in figure 3-13, where the curve $q_2 = \psi(q_1)$ has a positive slope (and ψ' is positive). The harvest is beneficial to hunting quality in the lower sections of the isocost curves in figure 3-13. Raising the harvest level increases the base level of wildlife service, and so reduces the expenditures c needed to reach a chosen level of service. In contrast, at the higher levels of harvest where ψ' is negative, timber marginal cost will exceed κ. Here the base level of wildlife service is declining with increased harvests, so the cost of maintaining the chosen level of wildlife service increases. The marginal cost of timber will include the mitigation expense to counter the damage to wildlife. Marginal cost will exactly equal κ only when slope ψ' is zero or the slope of curve $q_2 = \psi(q_1)$ is vertical (as in figure 3-13).

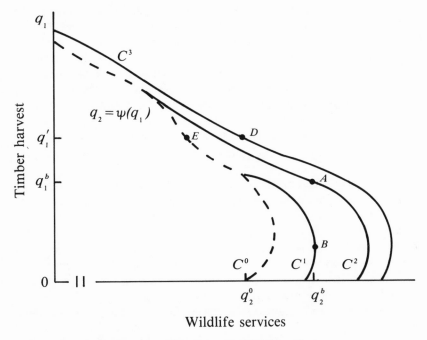

Figure 3-15. Separable costs of timber and wildlife services

Separable Costs Illustrated

The separate cost of an output has been defined above as the extra cost of including that output in the overall product mix. It is useful to focus on the costs associated with the decision to actively produce outputs—that is, to look at the costs of providing the increment in outputs above the level of production that would be incidental to the provision of other outputs. The incidental production of wildlife that results from timber management has already been noted. Notice now that in figures 3-13 and 3-15 (which reproduces a section of figure 3-13) there is also some timber production incidental to the least-cost provision of wildlife services. For example, the least-cost production of wildlife service q_2^b is C^1, at point B where the slope of the isocost curve C^1 is vertical.

The separable cost of providing the timber harvests in an output mix q_1 and q_2 is measured by determining the least cost of producing the overall output mix, and subtracting from it the least cost of providing the same level of wildlife service q_2^b. For example, in figure 3-15 the separable cost of the timber harvest in (q_1^b, q_2^b) is the cost difference $C^2 - C^1$ between points A and B.

Looking at the separable costs of providing timber and wildlife in the output mix q_1', q_2^0 marked as point D on isocost curve C^3 in figure 3-15, it can be seen that the wildlife output q_2^0 is the same level provided at C^0, the wildlife service from the land when there is no harvesting. The separable cost of the wildlife service in (q_1', q_2^0) is the cost difference between the points D and E. This is a cost greater than $C^3 - C^2$ but less than $C^3 - C^1$. Here it is the cost of mitigating the damage to wildlife services that would result from single-purpose timber management for the harvest volume q_1'. The separable cost of the timber harvest in (q_1', q_2^0) is found as the cost difference $C^2 - C^0$. With cost $C^0 = 0$, the total expenditure C^3 is the separable cost of timber.

These separable cost calculations illustrate two significant points relevant to potential cost-allocation schemes for forest products. First, separable costs do not uniquely assign the cost of activities to individual outputs. The cost of mitigating damage to wildlife services is in this case separable to both outputs. Second, the separable costs of the forest outputs here sum to more than the total costs of production. This can occur when products are substitutes to the extent that there are diseconomies to joint production. It is a result that is not convenient for cost-allocation schemes.

We turn now to the choice of output mix under economic multiple-use management.

7. The Choice of Product Mix

With the multiple-use benefits $B(Q)$ and forestwide costs of production $C(Q)$ determined, the problem becomes the selection of the best attainable product mix for each unit of the forest. The forestwide cost function $C(Q)$ now represents the sum of the cost functions for individual planning areas of the forest.

Budget-Constrained Product Choice

As a first step, consider the forest to be constrained to a particular budget level, choosing the best output mix on an isocost curve $C(Q) = \overline{C}$. The manager must solve the maximization problem

$$\underset{Q}{\text{Max}} \ \{B(Q) - C(Q)\}$$
$$\text{subject to: } C(Q) = \overline{C}$$

selecting Q^*, such that no other output mix on the isocost curve provides a greater net value. The solution is to produce at an output mix Q^* where the given isocost curve $\overline{C}$ just touches the highest attainable isovalue curve. The

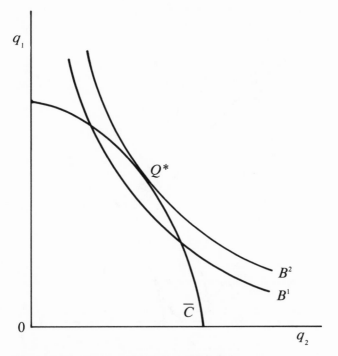

Figure 3-16. Cost-constrained product choice

expression Q represents the full list of outputs provided from all the planning units of the forest.

Slope Conditions for Constrained Optimality. At the optimal output mix, the ratio of marginal values for each pair of actively produced outputs must equal the ratio of marginal costs for that output pair.[21] If both q_i and q_j are produced, then

$$\frac{B_j(Q^*)}{B_i(Q^*)} = \frac{C_j(Q^*)}{C_i(Q^*)}$$

The output mix should be chosen so that the technical rate at which the forest can trade off production of one output for the other is just equal to the rate at which consumers are willing to make such trades. That is, at the solution, the slope of the isocost curve should equal the slope of the isovalue curve. The budget-constrained solution is at a point of tangency, where the given isocost curve just touches the highest attainable isovalue curve. Such solutions are illustrated in figures 3-16 and 3-17.

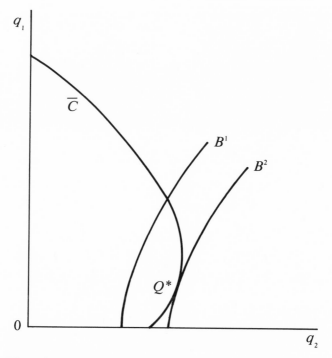

Figure 3-17. Cost-constrained product choice: negative marginal cost

In figure 3-16 the optimal production mix is at the tangency between the isocost curve $\overline{C}$ and the isovalue curve B^2. In figure 3-17, illustrating a more unusual case, the solution is again at the point of tangency between isocost curve $\overline{C}$ and isovalue curve B^2. The solution calls for the production of an output q_1 which has both a negative marginal value and a negative marginal cost of production. Output q_1 might be timber. Here timber is too expensive to harvest on its own merits, but some harvest is desirable because of the beneficial impact on the other amenity service.

In many cases it will not be desirable to provide all outputs on each land unit. In general, with output q_j produced and output q_i not produced in the optimal solution Q^*, we must have $B_i(Q^*)/B_j(Q^*) < C_i(Q^*)/C_j(Q^*)$. An output should not be produced if its marginal value is so low that the withdrawal of money from the production of other goods cannot be justified. Figures 3-18 and 3-19 illustrate two cases in which output q_2 should not be produced. To reach the highest isovalue curve B^2 requires production at the end of the isocost curve $\overline{C}$, at a so-called corner solution. Here the isovalue curves are always flatter in slope than the isocost curve at any possible

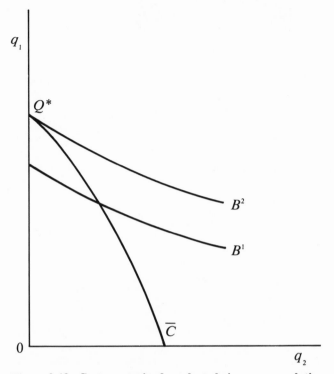

Figure 3-18. Cost-constrained product choice: corner solution

production mix. The marginal value of product q_2 is so low relative to the marginal value of q_1 that specialization in q_1 is preferred.

Corner solutions correspond closely to dominant-use land allocations. Such specialized solutions are particularly likely when isocost curves are convex, as in figure 3-19. Because of this convexity, the preferred output mix is likely to be dominated by one or the other service, depending upon relative marginal values. Balanced product mix solutions are not possible unless the isovalue curve is more convex than the isocost curve.

Marginal Cost Conditions for Constrained Optimality. For each actively produced output q_i^*, first-order conditions for optimality in the budget-constrained product choice require that we have marginal benefits in some common proportion δ to marginal costs, with

$$B_i(Q^*) = \delta C_i(Q^*)$$

That is, an extra dollar of expenditure must provide equal marginal benefits no matter which output it is allocated to.

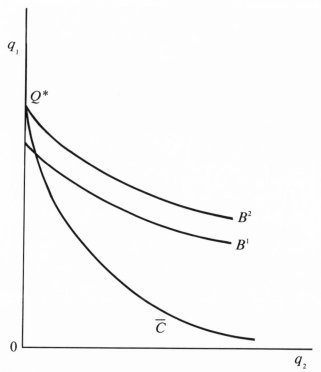

Figure 3-19. Cost-constrained product choice: convex isocost curve

For each output that is not to be produced, the marginal benefits must not be greater than δ times the marginal costs, with

$$B_i(Q^*) \leq \delta C_i(Q^*)$$

Alternative Budget Levels. Budget-constrained solutions may also be determined for a range of alternative budget levels. For example, we might generate various solutions along the locus of tangencies between isovalue and isocost curves, marked by the broken line in figure 3-20. Looking at the output solutions along this array of budget-constrained solutions in figure 3-20, it would be found that for the lowest budget levels the marginal value of each produced output would be less than its corresponding marginal costs. We would have $B_i = \delta C_i$, with $\delta < 1$. In a sense, we would be underspending at the low budget levels. An extra dollar of budget could be used to produce outputs that provide more than a dollar's benefits.

On the other hand, at the highest budget levels it would be found that the marginal value of each produced output was greater than the corresponding

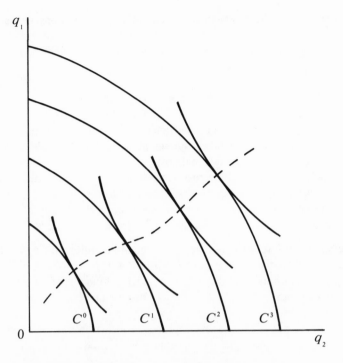

Figure 3-20. Locus of cost-constrained product choice solutions

marginal cost. We would have $B_i = \delta C_i$ with a value of $\delta > 1$. The forest would be overspending, in the sense that a one-dollar reduction in the budget need not reduce benefits by a dollar.

The economically optimal budget level is that at which the marginal value of each actively produced output just equals the marginal cost of production. Economic efficiency requires that outputs, and the associated budget level, should be adjusted until an additional dollar spent to increase the production of any one output provides exactly a dollar's increase in public benefits. Let us now look at the general economic multiple-use management problem more carefully.

Optimal Choice of Product Mix, Unconstrained

The general economic problem facing the land manager is to find the output mix for each unit of the forest that provides the greatest overall value of net benefits. With no budget constraint, the manager must solve

$$\underset{Q}{\text{Max}} \; \{B(Q) - C(Q)\}$$

selecting a product mix Q^*, such that no other output mix provides a greater net value.

A product mix Q^* is optimal if and only if, for every other output mix Q, the following incremental benefit-cost test is satisfied:

$$B(Q)-B(Q^*) \leq C(Q) - C(Q^*)$$

This is both a necessary and a sufficient condition for optimality. At an optimal solution, no further adjustment in output can provide incremental benefits that justify the incremental costs.

Marginal cost and separable cost tests are among the many direct applications of this incremental benefit-cost condition for optimality. Mathematical programming techniques provide practical means for applying this condition in a systematic search for optimal land-use allocations.

Marginal Cost Tests of Efficiency. Marginal conditions for optimality derive directly from the incremental benefit-cost test. By considering very small changes in each output alone, we can find the following first-order conditions, which must be met at Q^*, the net benefit-maximizing land-use solution.[22]

If the ith output is to be actively produced in the optimal solution Q^*, its marginal value should equal the marginal cost of production, with

$$B_i(Q^*) = C_i(Q^*)$$

Were this not true, either a small increase or decrease in the level of production of the output would lead to an increase in overall net benefits. For example, suppose that marginal benefits exceeded marginal costs. Then an increase in output would lead to higher net benefits. On the other hand, if marginal benefits were less than marginal costs, reduced production would increase net benefits.

If the ith output is not to be actively produced, the marginal benefit must not be greater than the marginal cost, with

$$B_i(Q^*) \leq C_i(Q^*)$$

If this were not true, introducing production of the output would lead to an increase in net benefits.

As illustrated earlier in figure 3-16, if a pair of outputs i and j are to be actively produced, the solution must be at a point of tangency of an isovalue curve and an isocost curve. The overall optimal product mix solution corresponds to a budget-constrained solution, with the budget at the optimal level.

Incremental and Separable Cost Tests. Incremental benefit-cost tests can be useful in investigating the economic efficiency of a selected product mix.

A product mix is economically efficient only if there is no adjustment in outputs that would lead to an improvement in net benefits. If there is an adjustment in output from Q to Q' such that the incremental benefits $B(Q')$ − $B(Q)$ exceed the incremental costs $C(Q')$ − $C(Q)$, that change in output is economically desirable. That is, the adjustment in outputs would lead to an increase in net economic benefits.

The separable cost test is an incremental benefit-cost test focusing on adjustments to the production of individual outputs. The separable cost of an output is the cost of including that output in the overall product mix. At an optimal product mix, the increment in benefits from active production of any output in the product mix must exceed the corresponding separable cost. For example, the selection of a product mix with active timber production would require that the benefit increment from the timber production exceeds the corresponding separable cost. Marginal costs tests provide a similar focus on adjustments in single outputs, although for a much smaller change in the level of production.

As a test for the optimality of a particular product mix, the separable cost test is limited, and is best viewed as one of many possible comparisons of incremental benefits to incremental costs that should be considered. The fact that all individual products pass the separable cost test is not sufficient to demonstrate optimality. There will be many product mixes for which the separable cost test is passed, with great differences in the net benefits among these alternatives. Of course, if it is found that the incremental costs of providing one output exceed the corresponding incremental benefits, then eliminating that output will certainly improve overall net benefits. But even this result should be viewed with some caution. An altogether different product mix might be found that both includes production of this output and provides even greater net benefits.

Below-Cost Timber Sales. One useful example of an incremental benefit-cost test is an application to the below-cost timber sales issue. Recently there has been much concern that some of the national forests are receiving less money from the sale of timber than it costs them to provide the timber. To conveniently illustrate this problem, suppose that there are two outputs q_1 and q_2, and let the benefit function $B(Q)$ be equal to $\beta(q_1) + A(q_2)$. Here $\beta(q_1)$ is the benefit from the timber supply q_1, and $A(q_2)$ the benefits from the supply of amenity services q_2.

We will say that timber sales are below-cost if the benefits from those sales are not sufficient to cover the costs directly associated with the sales. That is, the sales q_1^* are below cost if

$$\beta(q_1^*) < [C(q_1^*, q_2^*) - C(0, q_2')]$$

where $C(q_1^*, q_2^*)$ is the overall management cost with the sales program, while $C(0, q_2')$ would be the cost if there were no timber sales. Output q_2^* is the amenity level that results with the sales program. Output q_2' gives the amenity services that would result if the sales did not take place.

Can below-cost sales be consistent with economic efficiency? They can be, but only under limited conditions. Suppose the output mix (q_1^*,q_2^*) maximizes net benefits; then it must be true that the net benefits at this output mix are no less than the net benefits from any other mix of outputs. In particular, we must find this output mix better than production at $(0,q_2')$ with no timber program, and so

$$[\beta(q_1^*) + A(q_2^*) - C(q_1^*,q_2^*)] \geq [A(q_2') - C(0,q_2')]$$

This expression can be rearranged to focus more clearly on the timber harvesting decision, requiring

$$\beta(q_1^*) \geq [C(q_1^*,q_2^*) - C(0,q_2')] + [A(q_2') - A(q_2^*)]$$

That is, if the timber sale program is economic, the benefits of that program must exceed both its direct costs and the opportunity costs associated with the impact of timber sales on amenity services. Only if these opportunity costs $A(q_2') - A(q_2^*)$ are negative can we accept below-cost timber sales; a below-cost sale, in other words, is acceptable only if it improves the other services (perhaps future services) of the forest. (The extreme case of an economically desirable below-cost sale is illustrated in figure 3-17, where the timber benefit $\beta(q_1)$ is negative.)

Solving for Optimal Product Mixes in Practice

In practice, with complex production cost functions such as exist for the multiple-use forest, the selection of the value-maximizing production targets is best accomplished by mathematical programming techniques. It is likely that the full cost function will never be known. In general, only a few alternative output mixes will be formulated for consideration. Selection among a discrete set of production alternatives so as to maximize economic value may be accomplished quite easily by the use of linear programming—a method by which a systematic search can be made through many alternative management schemes until the best solution is discovered. From an initial candidate allocation of land use, the linear program can search for adjustments in the land-use activity that add to net benefits. When no further adjustments can be found, an optimal solution is considered to have been found. If the range of production possibilities has been adequately represented in the formulation of the problem, the solution should be optimal and analogous to that described above by the marginal cost conditions.

Supply Response to Changing Relative Marginal Value

The choice of an economically optimal product mix will depend upon the relative values of the products. By systematically altering the price level of a particular output in a mathematical program of forest production, supply curves could be generated for each product. As is usual, the supply of a product would be increased as its relative value increases. To say more about the nature of the supply relationships, it is necessary to consider the shape of the isocost curves and the jointness among products.

With convex isocost curves, small changes in relative prices can lead to sudden jumps in the desirable land use. Referring back to figure 3-19, it can be seen that a sufficient increase in the relative value of product q_2, making the isovalue curve more steeply sloped, would lead to a switch from the production of q_1 to the dominant-use production of q_2. Small changes in marginal values will lead to more moderate adjustments in the supply of products for which the isocost curves are convex.

Of course, in general the response to changing demand is not going to be confined to adjustments in product mix along a single isocost curve. The optimal budget level probably must also be changed as we seek to balance the marginal costs of each output to equal the marginal benefits. With jointness in production, changes in the production of any one output lead to changes in the marginal costs of producing the others. As a result, the optimal production level for every output is likely to change in response to changing demands for any single product.

It can be seen, then, that among the practical implications of the classification of products as substitutes or complements in production are the following. If the demand for one output increases, and in response the output of this product is increased, then the corresponding reduction in marginal cost of complementary outputs implies that the supply of these other products should also be increased. Conversely, with a substitute product, an increase in the demand for one output will lead to a reduction in the optimal supply of the other.[23]

8. Summary and Conclusions

The discussion in this chapter has focused on the conceptual nature of joint production and the choice of product mix, particularly the decision to select either balanced or dominant-use production. With fixed budget and resources, production of one output should be increased to the extent the extra value from this output exceeds the sacrificed value from the other services. With flexibility in the budget level, production of one output should be increased if its marginal value exceeds its marginal cost. With jointness in production, such marginal costs reflect more than the costs of increasing the one output without regard to other services. Marginal costs also properly reflect the costs of maintaining constant the production level of other goods and services.

There are many practical difficulties facing any attempt to implement economic multiple-use forest management. One difficulty which may not have been made apparent is the choice of the land units for which we develop production possibilities. What happens on one plot of land can affect the value of the services from other units of the forest. If too small a unit of analysis is considered, the interaction of the multiple-use services will not be meaningfully reflected. On the other hand, if very large areas of land are considered in an attempt to internalize the interactions between adjacent areas, then the detail in which production alternatives can be sensibly speci-fied may have to be sacrificed. The sheer number of combinations of treat-ments that could potentially be considered for even a moderately sized area of land suggests that in practice any planning effort is likely to fall short of the ideal.

A second noteworthy difficulty is the determination of the multiple-use product values. Such evaluation is especially troublesome for those services which are not traded in competitive markets; for these services the values must be estimated from indirect evidence of consumer choice. Although methods for valuing recreation are well established, the problem facing the Forest Service is rather more difficult than is usual. Many management actions will lead only to changes in the quality of the service. It is more difficult to value such changes in quality than to estimate the total value of the activity. Further, management actions on one land area will affect the valuation of other areas also under Forest Service management. Taking account of the interrelated demands for all areas of the forest could present a formidable task both for the estimation and use of values in management decisions.

Among the conceptual difficulties with the descriptive approach taken in this chapter is our obvious lack of consideration for future flows of harvests and services from the land. The static (timeless) description of multiple-use production is of great convenience, and indeed many of the basic elements that make multiple-use management challenging can be adequately illustrated in this framework. In theory at least, one could even think of our list of products as including harvests and services at all future dates. In many ways, the problem of choosing among production levels between time periods is no different than the problem of choice among products in the same time periods. However, because of the special nature of the link between produc-tion possibilities in successive time periods, there prove to be much more enlightening ways to describe the problems of forest management over time. In the next chapter we illustrate some aspects of multiple-use forest manage-ment in this dynamic setting.

Notes

1. For a history of Forest Service planning, see D. C. Iverson and R. M. Alston, *The Genesis of FORPLAN: A Historical and Analytical Review of Forest Service Planning Models,* General Technical Report INT-214 (Ogden, Utah, USDA Forest Service, Intermountain Research Station, September 1986).

2. USDA Forest Service, *Even-Age Management on the Monongahela National Forest* (Washington, D.C., 1970).

3. K. N. Johnson, *FORPLAN Version 1: An Overview* (Washington, D.C., USDA Forest Service, Land Management Planning Systems, February 1986); K. N. Johnson, T. W. Stuart, and S. A. Crim, *FORPLAN Version 2: An Overview* (Washington, D.C., USDA Forest Service, Land Management Planning Systems, August 1986).

4. We treat logging costs as if incurred by the forest. In fact, purchasers usually incur harvest costs, with these costs reflected in lower price offers. Under that circumstance, it is even more likely that potential purchasers would not offer positive prices.

5. See R. D. Willig, "Consumer's Surplus Without Apology," *American Economic Review* vol. 66 (1976) pp. 589–597.

6. We greatly oversimplify here, ignoring the type and quality of timber and also the location and capacities of individual mills.

7. Determining the equivalent of these revenues might require an adjustment to observed revenues, adding harvest costs incurred by the purchaser.

8. See K. G. Mäler, "A Method of Estimating Social Benefits of Pollution Control," *Swedish Journal of Economics* vol. 73 (1971) pp. 121–133; D. F. Bradford and G. C. Hildebrand, "Observable Preferences for Public Goods," *Journal of Public Economics* vol. 8 (1977) pp. 111–131.

9. For more discussion, see M. D. Bowes and J. V. Krutilla, "Multiple-Use Forestry and the Economics of the Multiproduct Enterprise," in V. K. Smith, ed., *Advances in Applied Micro-Economics,* vol. 2 (Greenwich, Conn., JAI Press, 1982) pp. 157–190.

10. See, for example, D. McFadden, "Costs, Revenue, and Profit Functions," in M. Fuss and D. McFadden, eds., *Production Economics: A Dual Approach to Theory and Applications* (Amsterdam, North-Holland, 1978).

11. Other writings on the economics of multiple-use forestry that the reader may wish to refer to are: R. H. Alston, "Economic Tradeoffs of Multiple-Use Management," in D. D. Hook and B. A. Dunn, eds., *Proceedings of the Symposium on Multiple-Use Management of Forest Resources* (Clemson, S.C., September 1979) pp. 162–185; G. R. Gregory, "An Economic Approach to Multiple-Use," *Forest Science* vol. 1, no. 1 (1955) pp. 6–13; P. F. O'Connell and H. E. Brown, "Use of Production Functions to Evaluate Multiple-Use Treatments on Forested Watersheds," *Water Resources Research* vol. 8, no. 5 (1972) pp. 1188–1198; P. H. Pearse, "Toward a Theory of Multiple Use: The Case of Recreation Versus Agriculture," *Natural Resources Journal* vol. 9, no. 4 (1969) pp. 561–575; D. C. Teeguarden, "Multiple Services," in W. A. Duerr, D. C. Teeguarden, N. B. Christiansen, and S. Guttenberg, eds., *Forest Resource Management* (Corvallis, Oregon State University Bookstores,

1982) pp. 276–290; and G. R. Walter, "Economics of Multiple-Use Forestry," *Journal of Environmental Management* vol. 5 (1977) pp. 345–356.

12. See J. M. Griffin, "The Econometrics of Joint Production: Another Approach," *Review of Economics and Statistics* vol. 59 (November 1977) pp. 389–397.

13. Perhaps they should be called isocost curves in output-space, to stress our narrowed focus on outputs. With single-product technologies, it is usual to focus on isocost curves as functions of input prices.

14. See, for example, the complicated tests in E. Burmeister and S. J. Turnovsky, "The Degree of Joint Production," *International Economic Review* vol. 12 (1971) pp. 99–105.

15. See R. E. Hall, "The Specification of Technologies with Several Kinds of Outputs," *Journal of Political Economy* vol. 81 (1973) pp. 878–892.

16. See J. C. Panzar and R. D. Willig, "Economies of Scope, Product Specific Scale Economies, and the Multiproduct Competitive Firm," Bell Laboratories, Holmdel, N.J., 1978.

17. W. J. Baumol, "Scale Economies, Average Cost and the Profitability of Marginal Cost Pricing," in R. Grieson, ed., *Essays in Urban Economics and Public Finance in Honor of William S. Vickery* (Lexington, Mass., D.C. Heath, 1976).

18. W. J. Baumol, E. E. Bailey, and R. D. Willig, "Weak Invisible Hand Theorems on the Sustainability of Prices in a Multiproduct Natural Monopoly," *American Economic Review* vol. 67 (June 1977) pp. 350–365.

19. The formulation is really appropriate only if outputs are equally valued across sites and if it is overall output from sites that is valued. Outputs q_I^A and q_I^B should not be viewed by consumers as being distinct products.

20. To be more general, we might allow the costs to depend directly upon harvest level.

21. Actively produced outputs are those provided at a level greater than would be incidental to the least-cost production of the other goods.

22. They are necessary conditions for optimality, but in general are not sufficient to guarantee an optimal solution. If the cost function is convex and the benefit function is concave, they are also sufficient conditions.

23. See E. W. Erickson and R. M. Spann, "Supply Response in a Regulated Industry: The Case of Natural Gas," *Bell Journal of Economics* vol. 2, no. 1 (1971) pp. 94–121.

4

Dynamic Models of Multiple-Use Management

1. Introduction

In looking at the economic theory of multiple-use forestry, we now turn to consideration of the dynamics of forest growth and the asset value of the forest stock.* The economic multiple-use management problem is the selection of a sequence of management actions so as to maximize the discounted value of net benefits from the flow of harvests and other resource services over time. In order to simplify a nevertheless complex discussion, our focus will be on the timing of timber harvests alone; no other management actions are explicitly considered. In choosing to harvest timber, one must be mindful of the value of maintaining a stock of standing timber. Standing stocks will provide for future harvests and will determine the flow of other multiple-use services over time. A timber harvest will alter, perhaps for an extended period, the flow of all services which depend upon the condition of the standing stocks.

The principles of public forestry in the United States have often been driven by a fear of timber shortages and the destabilizing effects that might arise from fluctuations in the supply of timber over time. A distrust of the private market as a supplier of timber arose in reaction to the rapid clearing of forestlands that came with the settlement of the West and as a response to the thefts of timber from remaining public forestlands.[1] These public forests, for all practical purposes, were common property, unprotected against trespass until after the Organic Administration Act of 1897. The Organic Act

*Portions of this chapter have been adapted, with permission, from M. D. Bowes and J. V. Krutilla, "Multiple Use Management of Public Forestlands," in A. V. Kneese and J. L. Sweeney, eds., *Handbook of Natural Resource and Energy Economics,* vol. 2 (Amsterdam, The Netherlands, North-Holland, 1985); copyright Elsevier Science Publishers B. V., 1985.

authorized the protection and active management of previously set-aside forest reserves, and is the act to which the U.S. Forest Service traces its origin.

The early philosophy of public forest management (associated strongly with Gifford Pinchot, the first chief of the USDA Forest Service in 1905) became known as conservationism. It was based on biological and techno-cratic rather than economic principles. The goals were protection of the public forests and the promotion of a high perpetual level of services from these lands. Concern for protection, stability, and high levels of yield led eventually to a widespread acceptance of some rule-of-thumb policies for public forest management. First, there is the policy of maximum sustained yield (MSY) management. Lands under MSY management for timber must be capable of prompt regeneration and continued management and are to be cut at an age that ensures a sustained average flow of harvests over time at the maximum biological potential. MSY management is intended to ensure the long-run protection of managed lands while providing for high levels of timber harvest. Second, there is the policy of even-flow management. Under an even-flow policy, current timber harvest targets must be at a level that can be constantly maintained over time. The even-flow policy reflects concern for the stability of output flows. The goal of the "fully regulated" forest combines these two principles of management, with a regulated forest having a sustained and even flow of harvest over time, and with this flow at the maximum potential biological yield.

The maximum-sustained-yield even-flow philosophy has long been criti-cized as devoid of economic rationale. Indeed it has little to say about the wise investment of scarce appropriations among lands of varying productiv-ity. Certainly such a philosophy does not indicate how one should respond to changing patterns of demands for the various products and amenity serv-ices of the forest, which in addition to timber include water flow, wildlife, range, wilderness, and other recreational services dependent upon the con-dition of the standing vegetation. This philosophy can result in the sacrifice of potential consumption in times of plentiful timber stocks, and can lead to uneconomic broadening of the timber land base and intensification of timber management.

While the Forest Service has always had a stated concern for the nontimber resource services, it was not until after World War II, with the rapid increase in demand for both timber and outdoor recreation and resulting political pressures on behalf of such single purposes, that the need for explicitly balanced operating criteria became apparent. Not surprisingly, attempts to extend the maximum-sustained-yield even-flow philosophy to all outputs have been unsatisfactory. Now, under the Multiple-Use and Sustained-Yield Act of 1960 and the Forest and Rangeland Renewable Resources Planning Act of 1974 (RPA) as amended by the National Forest Management Act of

1976 (NFMA), the Forest Service has been given the legislative mandate to manage the forest so as to maximize the benefits from sustained-yield multiple-use production, with consideration given to the relative values of all resources and subject to preserving the productive potential of the land. Despite considerable ambiguity, most would agree that the legislation at least now accommodates economic concerns and requires that a considerable amount of attention be paid to the economic efficiency of forest plans.

In response to the RPA and NFMA legislation, the Forest Service has begun an ambitious program of planning the management of some 190 million acres of National Forest lands by using linear programming methods. Large linear programming models are currently in use or under development on many forests, where they are to be applied in the evaluation of options for scheduling harvests and land use over time. Planning choices still seem to be heavily constrained by policy, however. The models often contain even-flow restrictions and the treatments considered generally reflect prior restrictions on minimum harvest ages as well as prior standards for the intensity of harvest activity. A harvest-age restriction, designed to ensure a high average level of yield, is now often justified as a means of protecting amenity values through long rotation cycles. Standards, although often intended to protect amenity values, tend to limit the possibility for advantageous specialization of uses on different areas of the forest. The planning effort has served to highlight the need for clarification of the inherent contradictions between the biological-technocratic principles of which the Forest Service is not entirely free and those principles of allocative efficiency implied by the balancing of supply and demand.

In this chapter we compare the economic multiple-use management solution to maximum sustained yield management. While MSY management may result in a greater flow of multiple-use outputs in the longer run, it does so at the cost of current consumption by holding large amounts of potential wealth in timber stocks and other capital resources. Economic multiple-use management results in a greater net present value from the flow of multiple-use outputs and services.

Economic multiple-use management will also be compared below to the single-purpose management of land for timber. The introduction of multiple-use values does often lead to a change in the optimal level and timing of timber harvests, but not always in a manner that might be anticipated. In some cases, consideration of multiple-use values may lead to higher levels of harvest despite nontimber uses which seem incompatible with timber harvesting.

Also considered in chapter 4 are the compatibility of land uses, the desirability of specialized use of lands for single-purpose management, and the importance of diversity of stocking conditions. Our intuitive sense of the compatibility of uses can be strained once the timing of actions is considered.

Further, the various multiple-use services typically can be meaningfully defined only by addressing the patterns of vegetation cover over a fairly wide forest area. The harvesting decision for a single timber stand must reflect this interdependence among forestland areas.

The chapter is divided into three main sections. The section on Single-Stand Models of Harvest Timing (section 2) introduces the basic concepts of dynamic models and links the static approaches of chapter 3 to the more general discussion of multiple-use management provided in the last two sections of this chapter. Discussion of the Faustmann model focuses attention on the age at which single-aged stands of trees are to be optimally harvested, and the Faustmann formula provides the solution to this optimal rotation-age problem.[2] In a later modification of the Faustmann problem that reflects multiple-use values, Hartman assumes that various nontimber services of a stand can be related to the stand age.[3] In section 2 the Faustmann and the Hartman models are described in detail and a graphic construct is used to illustrate the long-run tradeoffs between timber harvests and nontimber services in these single-stand models.

The section on Multiple-Use Management of Related Forest Stands (section 3) provides a further generalization of the Faustmann-type models. Much of what makes multiple-use forestry interesting and difficult arises from the interdependence of the harvest decision across stands. While the analysis of single-stand models can provide much economic insight, many of the conclusions from these simple models prove to be misleading if considered carelessly. We provide a fairly general model which takes the interdependence of timber stands into account. This multiple-stand model seems well-suited to describing at least the basic elements of true multiple-use management for combined timber, water-flow, dispersed recreation, and wildlife services.

It proves difficult to provide a complete analytic description of the harvest solution for our multiple-stand model. The model is most useful for illustrating the shortcomings of the traditional descriptions of multiple-use management. In the section on Illustrations of Multiple-Stand Harvesting Solutions (section 4) a few examples of multiple-use harvesting solutions are provided. These show that consideration of the link among stands leads, as might be expected, to richer and more intuitively appealing harvest decisions than are found with single-stand models. Many readers may find it appropriate to skim the theoretical material in section 3 and proceed to the simpler treatment in section 4.

The harvest timing models described in this chapter are not meant to be taken as practical approaches to the solving of large-scale forest management problems. Linear programming models such as the FORPLAN models used for national forest planning,[4] if appropriately structured, do provide such practical tools for multiple-use management (FORPLAN models for two forests are described in detail in chapter 10). The discussions presented here in chapter 4 can be viewed as analyses of two stylized versions of the FORPLAN models. The discussion of the Faustmann-type single-stand models addresses the basic economic features underlying most earlier FORPLAN models, which considered the choice of management over time for areas of the forest that were homogeneous in timber-growing condition, stock, and accessibility. Each such land area was treated as nonjoint or unrelated in the production of multiple-use outputs.

Our discussion of multiple-stand models addresses the economic features of the newer FORPLAN models which consider the management of contiguous geographical zones of the forest. Within such forest zones (or planning units) there are separate stands, each of which may differ in stock conditions and productivity. The management choices for individual stands within the planning units are treated as related, linked perhaps because the stands share a road network or because of their joint role in the provision of nontimber services. It is our opinion that a FORPLAN model will be of little use if, as in earlier versions, it is applied to choose management prescriptions for homogeneous timber stands without concern for spatial location or for the linkage of these stands to adjacent stands within the forest.

2. Single-Stand Models of Harvest Timing: Faustmann-Type Models

Faustmann-type models are by now the traditional approach to describing the economics of timber management. The Faustmann problem is to maximize the present value of net receipts from a single-aged timber stand by the selection of a sequence of harvest dates. The multiple-use values introduced into the Faustmann model by Hartman lead to a change in the optimal harvest timing, with the extent and direction of the change depending upon the nature of the services and their value relative to the timber harvests. Linear programs can extend these Faustmann-type models, providing practical means for determining optimal harvest scheduling for several stands under more general conditions of time, varying prices, and costs, and with allowance for a greater choice in management actions over time.

The Faustmann Model

The Faustmann model describes the optimal economic management of a single-aged timber stand under conditions of unchanging productivity and prices over time.[5] Beginning with a single unstocked stand, a manager incurs per-acre regeneration costs C. The timber grows according to a function $V(T)$ which gives the salable volume per acre for a stand aged T. At harvest the manager receives $PV(T)$, with the price P given net of any harvest costs. The land is then regenerated again, and the cycle repeats perpetually. The manager may choose only the age at which the stand is to be harvested. With productivity, prices, costs, and the interest rate assumed to be unchanging over time, this optimal harvest age will be the same in each rotation cycle.[6] The Faustmann problem, then, is to find the rotation length T that maximizes the present value of net receipts from the current and all future harvest cycles.

The present value of revenue from a sequence of harvest cycles each of length T can be given as

$$\lambda(T) = \sum_{k=1}^{\infty} [PV(T)-C]e^{-kiT} - C$$

and this expression for the present value can be simplified to

$$\lambda(T) = [PV(T)e^{-iT} - C]/[1 - e^{-iT}] \qquad (4\text{-}1)$$

where i is a market-determined rate of interest and e^{-iT} is a factor which discounts net revenues at time T to their present value.[7] The practice of discounting future values can be thought of as a means of reflecting the forgone opportunities for selling timber and land and investing the proceeds in other assets that yield at the market rate of return.

The Faustmann problem requires the selection of a rotation age to maximize the present value (equation 4-1). We find the maximum value λ^*, with

$$\lambda^* = \underset{T}{\text{Max}} \ \{\lambda(T)\}$$

by choosing a rotation age such that the derivative $\lambda'(T)$ is equal to zero; that is, the solution requires the selection of a rotation age—call it T_F—such that we meet the first-order condition

$$PV'(T) = i[PV(T) + \lambda^*] \qquad (4\text{-}2)$$

Interpreting the Faustmann Solution. The optimal harvest rotation age should be that age at which the benefit from a marginal delay in the harvest date is just equal to the opportunity cost of this delay in the harvest. The present value λ^* may be viewed as the value of an acre of land under forestry management. It is the greatest amount that could be offered for the purchase of an acre of bare land to be used for timber management purposes. The term $PV(T) + \lambda^*$ is the combined value of the land and timber stock at harvest. $PV'(T)$, the derivative of $PV(T)$ with respect to T, gives the growth in the value of these assets over time. This value $PV'(T)$ is the benefit from a marginal delay in the harvest date. The opportunity cost of a delay in the harvest is $i[PV(T) + \lambda^*]$, the potential interest income that could be earned if the land and timber were sold and the proceeds invested to yield the market rate of return.

Put another way, timber should be held uncut until the rate of growth in the combined asset value of the timber and land just equals the market rate of return on investments. That is, one should not harvest until $PV'(T)/[PV(T) + \lambda^*] = i$. This optimal harvest rule is consistent with the usual economic principles of asset management.

The Initial Stand-Age Assumption. No difficulty would have been introduced if it had been supposed that the initial stand was already stocked with growing timber. The optimal harvest-age solution would prove to be the same. If the current stand age already exceeds the Faustmann age, the stand should then be cut immediately, with subsequent rotations following the Faustmann harvest cycle.

The Decision to Manage. There is the possibility that the greatest present value of land under any timber rotation is negative. Under such conditions it would be better to leave the land unmanaged. More generally, if there exists a set of possible management choices for the land, timber management should be selected only if it provides the greatest present value among the alternative uses of the land.

If the land were currently stocked, possibly an initial harvest might be profitable while continued timber management would be uneconomic. We would then choose to cut once and abandon management, with no regeneration. Consideration of nontimber values could make the decision to leave lands unregenerated a less than attractive solution.

Illustrating the Faustmann Solution. One useful method for illustrating the Faustmann solution is given in figure 4-1. From the present value definition (equation 4-1) above, we have at the Faustmann solution

$$PV(T) + \lambda = [\lambda + C]e^{iT} \qquad (4\text{-}1')$$

while from the first-order condition (equation 4-2) we find that present-value maximizing solution must satisfy

$$PV'(T) = i[\lambda + C]e^{iT} \qquad (4\text{-}2')$$

These two equations can be solved for the two unknowns, the maximum land value λ and the optimal rotation T.

The right-hand side of equation (4-1') gives the value of assets one could hope to have at time T if the initial monetary investment in land and regeneration had been put into assets yielding the market rate of return. The left-hand side gives the actual value of the investment in timber production at time T. Equation (4-1') says that at an optimal solution we must be indifferent in choosing to invest in forestry or other alternatives. The two sides of equation (4-1') are separately drawn in figure 4-1 as functions of stand age T for an assumed land value λ^*. That is, in figure 4-1 an optimal rotation must be at an age where the two curves intersect. Equation (4-2') indicates that at an optimal solution, these two curves must in fact be equal slopes. With condition (4-1'), this means that the Faustmann rotation age must be at a point of tangency between the two curves, a point where the curves are just touching. Figure 4-1 illustrates the solution for the maximum land value λ^* and the Faustmann rotation T_F.

In developing figure 4-1, we could have begun by guessing at λ, the value of land. The optimal solution requires us to find the highest such land value

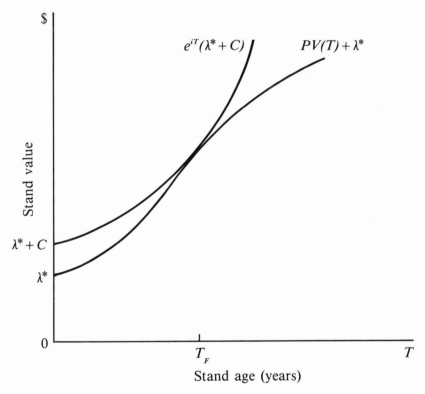

Figure 4-1. The Faustmann solution: optimal rotation age and land value

at which the two curves from equation (4-1′) still intersect. If figure 4-1 were redrawn for a value of λ less than λ^*, the two curves would be found to intersect twice. In searching for higher possible values of λ, eventually at values of λ greater than λ^*, it would be found that the two curves diverged, never intersecting. Land value λ^*, which leads to the two curves being tangent, would be the highest possible value for λ at which the two curves intersect.

The unknown land value λ^* and the optimal harvest age solution T_F are determined by the two conditions (4-1′) and (4-2′) together. The land value λ^* is the highest possible value that could be paid for land such that investors would still find investment in forestry as attractive as other market investments. The Faustmann rotation age T_F is the rotation age that will allow the forest investor to afford to use land selling at a price λ^*.

Effect of Changed Parameters. To illustrate other characteristics of the Faustmann solution, it is most convenient to rewrite equation (4-2), substituting explicitly for the land value λ by using equation (4-1). We then find yet another version of the Faustmann condition:

$$PV'(T)/(PV(T)-C) = i/(1-e^{-iT}) \qquad 4\text{-}2'')$$

The right-hand side of equation (4-2″) is a downward-sloping function of the rotation age T. The value of the function $i/(1-e^{-iT})$ is always greater than i, although it is not much greater than i at high stand ages T. The curve $i/(1-e^{-iT})$ in figure 4-2 is evaluated for an interest rate of $i = .04$. The left-hand side of equation (4-2″) may be interpreted as the relative growth rate in net harvest revenues; for illustrative purposes, in figure 4-2 the relative growth-rate curve is based on Douglas fir yields. The harvest age solution is the age T_F at which these two curves intersect.[8]

The comparative statics of the Faustmann solution are easily illustrated with the aid of figure 4-2. With regeneration costs, a higher (constant) price level P leads to a shorter rotation-age solution. To see this, it should be noted that a higher price level shifts the relative timber-value growth curve downward. The result is an intersection with curve $i/(1-e^{-iT})$ at a shorter rotation age. Similarly, a higher regeneration cost C shifts the relative value growth-rate curve upward, leading to a longer rotation age. A higher interest rate raises curve $i/(1-e^{-iT})$ and leads to a shorter rotation age.

If there is no regeneration cost, the Faustmann solution age is at T_0, where $V'(T)/V(T) = i/(1-e^{-iT})$ (see figure 4-2). The solution T_0 is then independent of changes in the price level. With regeneration costs, the Faustmann rotation age must be above the age T_0, approaching this age as a lower bound as price rises relative to cost. If regeneration costs are so high relative to

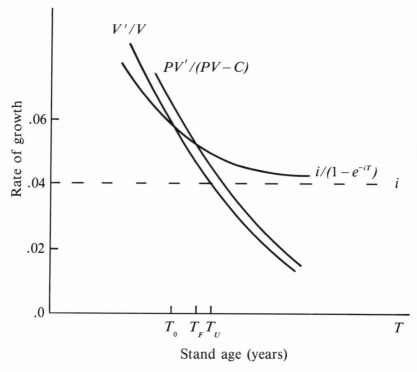

Figure 4-2. The Faustmann rotation age

price that the maximum land value λ^* just equals zero, then the rotation age is at T_U, where $V'(T)/V(T) = i$. This solution gives the upper bound on the Faustmann rotation-age solutions.[9]

A Variation on the Faustmann Problem—Rising Prices. The Faustmann problem is often dismissed as unrealistic because of the simplistic assumption that future conditions are unchanging. In fact, there is much to learn from the method, and the problem formulation is easily generalized to more realistic conditions. In particular, much the same form of the optimal harvesting rule will apply even if timber prices are not constant over time.

Suppose that timber price varies over time according to a function $P(t)$. Under these changing price conditions, each successive harvest should generally be at a different age, and the value of land will itself vary over time. Let $\lambda^*(t_k)$ represent the maximized value of land at time t_k, the time of the kth harvest; that is, $\lambda^*(t_k)$ is the present value of the net revenues from all subsequent harvest rotations, with these harvests at optimal rotation ages. The harvest problem is to select each rotation age T_k (for $k = 1, 2, \ldots$) so as to maximize the present value of future net receipts. It is easily shown that the optimal harvest ages T_k must each satisfy the following rule:[10]

$$P(t_k)V'(T_k) + \dot{P}(t_k)V(T_k) + \dot{\lambda}*(T_k) = i[P(t_k)V(T) + \lambda*(T_k)] \quad (4\text{-}2''')$$

This harvest rule is similar in spirit to the Faustmann harvest rule (equation 4-2). The left-hand side again measures the benefits from a slight delay in the harvest date. These benefits now include not just the value of incremental timber growth PV', but also the appreciation in value of timber stocks resulting from rising prices and the appreciation in the value of the land itself.

It can be shown that with exponentially rising prices $P(t) = e^{\alpha t}P$ (with growth rate $\alpha < i$), the harvest age solution will be initially longer than the Faustmann rotation that would result for a price constant at P. The optimal rotation ages then gradually decline over time, approaching the previously defined T_0 as a lower bound.[11]

Maximum Sustained Yield. The maximum sustained yield of timber is provided by selecting a rotation age that maximizes the average annual flow of harvest volume $V(T)/T$. This requires a rotation age T_M which satisfies the first-order condition

$$V'(T)/V(T) = 1/T$$

The solution is the age at which the marginal increment to value $V'(T)$ has declined to just equal the average yearly harvest volume, $V(T)/T$. The MSY rotation-age solution T_M is known in the forestry literature as the age of "culmination of mean annual increment."

For comparison, the Faustmann solution requires (see equation [4-2]) a harvest age for which $PV'(T) = i[PV(T) + \lambda*]$, or

$$V'(T)/V(T) = i + i\lambda*/PV(T)$$

With $V'(T)/V(T)$ a decreasing function of T (as expected), it is apparent that the Faustmann economic harvest rotation T_F is generally shorter than the MSY rotation T_M. This is because the value $1/T_M$ is almost always less than the interest rate i. The longest possible Faustmann age solution (T_U) will occur at $V'(T)/V(T) = i$, when timber prices are so low (or costs so high) that $\lambda*$ is zero. With any higher prices or lower costs, the Faustmann rotation must be shorter than T_U. On the other hand, the MSY rotation age T_M is greater than T_U except in the most unusual circumstance that $1/T_M$ is less than the interest rate. If, for example, the interest rate were .04, then the

MSY solution would have to be less than age 25 ($^1/_{25}$ = .04) in order for an economic rotation to be longer. A more typical MSY harvest would be at a stand age of 100 years for North American softwoods. Only with remarkably fast-growing trees (perhaps tropical species) or with very low interest rates can the Faustmann harvest age T_F exceed the MSY-yield harvest-age T_M.

Maximum Sustained Yield of Rent.   Others have argued that the MSY solution can never exceed the Faustmann rotation age. That conclusion is usually based on a somewhat different view of maximum sustained yield. For example, Samuelson describes a forest as maximizing the average flow of yearly net income (rather than volume).[12] While forests do not typically try to maximize the net income flow, it is in some ways more logical that they do so, rather than maximize volume. If we allow for the possibility of increasing yields through intensive management, the goal of maximizing the flow of volume becomes rather absurd. It would require implementation of every possible activity that increases timber yields—clearly not a practical approach to management. In much the same manner, sustained yield management policies require the use of excessive amounts of another valuable input to production, time.

Suppose the forest does choose to maximize the average yearly flow of net income $[PV(T)-C]/T$. This would require a rotation age T_m that satisfies the condition

$$PV'(T)/[PV(T)-C] = 1/T,$$

with the increment in value from delay just equal to the average yearly net harvest value. The result can be described as the maximum sustained yield of rent (MSY_R) solution.

For comparison, the Faustmann economic solution (see equation [4-2″]) requires

$$PV'(T)/[PV(T)-C] = i/(1-e^{-iT})$$

With a positive interest rate and $PV'(T)/[PV(T)-C]$ a downward-sloping function of age, the maximum sustained rent rotation solution T_m is always greater than T_F, the economically optimal harvest age under timber management. This conclusion may be understood by noting that the value $i/1-e^{-iT})$ is always greater than $1/T$. An illustration makes this conclusion more understandable.

Illustrating Maximum Sustained Yield Solutions.   In figure 4-3, illustrating maximum sustained yield solutions, the curves $i/(1-e^{-iT})$ and $1/T$ are drawn as functions of harvest age T. The volume growth rate $V'(T)/V(T)$

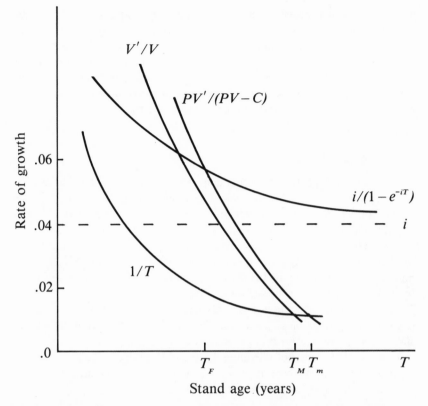

Figure 4-3. Faustmann rotation versus maximum sustained yield rotations

and net value growth rate $PV'(T)/[PV(T)-C]$ are also shown. These growth rates decrease with stand age; with regeneration cost C greater than zero, the value-growth curve must be above the volume-growth curve (as drawn). It is apparent in figure 4-3 that the Faustmann solution T_F is at a younger age than the age of maximum sustained rent T_m. The maximum sustained volume yield solution T_M is also marked in the figure. This MSY solution T_M is always at a somewhat lower age than the MSY_R solution T_m and generally above the Faustmann age T_F, as drawn.

It is often suggested that a maximum sustained yield policy amounts to ignoring interest rates, or, more accurately, to ignoring the returns that could be gained from forgone alternative investments that yield at the market rate of interest. This conclusion may be understood by noting that, as the interest rate i approaches zero, the value $i/(1-e^{-iT})$ approaches the value $1/T$. With interest rates close to zero, the Faustmann solution would approach the MSY_R solution. While the higher average yields or value flows that do result under maximum sustained yield policies may seem appealing, it must be realized that these yields come at some sacrifice. The higher yields are attained by building up larger amounts of timber stock than is economic. Selling some of this stock earlier, and investing the proceeds at the market rate of return (whether in economical timber management or elsewhere), would result in an increase in net economic value that more than offsets the somewhat lower flow of timber or net timber value over time.

The Hartman Model

The conventional Faustmann model does not reflect the value of those services provided directly by the standing forest. Hartman, in reacting to Samuelson's paper, provided a generalization of the Faustmann harvest problem that did account for the various multiple-use services of a single stand. The analysis again considers the management of an even-aged stand that is initially clear of timber. The problem becomes the selection of a rotation period so as to best provide the combined net benefits from timber and other multiple-use services of the standing timber.

Assume that the flow of net benefits from the nontimber services can be expressed as a function of stand age. The net benefit flow from an acre of standing stock of age n can be represented by the term $a(n)$. This measure reflects any direct costs of providing the amenity services. The present value from the net amenity benefit of the stand is the discounted sum of these flows over all future harvest cycles:

$$\psi(T) = \int_0^T [a(n)e^{-in}]dn/(1 - e^{-iT}) \qquad (4\text{-}3)$$

The Hartman problem is to select a rotation period T that maximizes the combined present value from the timber harvests and the amenity services of the standing stock, with

$$\phi^* = \underset{T}{\text{Max}} \{\lambda(T) + \psi(T)\}$$

$$= \underset{T}{\text{Max}} \left\{ \left[PV(T)e^{-iT} + \int_0^T [a(n)e^{-in}]dn - C \right]/(1 - e^{-iT}) \right\}$$

where ϕ^* is the overall asset value of an acre of land and $\lambda(T)$ is the timber value as defined for the Faustmann problem in equation (4-1). The overall land value now reflects both the net present value from future harvests and the amenity services of the land.

The rotation-age solution, which we call T_H, must satisfy the first-order optimality condition

$$PV'(T) + a(T) = i[PV(T) + \phi^*] \tag{4-4}$$

Just as for the Faustmann solution, the optimal rotation should be at that age at which the increase in value from a marginal delay in the harvest date equals the opportunity cost of that delay. In this multiple-use formulation, delaying the harvest provides the benefit of increased timber value growth and in addition provides $a(T)$, a flow of amenity value over the period of delay. Timber should be held uncut until the rate of growth in the full asset value of the land and timber just equals the rate of return i available on comparable alternative investments.

Because of the varied mix of services they may represent, there is no a priori reason to expect the nonharvest values $a(n)$ to be monotonically increasing or decreasing with stand age. It is quite possible that one may find several ages meeting the first-order conditions, some of which are local minima or maxima. With $a(n)$ sufficiently large and increasing with stand age, we may find no age at which the first-order conditions are met, and should choose never to harvest.

Characteristics of the Hartman Harvest-Age Solution. If the maximum land value λ is less than zero, the land should be left unmanaged. Consideration of the full mix of multiple-use values makes the likelihood of such negative returns to management less than if timber values alone were considered. Not surprisingly, we would be less willing to harvest and abandon an area if it provides positive amenity services when stocked. It is interesting that the amenity values may be sufficient to justify perpetual harvest management on land which would appear to be not economically managed when timber alone is considered. In fact, even recreational values, supposedly diminished by timber management, may justify restocking, yet then be neither large enough nor increasing sufficiently with stand age to justify shifting from a typically short timber harvest rotation. This suggests a danger in making an arbitrary allocation of fully joint regeneration costs to the timber output alone.

In contrast to the Faustmann problem, the age of inherited stocks may matter in the Hartman problem. The harvest-age solution is unaffected unless the current age exceeds the Hartman age. However, if the inherited stand does exceed that age, it may be preferable to further delay harvest, or never to harvest. With timber value alone, the solution was to harvest such older stands immediately; the declining timber growth rate penalized any further delay. Here, with an initially clear stand, discounting may weigh against waiting for the high amenity values from older growth, and a short timber rotation may be called for. Yet if we inherit an older stand, the existing high flow of amenity values may be sufficient to justify preservation, especially if such old-growth lands are relatively scarce and so highly valued. For it to be economic to maintain an older stand uncut, the present value of the subsequent flow of amenity values must certainly exceed the net value of an immediate harvest.

In general, the Hartman rotation age will be somewhere between the Faustmann age and that age which would maximize the present value of the returns from the amenity services alone. The solution age will depend both on the total amenity benefits of a harvest cycle relative to the net timber receipts and on the separate relative growth rates in the amenity and timber values. When the amenity value of stocks rises with stand age, the Hartman rotation age T_H will be greater than the Faustmann age. If the amenity value flow is large and increasing with stand age, it may be optimal to leave the stand unharvested forever. In many areas where forage and increased water flow are important we might anticipate declining amenity values with stand age. When amenity values decline with stand age, the Hartman rotation T_H will be shorter than the Faustmann rotation. If amenity values are completely insensitive to stand age, the timber rotation will not be affected by these values (no matter how large the values).

One surprising implication is that introducing multiple-use values into the Faustmann model can lead to an increase in the average supply of timber. First, note that the higher average level of stocking that is often needed to maintain amenity services also allows for a greater sustainable harvest. That is, if the multiple-use harvest rotation is longer than the Faustmann rotation (amenity values rising with stand age), then the optimal rotation age may move closer to the age of maximum sustained timber yield. Second, as we have noted, multiple-use values may justify harvesting on some lands that would be uneconomical to manage for timber alone. There may be a positive overall net value from land that was not economic when timber values alone were considered. Under such circumstances, the supply of timber would again be increased. It is interesting that those amenity services which are in a sense competitive with timber harvesting in the short run (immediately decreased by timber harvests) may justify higher average levels of harvesting in the long run.

The comparative static results are somewhat more complicated for the Hartman solution than for the Faustmann solution. A higher timber price level can be shown to decrease (increase) the rotation age if the Hartman solution T_H is above (less than) age T_0, where T_0 is the Faustmann solution when there are no regeneration costs. An equal, proportionate increase in $a(n)$ for each age n lengthens (shortens) the rotation if the Hartman rotation is longer (shorter) than the Faustmann rotation. A proportionate increase in both the timber price and all amenity values leads to a decrease in the optimal rotation. It is a little surprising to find that an increase in the interest rate does not necessarily lead to a shorter rotation. With timber value alone considered, the higher rate raised the opportunity cost of delaying the harvest and a shorter rotation resulted. Here we have a potentially offsetting effect. The greater rate of discounting tends to raise the relative growth rate in the amenity value of the forest stand, $a(T)/\int_0^T(a(n)e^{-in})dn$. If the cumulative flow of amenity benefits is fairly high relative to the timber share, a higher interest rate can lead to longer rotation ages.

In general, because the amenity values of the forest may represent a varied mix of services, it will be difficult to guess whether the multiple-use values of timber stands are increasing or decreasing with stand age. While the aesthetic condition of the forest may improve with greater stand ages, other services such as water flow and even wildlife habitat quality might be favored by rather short rotations. The combined effect of several multiple-use services may offset any advantage gained from altering the timber rotation. This conclusion has led to some debate as to how sensitive the optimal rotation age might be to the inclusion of multiple-use values.

Indeed, it has been suggested that multiple-use values will have little impact on the optimal rotation age even when these values are uniformly increasing or decreasing with stand age. On the basis of their analyses of what they felt were reasonable (although hypothetical) multiple-use values, Calish, Fight, and Teeguarden concluded that there would be little sensitivity of the management solution to the introduction of stock-related amenity values.[13] In their analysis, the timber stand was a productive Douglas fir stand. Now, one should certainly be mindful that the opportunity cost of forgoing a timber harvest on a highly productive timber stand is likely to be far greater than the potential gain in amenity services from preserving the standing stock—except in special circumstances. Nevertheless, there are several reasons why one should be cautious in accepting a blanket conclusion that amenity service values do not matter. First, note that many public forestlands have very low timber productivity. Multiple-use values may be a dominant factor in determining the appropriate management of these poorer lands. As noted above, on lands of very low timber productivity, timber management might not be justified at all were it not for the multiple-use values. Further, when poor timber stands are inherited with older growth already supplying considerable amenity value, harvesting may not be desirable at all.

Perhaps more important, it must be recognized that very little is known about the multiple-use values of a single stand. It is not intuitively easy to understand how the amenity benefits of a single stand should be measured. Usually the multiple-use value from a single stand will depend upon the condition of neighboring areas of the forest. To meaningfully define the

values $a(n)$, they must probably be thought of as increments to value resulting from management of this one stand, given that the management of
associated areas of the forest is somehow already determined. The net amenity values from a stand will depend upon the nature of the management
selected for this stand and the surrounding area. Each alternative for management is likely to lead to a different pattern of amenity flows over time.

Further, the impact that one stand may at times have upon the services
available from a much wider area of forestland should not be underestimated.
A single highly visible harvest can reduce the amenity benefits of a considerable land area for many years. An old-growth stand among younger stands
might take on particularly great value as a critical element in wildlife habitat.
The value from improving the distribution of stand ages in the wider forest
area can at times be sufficiently great as to justify dramatic adjustments in
the optimal timber-harvesting strategy for the single stand. Scarcity can lead
to high values. It is doubtful whether it really makes sense to look at the
single stand in isolation if diversity of conditions across stands matters. We
will return to this point later in the chapter as we consider the simultaneous
management of several adjacent stands in a forest.

What proves to be of great importance to the harvest-age decision is the
rate of change in amenity values as the harvest age is delayed. As it turns
out, even very low absolute levels of stock amenity values can have a
profound effect on rotation-age solutions if these values change sharply with
stand age. On the other hand, very high levels of amenity values may have
little impact on the timber management decision if these values are not
especially sensitive to the stand age. The natural focus on relative level of
total benefits from timber versus amenity values can prove to be misleading.
Finally, it should be pointed out that the focus on the change in rotation age
can be a misleading measure of the impact of multiple-use values. A small
change in the rotation-age solution might have a nontrivial impact on the
value from the multiple-use services.

Variations in Harvest-Age Solutions

To illustrate some of the more interesting variations in the multiple-use
harvesting solution, we turn to one of the very few studies of optimal
multiple-use rotations, the investigation by Calish, Fight, and Teeguarden of
the relation of several nontimber services to the age of a single stand set
within a larger forested area.[14] The indices of multiple-use production they
employed were based on suggestive but limited empirical data. The indices
suggest that for certain game species, such as deer, the flow of service may
initially be an increasing function of stand age and then begin to decline with
higher stand ages. For water flow, the indices suggest that the current flow
of value may steadily decrease with stand age. Aesthetic value, according to

the indices, is likely to steadily increase with stand age. Some variations on the multiple-use values used in the Calish, Fight, and Teeguarden study provide the basis for the illustrations that follow.

The Amenity Values. Assume that a water-flow value function might be like that in figure 4-4. The cumulated present value from water flow over repeated harvest cycles would then be a decreasing function of the rotation age, as in figure 4-5. An aesthetic value function might be like that in figure 4-6. The present value from these aesthetic services over repeated harvest cycles would be an increasing function of the rotation age, as in figure 4-7. A summed value flow from water flow, aesthetic conditions, and hunting services (not illustrated) would appear as in figure 4-8. The combined value of these three services over repeated rotation cycles is given in figure 4-9.

Harvest-Age Solutions. We now turn to look at optimal harvest-age solutions based on the amenity values represented in figures 4-4 through 4-9.

The optimal harvest rotation for timber and water flow is illustrated in figure 4-10. The water-flow value function ψ from figure 4-5 is reproduced, along with a timber value function λ based on the yields and regeneration costs used in figure 4-1. The combined net present value from these two services is represented by ϕ, the solid line (with $\phi(T) = \lambda(T) + \psi(T)$). Maximizing the combined multiple-use value leads to the selection of the rotation period T_H. This rotation is here slightly shorter than the Faustmann rotation T_F. The shorter rotation is expected, since water-flow value is declining with stand age. Despite the rather small influence of the water-flow value on the optimal rotation, there is a significant effect on the level of service provided. In comparison to the Faustmann solution, the multiple-use rotation is seen to provide perhaps a 50 percent increase in the net present value from water flow.

The optimal rotation for timber and aesthetic services—a timber-aesthetics solution—is illustrated in figure 4-11. The aesthetic service value ψ from figure 4-7 is shown, along with the timber present value function λ. The combined value ϕ is maximized by the selection of rotation T_H, which is slightly longer than the Faustmann rotation T_F. The longer rotation is expected, since the aesthetic value flow increases with stand age.

It seems fair to say that in this solution the aesthetic values have little effect on the optimal solution. However, it should be noted that the amenity value function is continuing to increase with age, and that the overall land value ϕ is itself beginning to rise with age again. It is possible that if these curves were extended further we would find it best to forgo harvesting this stand. With a small increase in the amenity value this would certainly be the case. Finally, note that if such a stand were inherited at an age greater than T_H, it would probably be preferable to maintain the existing high amenity services rather than to begin a cycle of periodic harvesting.

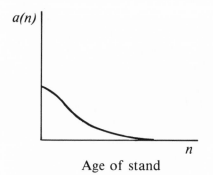

Age of stand

Figure 4-4. Value flow: water flow

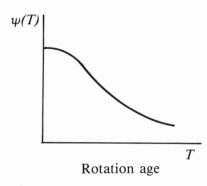

Rotation age

Figure 4-5. Present value: water flow

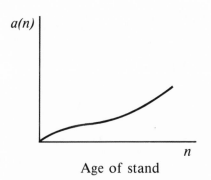

Age of stand

Figure 4-6. Value flow: aesthetic

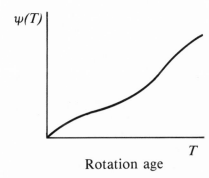

Rotation age

Figure 4-7. Present value: aesthetic

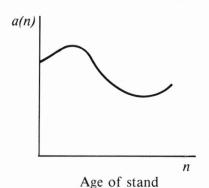

Age of stand

Figure 4-8. Value flow: all amenities

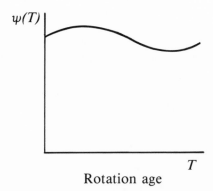

Rotation age

Figure 4-9. Present value: all amenities

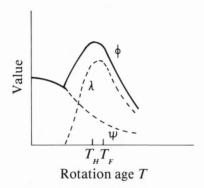

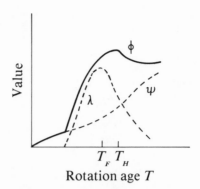

Figure 4-10. Multiple-use rotation: timber and water flow

Figure 4-11. Multiple-use rotation: timber and aesthetic value

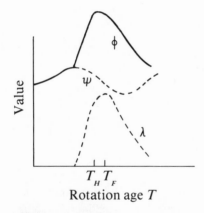

Figure 4-12. Multiple-use rotation: combined values

Key to figures 4-10 through 4-15:

λ = Present value: timber

ψ = Present value: amenities

ϕ = Present value: all outputs

$$(\phi = \lambda + \psi)$$

A combined multiple-use solution is presented in figure 4-12, where the combined stock amenity values from water flow, deer hunting, and aesthetics are represented by curve ψ from figure 4-9. The total value from timber and these other multiple uses is represented by curve ϕ. The multiple-use rotation T_H which maximizes the overall value is seen to be virtually the same as the Faustmann rotation T_F. This result is far from surprising, given the rather flat shape of the stock service value function ψ. The combined value from the amenity services of the stock is not particularly sensitive to stand age (although the composition of these services would certainly change with the selected rotation).

Curvature of the Value Functions and Harvest Solutions. The relative insensitivity of the harvest-age solutions to multiple-use values in the above

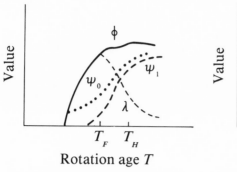

Figure 4-13. Multiple-use rotation: modified aesthetic value

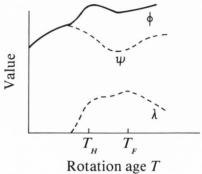

Figure 4-14. Multiple-use rotation: modified timber value

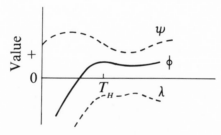

Figure 4-15. Multiple use and the decision to manage

examples is largely a result of the nature of the timber value function λ. This value function is sharply peaked, with the present value falling off rapidly for any adjustment to the rotation age away from the Faustmann solution. Unless the amenity services of the stock exhibit an offsetting sharp change in value with age, there will be little effect from the nontimber service values on the overall harvest solution. It may be noted that simply raising the amenity values by a constant amount at all rotations will have no effect on the optimal harvest-age solution. In figures 4-13 and 4-14 we illustrate the importance of relative change in value.

A modified aesthetic value solution is presented in figure 4-13, where a modified aesthetic value function ψ_1 is combined with the previous timber value function λ. Compared to the aesthetic value function from figure 4-7 (marked in figure 4-13 as ψ_0), the new function ψ_1 has a lower value at every rotation age. This does not mean that it plays less of a role in the harvest timing decision. Notice that at intermediate rotation ages the new function ψ_1 rises much more steeply than the old one. As a result, the combined timber and aesthetic value is now maximized with the choice of a harvest rotation T_H, which is significantly longer than the optimal timber rotation

T_F. We would argue that there is insufficient empirical evidence to allow us to determine which aesthetic value function—that in figure 4-7 or that in 4-13—is more reasonable.

A modified timber value solution may also be offered. The sharp peak of the timber value function is by no means to be generally expected. If steady or increasing rates of timber value growth continue over several years, the present value function for timber will have a broad top, and multiple-use considerations will be more likely to determine the choice of rotation. For example, if a stand undergoes periodic thinnings, the stand value is then likely to be fairly insensitive to the timing of the ultimate regeneration harvest. In figure 4-14, a modified timber value function λ is drawn which exhibits little sensitivity to the chosen rotation. Here it is seen that the combined multiple-use value from water, deer habitat, and aesthetics (ψ from figure 4-9) can indeed have a dramatic effect on the choice of rotation. It should be added, however, that there is little opportunity cost to selecting the wrong rotation age in this case, with both timber and amenity values fairly constant over a wide range of rotation ages.

Uneconomic Timber Values. One final aspect of the multiple-use solution is illustrated in figure 4-15. Multiple-use values may justify the harvesting of lands which are not economically managed for timber alone. Suppose that because of high regeneration, management, or harvest expenses the present value from timber λ were negative at any rotation age, as in figure 4-15. Were this timber value alone to be considered, the land should not be managed. However, when the multiple-use values of the standing timber are considered, the overall present value of the land ϕ is positive and management is justified, with an optimal rotation age T_H. An example of exactly this case is provided in chapter 5, where it is shown that water-flow values justify otherwise uneconomic timber management in the Colorado Rockies.

Before going on, two matters should be made quite clear. First, multiple-use values will not always justify the harvesting of uneconomic timber stands. Second, it should be realized that we have provided a very limited description of the forest management problem, allowing the manager only the choice of the age at which the full stand is harvested and the decision of whether to initially regenerate the stand. There will in general prove to be a variety of options for managing lands for amenity services alone (perhaps with some limited low-cost timber clearing). Such options may well provide higher overall present values than would any timber management regime.

The Present-Value Production Possibility Curve

An alternative and useful means of demonstrating the multiple-use harvest solution requires development of a curve analogous in some ways to the isocost curves of chapter 3. Such a "production" possibility curve will illustrate the possible tradeoffs between the present values of timber and stand amenities for a given land area and regeneration expenses. Depending upon the choice of rotation age, various combinations of λ (net present value from timber) and ψ (net present value of the amenity service flow) are possible. The production possibility curve shows the maximal combinations of present values from these two output groups; that is, it represents those combinations of λ and ψ such that no higher value of one is attainable without sacrificing some of the other.

Much of the discussion in chapter 3 could be reinterpreted in terms of present values from output benefits and present values of costs. We shall describe only the compatibility of outputs over time, demonstrating that an overall compatibility in use between services may exist despite an apparent incompatibility at any point in time.

The Timber/Water-Flow Production Possibility Tradeoff. The tradeoff between the present value from timber receipts and the present value of water flow is developed in figure 4-16.[15] Quadrant II of the figure reproduces (with

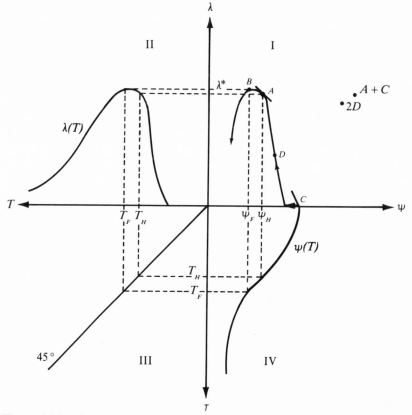

Figure 4-16. The tradeoff between timber and water-flow values

a reversal of the axes) the timber present-value function $\lambda(T)$ from figure 4-10. Note that the rotation period T increases to the left. The Faustmann rotation age T_F which provides the maximum present value of timber receipts is marked. In quadrant IV, $\psi(T)$ (also from figure 4-10) represents the present value from water flow as a function of the rotation length. The optimal rotation for water flow alone would require keeping the stand clear of timber growth. Quadrant I thus illustrates the various possible combinations of timber value λ and water-flow value ψ.

For example, if the Faustmann rotation T_F were chosen there would be value λ^* from timber (read from quadrant II). With this same rotation period T_F there would be value ψ_F from water flow (read from quadrant IV). Point B on the value possibility curve in quadrant I is the result of selecting the Faustmann rotation. The arrow on the present value curve indicates the direction of change resulting from increased rotation length. Shorter rotations, for instance, provide higher water value and lower timber value. Long rotations result in low values from both timber and water flow.

Maximizing the overall sum of the present values $\lambda + \psi$, subject to the present-value possibilities of quadrant IV, is achieved by choosing to produce at point A in quadrant IV. Point A indicates the tangency of the highest isovalue line $\lambda + \psi = \phi$ to the value possibility curve. This solution corresponds to selecting a rotation age T_H somewhat shorter than T_F, the Faustmann age.

The Response to Price Changes.   By parametrically varying the price level of the amenity services, the tradeoff curve can be used to illustrate the sensitivity of the harvest-age solution to errors in value estimation. Suppose our estimates of the water-flow values might be too low by a factor of two. Maximizing the total value $\lambda + 2\psi$ would reflect this higher value, and is seen to result in the choice of a solution at point C in figure 4-16. With the doubled water-flow value, the land should be kept clear and no timber value produced. This is quite a drastic change in management. In contrast, lowering the water-flow value can be shown to have little effect. Lower water-flow values result in a multiple-use solution somewhat closer to the Faustmann solution B, but not much different from the original solution A.

The frontier of the timber/water-flow value tradeoff curve in figure 4-16 is roughly convex to the origin between points A and C. As in chapter 3, with a convex production possibility frontier we anticipate specialization in land use to be optimal, and also expect drastic adjustments in the optimal land use in response to a sufficient price change. With a concave production possibility surface, more gradual responses to price changes would be seen and more balanced mixes of product values might be optimal.

Specialization and Convexity of Production Possibilities.   Convexity of the frontier of the production possibility curve (convex or bowed toward origin) may be thought of as reflecting some incompatibility in the production of present value from the two services. Incompatibility of product-value flows is defined here in the following way. Suppose there were two land units, each with identical convex production possibility frontiers. In that case, a balanced mix of present values from the two outputs is best produced by having each unit separately specialize rather than by producing an identical mix of services from each unit.

Consider the output mix $A + C$ in figure 4-16 that results from the separate specialization of the two units, one producing A while the other produces C. The output $A + C$ cannot be matched if the two areas each produce an identical output mix. For example, if D were produced on each area, the total production $2D$ would be less than $A + C$. Note that specialization here means the selection of a rotation period best-suited to one particular output or service, such as the Faustmann rotation for timber. Despite such specialization, significant value from other products can result on each land unit.

The likely classification of products as compatible or incompatible is by no means immediately apparent. In figure 4-17, present-value curves for

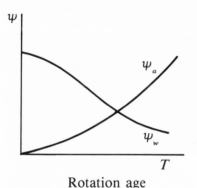

Rotation age

Figure 4-17. Present values

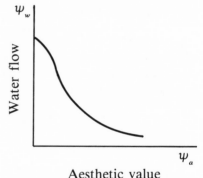

Aesthetic value

Figure 4-18. Convex tradeoff curve

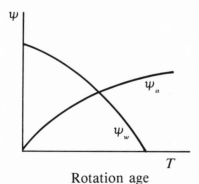

Rotation age

Figure 4-19. Present values

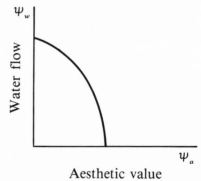

Aesthetic value

Figure 4-20. Concave tradeoff curve

water flow and aesthetics are labeled ψ_w and ψ_a, respectively. The two services, water flow and aesthetic condition, would seem to be incompatible. One service is favored by very short rotations, the other by long rotations. In figure 4-18, the present-value tradeoff curve for these services is shown to be highly convex, and the services are indeed incompatible.

However, suppose the present-value flows from these two services were slightly different, as represented in figure 4-19. Water-flow value ψ_w is still favored by short rotations and aesthetic value ψ_a is favored by long rotations, but the curvature of the value functions has changed. In figure 4-20 the corresponding present-value tradeoff curve for these services is highly concave, and the products are compatible. They are compatible in the sense that we would choose to produce these services together, rather than specialize in the production of the services on separate land areas. Concavity of the tradeoff curve results because the underlying value curves are now themselves concave.

Extending the Production Possibility Approach. The production possibility curve described here was developed in the context of the Faustmann-type problem, with a focus on the rotation age for the single stand. The rotation age provided the only means for altering the mix of services. The production possibility approach is not restricted to such a simple view of management. More generally, one could consider the output from larger units of several related stands to be managed over time under a variety of possible harvest and nonharvest activities. One could also allow for changing prices and costs over time. Much as in chapter 3, a cost-function view of the production possibilities could also be developed, illustrating the tradeoffs at several (present-value) levels of cost.

Although extending the graphic approach might be useful, it can become rather complex. Nevertheless, displaying the present values available from each of the multiple uses focuses attention on the overall conflict or compatibility between outputs. Doing so may aid the planner in formulating improved sequences of management activities.

Linear Programming Models for Multiple-Use Management

A variety of linear programming models for timber harvest scheduling that provide practical analogs to the Faustmann model are described by Johnson and Scheurman.[16] The initial FORPLAN models used for the forest planning effort of the USDA Forest Service were multiple-use extensions of these stand-harvesting models.[17] In its simplest form, with a focus on the harvest timing for even-aged timber stands, the FORPLAN model is the analog to the Hartman model. In a linear program timber prices, costs, and the amenity values of stands can be easily allowed to change over time. In addition, a wide variety of management regimes (differing in more than harvest date) may be considered for each land area. Further, the linear program is not inherently limited to a focus on management of even-aged stands—a limitation that so restricts the Faustmann-type formulation.

The essential features of a linear programming model for forest planning can be simply presented. For each land unit in the forest there are a variety of possible management "activities." These activities correspond to a sequence of treatments and harvest dates over time. Choosing a management sequence for a land unit results in a particular time flow of timber harvests, multiple-use services, and management costs. An economic objective would call for the allocation of each land class to management sequences that would maximize the present value from the overall stream of goods and services over time.

The model can be written, in its simplest form, as

$$\text{Max} \left\{ \sum_{\iota=1}^{L} \sum_{s=1}^{S_\iota} [\phi_{\iota s} x_{\iota s}] \right\}$$

subject to the area constraint:

$$\sum_{s=1}^{S_\iota} x_{\iota s} = A_\iota \qquad \text{for } \iota = 1, \ldots, L,$$

where $x_{\iota s}$ is the acreage of land unit ι assigned to management activity s, A_ι is the acreage in land unit ι, L is the number of land units, S_ι is the number of management activities considered for land unit ι, and $\phi_{\iota s}$ is the present value per acre from the flow of outputs and costs when land unit ι is managed under activity s.

Specifically, ϕ the net present value from land is defined by

$$\phi_{\iota s} = \sum_{t=1}^{N} [p_{\iota st} h_{\iota st} + a_{\iota st} - c_{\iota st}] \delta^t$$

where N is the number of planning periods, $p_{\iota st}$ is the received price for timber in time t from unit ι under management s, $h_{\iota st}$ is the volume per acre harvested in time t from unit ι under management s, $c_{\iota st}$ is the cost per acre in period t from unit ι under s, $a_{\iota st}$ is the amenity service flow at time t from unit ι under management s, and δ^t is the discount factor for time t.

In the earliest applications of FORPLAN, a land unit was typically a timber stratum, made up of all lands of similar productivity, timber species, and timber age. The individual stands within such homogeneous lands might be distributed widely across the forest. There was little attention paid to spatial location of the individual stands within the timber stratum. When the land units in the linear programs represent single-aged timber classes, it is easy to see that the linear programming model is closely related to the Faustmann-type harvest rotation problems. Indeed, with prices and costs constant and with management choices corresponding to all possible sequences of harvest rotation age, this linear program would be equivalent to the Hartman problem. It would differ only in that the linear program offers a discrete set of harvest-timing options, rather than having the continuous choice in harvest age of the Faustmann-type models. The solution to the linear program would give a sequence of harvest dates meeting the same first-order conditions described for the Hartman problem.[18]

The essential weakness of all Faustmann-type models of multiple-use management, including linear programming models based on homogeneous land classes, is their lack of attention to the production linkage between individual stands. Although the early FORPLAN applications did schedule the harvest for several areas at once, there was no accounting for the inherent linkage between stands in the production of multiple-use services. In general, treatment of one stand is likely to affect the amenity services provided from adjacent units of the forest.

Consideration of the spatial location of stands is also critical to the accurate determination of the costs of harvest access. The early FORPLAN models carried no spatial information about the size of individual stands within the land class, and no information on the transportation network or on the likelihood of a harvest on the lands adjacent to these stands. It is very hard, if not impossible, to relate amenity flow to the age of a homogeneous forest land stratum that might be widely dispersed across the forest in an unknown (within the model) manner.

A more appropriate treatment of multiple use in a linear programming framework may be accomplished if geographical management zones are identified. Define a management zone as a small watershed encompassing perhaps several stands of timber (and nontimbered lands). The size of such an area and the natural features which serve to bound it should allow better representation of the amenity benefit flows and road-access conditions. We should be able to relate amenity value flows to the overall conditions of a well-defined geographical zone of the forest. In fact, the latest applications of FORPLAN do try to account for the spatial location of timber stands, and associate road construction and nontimber amenity values with the management emphasis chosen for geographic forest zones (see chapter 10 for details).[19]

Summary Comments on the Faustmann Methods

The multiple-use version of the Faustmann model provides an elegant but incomplete description of the forestry management problem. In a more general view of forest management there would be a variety of activities and harvest practices considered. The assumptions of unchanging prices, costs, and productivity are also most unlikely to hold in fact. Despite the simplifications, the essential economic features of timber stock management are clearly developed in the multiple-use Faustmann problem. Land should be managed so as to maximize the net present value from the flow of harvests and resource services of the forest. Timber should be held uncut as long as the value of the land and timber is increasing at a rate greater than the rate of return on alternative investments. The timber should be harvested at that date when the growth in asset value of the stand has declined to just equal the market rate of return. A maximum sustained-yield policy for providing timber does not reflect the opportunity costs of holding timber stocks.

For the multiple-use version of the Faustmann model, the focus on the single stand is a considerable weakness. The nontimber values are intended to reflect the incremental value from the timber stock on the single stand, with consideration given to its setting within the larger forest area. The difficulty arises in the assumption that $a(n)$, the amenity value flow from a stand of age n, will take the same value in each subsequent rotation. In fact, the condition of the adjacent forestlands might be quite different the next time the stand reaches any particular age n. As a result, $a(n)$ might take very different values over successive rotations.

The multiple-use version of the Faustmann model can be best viewed as a description of the "steady state" forest for which conditions are constant over time. Under more general conditions, one would have to simultaneously consider the optimal management of several related stands in order to properly reflect multiple-use values. Although linear programming models do simultaneously schedule treatment across all forest stands, they do not necessarily reflect any inherent production jointness between stands. The structure of the early FORPLAN linear programming models used for national forest planning precluded any accounting for the link between stands in providing multiple-use services. Newer versions of FORPLAN attempt to schedule harvests within a geographic zone of the forest, with some attention being given to the impact of these harvests on the overall multiple-use services of the forest zone. In section 3 of this chapter we will describe a simple extension of the Faustmann-type model that illustrates, in a stylized manner, the basic features of multiple-use harvest scheduling for several related timber stands within one geographic zone.

3. Multiple-Use Management of Related Forest Stands

In considering a discrete-time model of multiple-use management which takes into account the interdependence among stands, we do not mean to suggest a practical method of forest management; rather, we mean to provide a framework for describing some of the important elements of the economics of multiple-use management that are embedded (to some extent) in the newer forest-planning models. Many of the interesting features of multiple-use management become obscured because of the structure of simpler presentations. Unfortunately, the introduction of even limited realism into our model makes for a complicated problem that is not amenable to complete analysis.

As before, the focus is on the optimal scheduling of harvests. A forest planning unit is assumed to be made up of a number of timber stands, each of which may be in a different age class. We look at two broad sets of demand-based values, the value from timber stumpage sales and the value received from the services of the land and its stock of vegetation. The demands are assumed to be unchanging over time and known with certainty. The productivity of the forest is similarly assumed to be known and unchanging, and for convenience this productivity is taken to be uniform across the planning unit.

Amenity values are taken to be related to the overall conditions of the forest unit. In particular, we assume that the amenity values depend, in some perhaps complex manner, upon the mix of ages across the set of timber stands. In this manner the amenity services are related to the overall pattern of diversity resulting from the treatment of the individual stands. The chosen sequence of harvests over the stands results in a flow of timber-sale revenues and a changing pattern of timber stocks on the forest unit. The future flow of amenity values depends upon this progression of the forest condition. As a result of the dependence of amenity values on mix of stand ages, the optimal harvest decision for any single stand must also depend upon the current mix of age classes across the planning unit. Any harvest may then be as much motivated by the goal of improving the future distribution of age classes as by the immediate harvest revenues.[20]

The rather simple view of the management problem presented here proves sufficiently rich for expository purposes and allows for easy comparison of our model to the traditional Faustmann model, with its focus on harvest timing for the single stand. A choice among a richer set of treatments would no doubt indicate much more flexibility in adapting to multiple-use demands than is apparent when we consider only the timing of harvests. Our discussion begins with a description of a general model and theoretical results. Some illustrative examples follow (in section 4) that more clearly highlight the essential conclusions.

The Multiple-Use Values

Amenity Values. The net amenity values of the forest unit will depend upon the mix of stocks remaining after the harvest. We will represent the flow of amenity values from the forest site at time t by the function $A(Z_t)$. The timber stock vector Z_t represents the stocks held following the harvest at time t, and is defined as

$$Z_t = [z(0,t), z(1,t), \ldots, z(N_t,t)]$$

Each term $z(n,t)$ gives the acreage of land stocked with timber of age n held during time period t (term $z(0,t)$ is the acreage of cleared land). N_t is the age of the oldest timber held at time t. The function $A(Z_t)$ is assumed not to change over time, although changing stock conditions may lead to a change in amenity value in each time period.

The amenity function $A(Z_t)$ should be viewed as a sum of willingness-to-pay values over all individuals who value the condition of the standing stock on this land area.[21] The interests of the set of such individuals may be diverse. Holders of grazing rights might prefer the area if it were predominantly clear; hunters might prefer to have some clearings in a mix of older stands; other recreational users might prefer the area to be predominantly in the older age classes.

Changing the proportion of the forest acreage held in each age class alters the flow of amenity value. To make our notation fairly consistent with the Hartman version of the Faustmann model, we will represent the value of a marginal increase in the holding of stock aged n at time t by the expression $a(n,t)$; that is, we define

$$a(n,t) = \partial A(Z_t)/\partial z(n,t)$$

This marginal value is the sum of the amounts individuals would be willing to pay for a marginal increase in the acreage of stock aged n. Note that (despite our notation) the marginal flow of value from a stand aged n may now depend upon Z_t, the full mix of stocks across the planning area. A nonlinear dependence of multiple-use value on the mix of stand ages makes the harvesting decision richer than for the earlier Hartman problem.

Harvest Value. The timber harvest value is represented by the function $\beta(H_t)$, where H_t is the total volume of harvest at time t. The harvest in time t, H_t is defined by

$$H_t = \sum_{n=1}^{Nt} V(n)h(n,t)$$

where $V(n)$ is the merchantable volume per acre from stands of age n,[22] and $h(n,t)$ represents the acreage of stock in age class n harvested at time t.

The function $\beta(H_t)$ is a willingness-to-pay measure of value based on the demand for timber. The function $\beta(H)$ is assumed not to change over time. Harvest costs are presumed to be borne by the purchaser and so are reflected in this value function. The marginal value $\beta'(H_t)$ of an increment in the harvest volume supplied is the net price at time t that would be offered for timber in a competitive market, and will be represented by $P(t)$. Unless otherwise stated, we assume that $\beta''(H) < 0$ so that the price of timber declines with an increase in supply. If the level of harvest from our site does not influence the timber price ($\beta''(H) = 0$), then $\beta(H_t)$ represents the net revenues from timber sales on the planning unit. Otherwise, $\beta(H_t)$ includes a consumer surplus component in addition to these revenues.

Costs. The costs associated with the regeneration of timber are given by a constant C times the acreage harvested, and are represented by the function $C(h_t)$, with

$$C(h_t) = C \cdot [\sum_{i=1}^{N} h(t,n)]$$

where h_t is a vector listing the acreage harvest from each age group:

$$h_t = [h(1,t), h(2,t), \ldots , h(N_t,t)]$$

The Forest Unit Production Equations

The stock of timber held at time t, before harvest, is represented by vector X_t, with

$$X_t = [x(1,t), x(2,t), \ldots , x(N_t,t)]$$

where $x(n,t)$ is the acreage of stock in age class n held at the beginning of time period t. In each time period, the manager must decide whether to harvest an age class immediately or to leave the stock standing. The harvested lands are regenerated and by time $t + 1$ are stocked with timber in age class 1. Timber unharvested at time t will grow in age, and is reconsidered for harvest at time $t + 1$.

The production system at time t is represented by the set of equations:

$$
\begin{aligned}
x(1,t+1) &= z(0,t) = \sum_{n=1}^{Nt} h(n,t) \\
x(2,t+1) &= z(1,t) = x(1,t) - h(1,t) \\
&\quad \vdots \\
x(N_{t+1},t+1) &= z(N_t,t) = x(N_t,t) - h(N_t,t)
\end{aligned}
\tag{4-5}
$$

That is, the stocks Z_t held during period t depends upon the harvest from stocks X_t held at the beginning of time t. The stocks X_{t+1} held to begin the period $t+1$ result from the aging of stocks Z_t held during period t.

Harvests must be non-negative and cannot exceed the available stock, as reflected in the constraints

$$0 \leq h(n,t) \leq x(n,t) \qquad \text{for } n = 1, \ldots, N_t \qquad (4\text{-}6)$$

The Multiple-Use Management Problem

The overall problem facing the land manager is to select a sequence of harvests and stocks over time $\{h_t\, X_t\}$ so as to maximize the net present value from all current and future flows of harvests and resource services of the area.[23] The problem can be stated as

$$\Phi(X_0) = \underset{\{h_t\}}{\text{Max}} \left\{ \sum_{t=0}^{\infty} \delta^t [A(Z_t) + \beta(Ht) - C(h_t)] \right\}$$

where the maximization is constrained by the production equations (4-5) and (4-6).[24] The discount factor δ^t represents $1/(1+r)^t$, with r the (discrete time) rate of interest for discounting future consumption flows. The term $C(h_t)$ gives the regeneration costs associated with newly harvested lands. To simplify, we do not allow land to be left unregenerated.

This multiple-use management problem can be presented as a sequential maximization. Let $\Phi(X_t)$ represent the maximum present value attainable from land with given current stocks X_t. In each time period we must solve the problem

$$\underset{h_t, X_t}{\text{Max}} \left\{ A(Zt) + \beta(H_t) - C(h_t) + \delta\Phi(X_{t+1}) - \Phi(X_t) \right\}$$

subject to the production constraints (4-5) and (4-6).

The term $\Phi(X_t)$ may be interpreted as the overall (multiple-use) asset value of the forest management unit with stock X_t. The sequential formulation of the management problem can be interpreted as follows. In each period choose that pattern of harvests across the land area which maximizes the net flow of value plus the appreciation in the asset value of the land unit and its stock.

The advantage to the sequential formulation of the problem is that it focuses attention on the immediate harvest decision. It is sensible to consider most carefully those projects to be implemented in the near future. In practice, however, one might choose to use other approaches to solve the management problem.

The Optimality Conditions

The first-order conditions describing the optimal choice of harvests can be expressed in a fairly clear and concise form. For each age class we must choose either to harvest or allow further growth, selecting that activity with the highest return. Let us represent the value of a marginal increase in the acreage of age class n in time t by $\phi(n,t)$, so that

$$\phi(n,t) = \partial\Phi(X_t)/\partial x(n,t)$$

This marginal value is a function of the stocks held at time t.

The first-order necessary conditions to be met by the optimal harvests can be given as

$$\phi(n,t) = \text{Max}\{[P(t)V(n) + \phi(0,t)]; [a(n,t) + \delta\phi(n+1,t+1)]\} \quad (4\text{-}7)$$

with

$$h(n,t) = x(n,t) \qquad \text{if} \quad \phi(n,t) = P(t)V(n) + \phi(0,t), \text{ and}$$
$$\phi(n,t) > a(n,t) + \delta\phi(n+1,t+1)$$

$$h(n,t) = 0 \qquad \text{if} \quad \phi(n,t) = a(n,t) + \delta\phi(n+1,t+1), \text{ and}$$
$$\phi(n,t) > P(t)V(n) + \phi(0,t)$$

$$0 < h(n,t) < x(n,t) \quad \text{if} \quad \phi(n,t) = P(t)V(n) + \phi(0,t), \text{ and}$$
$$\phi(n,t) = a(n,t) + \delta\phi(n+1,t+1)$$

where we have used $\phi(0,t)$ to substitute for $a(0,t) + \delta\phi(1,t+1) - C$, reflecting the assumption that land cleared at time t is regenerated and grows to age one during the period after harvesting. (Definitions for the terms in equation [4-7] appear in table 4-1.)

The optimality conditions can be interpreted as follows. For each age class, the landholder should select that treatment (harvest or growth) that

Table 4–1. Definitions of Terms in Equation 4–7

$x(n,t)$	=	acreage of stock aged n held to begin time period t
$h(n,t)$	=	acreage harvest at time t from stock aged n
$z(n,t)$	=	acreage of stock aged n held during time t
$V(n)$	=	per-acre volume of timber for stock aged n
$P(t)$	=	value of a marginal increase in harvest volume at time t
$a(n,t)$	=	amenity value of a marginal increase in land with stock aged n during time t
$\phi(n,t)$	=	the marginal value of land with stock aged n at time t
C	=	per-acre cost of regenerating harvested land
δ	=	$1/(1+r)$ the discount factor

provides the highest marginal return. Under an optimal treatment schedule, the rate of return from a marginal unit in stock (of any age) will just equal the return available on alternative market investments. For example, timber should be retained uncut as long as the marginal benefits $a(n,t)/\delta$ + $\phi(n+1,t+1)-\phi(n,t)$ from amenity flow and stand-value appreciation equal $r\phi(n,t)$, the potential return expected from investing the value of the stocked land at the market rate. Stocks are to be held uncut as long as the flow of amenity value and the appreciating asset value of the land and stock continue to yield the required rate of return. The stocks are to be harvested at that date when further delay in the harvest would result in an insufficient rate of return over the period of delay.

Suppose a stand is optimally harvested at time t when it has reached age T. Further assume that we would be indifferent to a very short delay in the date of this harvest. By equation (4-7), the optimal harvest date then satisfies

$$\phi(T,t) = P(t)V(T) + \phi(0,t) = a(T,t) + \delta[P(t+1)V(T+1) + \phi(0,t+1)],$$

or

$$r[P(t)V(T)+\phi(0,t)] = [\phi(0,t+1)-\phi(0,t)] + V(T+1)[P(t+1)-P(t)]$$
$$+ P(t)[V(T+1)-V(T)] + a(T,t)/\delta \qquad (4\text{-}8)$$

That is, at the optimal harvest date the rate of appreciation in the total asset value over the period of delay should have declined to just equal r, the rate of return available on alternative investments. The appreciation in the asset value now reflects the potential capital gains on the land and timber stocks, as well as the timber volume growth and the flow of amenity value over the period of delay. Compare this harvest-age condition to the virtually identical (although continuous-time) condition in equation (4-2''') describing the optimal harvesting of single stands when prices change over time. Despite similarities, the dependence of the marginal values in equation (4-8) on the full mix of stocks makes for a more complex harvest schedule.

An Alternative View of the Harvest Timing Conditions. The conditions for the optimal choice of harvest age, as given by equation (4-7), can be written in an alternative form that focuses attention more clearly on the harvest timing. The first-order conditions indicate that land regenerated at time k should be optimally harvested at the age T that solves

$$\phi(0,k) = \underset{T}{\text{Max}} \left\{ P(T+k)V(T)\delta^T + \sum_{n=0}^{T-1} a(n,k+n)\delta^n \right.$$
$$\left. - C + \phi(0,T+k)\delta^T \right\} \qquad (4\text{-}9)$$

That is, the value of a small increment of clear land at time k equals the maximum present value from the resulting incremental flow of harvest and amenity services up to the harvest date, plus the present value of this land when it is cleared at the end of the harvest cycle. This form of the optimality conditions emphasizes the analogy to the Hartman model, and the differences. The differences arise because the timber price $P(t)$ and the amenity flow values $a(n,t)$ are now initially unknown and depend upon the selected harvest pattern for the whole area. Further, the land value at the initial time $\phi(0,k)$ can no longer be assumed to equal the land value $\phi(0,T+k)$ at the harvest date. By the harvest date, the forest may look very different than it did initially. As a result, the next harvest rotation may differ from the current rotation on this particular age class within the forest planning area.

Solving the Harvest Scheduling Problem.   The actual solution to the management problem will usually require an iterative search procedure which takes advantage of the recurrence relation in the conditions in equation (4-7).[25] In some cases a solution can be very simply found. Suppose the value functions $A(Z)$ and $\beta(H)$ were linear, so that the marginal values $P(t)$ and $a(n,t)$ were given constants. Then the equation (4-7) is trivially solved by a backwards recursion once we specify a terminal date N and the values $\phi(n,N)$ of stock held at the terminal date. More generally, an iterative procedure is required because the marginal values for timber and stock services depend upon the selected pattern of harvesting across the forest area.

Before providing illustrative solutions to a simplified version of the general multiple-use problem which can be easily solved by dynamic programming methods, we will discuss those features of the multiple-use harvesting solution that may be inferred from the optimality conditions for harvesting.

The Multiple-Use Harvest Solution

In presenting some results describing the pattern of harvests and stock holding under economic multiple-use management, our focus is largely on the long-run solution. Under the assumptions that demands and forest productivity are constant over time, we anticipate that the forest area would eventually settle into a stable pattern of supply. One example of such a steady state is the even-flow forest. On a forest of this kind there is an equal acreage in each age class, and this distribution of age classes remains identical in each period as the oldest age class is harvested each year. The multiple-use forest need not converge to an even-flow condition; rather, periodic cycles with fluctuating harvest levels may be preferred. Further, if an even-flow state is desirable, it may not resemble that state typical for timber management solutions. We may choose to permanently maintain some areas as clearings or old-growth habitat while the remaining area is managed under a more typical even-flow harvest cycle. The Hartman problem is shown to be an appropriate description of the economically managed multiple-use forest in an even-flow steady state.

The following discussion is complex, but many of the basic results can be anticipated from an understanding of the optimal management of economic assets over time.[26] The approach taken is based on the harvest-age condition given in equation (4-9). For convenience the continuous-time version of this optimality condition will be used.

The Harvest Timing Condition. With continuous time, an expression analogous to the harvest timing condition in equation (4-9) is given by

$$\phi(0,k) = \underset{T}{\text{Max}} \left\{ P(T+k)V(T)e^{-iT} + \int_0^T [a(n,k+n)e^{-in}]dn \right.$$
$$\left. + \phi(0,T+k)e^{-iT} - C \right\} \tag{4-10}$$

Optimality of a harvest at stand age T requires

$$\dot{P}(t)V(T) + \dot{\phi}(0,t)$$
$$+ \left\{ P(t)V'(T) + a(T,t) - i[P(t)V(T) + \phi(0,t)] \right\} = 0 \tag{4-11}$$

where t is the harvest date. Equation (4-11) is the continuous-time version of the harvest timing condition in equation (4-8), and has the same interpretation. It can also be seen as a more general version of the harvesting condition in equation (4-4) from the multiple-use version of the Faustmann problem. It differs from that equation by the inclusion of the first two terms, which represent the net capital gains from the possible changes in timber price and land value over time (compare to equation [4-2′′′]).[27]

Steady State Solutions. Define a forest area as being in a steady state condition if the age distribution and harvest repeat cyclically. One such steady state would be the even-flow condition in which the age distribution and the harvest mix remain constant over time. In a steady state, the stumpage price, the marginal value of the amenity services of stocks, and the marginal value of the land itself repeat in value cyclically with a period T equal to the harvest rotation interval. As a result, the marginal value of land held at time t can be expressed as

$$\phi^*(t) = \left\{ P(t)V(T)e^{-iT} + \int_0^T [a(n,t+n)e^{-in}]dn - C \right\} / \{1 - e^{-iT}\} \tag{4-12}$$

For a steady state solution to be optimal, the harvest optimality condition
in equation (4-11) must also hold; that is, we should be indifferent to a slight
delay in the harvest date for a small area of land. Now substitute $\phi^*t)$ into
equation (4-11) for $\phi(0,t)$. Take the derivative of equation (4-12) with respect
to time to find the value $\dot{\phi}^*(t)$, and substitute this into the optimality condition of equation (4-11). The result characterizes those steady states that are
consistent with optimality. In particular we find that a steady state must
satisfy

$$\frac{\{\dot{P}(t)V(T)+\int_0^T[\dot{a}(n,t+n)e^{-in}]dn\}}{[1-e^{-iT}]} =$$

$$-\{P(t)V'(T)+a(T,t)-i[P(t)V(T)+\phi^*(t)]\}$$

(4-13)

Note the correspondence between the right side of equation (4-13) and
equation (4-4), the condition for optimality in the Hartman problem. Suppose that the harvest cycle T is equal to the Hartman age T_H that would be
optimal with the timber price $P(t)$ and the sequence of amenity values
$a(n,t+n)$. Observe that as prices may be changing throughout the cycle, the
currently appropriate Hartman age will also be changing over time. The right
side of equation (4-13) would then be equal to zero. The right side of
equation (4-13) will be positive if the harvest rotation cycle is greater than
the Hartman age. It will be negative if the harvest cycle is shorter than the
Hartman age.[28] Using these conclusions, we can characterize those steady
states which are consistent with harvest optimality.

Even-Flow Steady States. Quite clearly there is at least one possible steady
state—the even-flow steady state with the harvest at the Hartman rotation.
Further, there can be no even-flow steady state except at an age which solves
the Hartman problem. This is consistent with the area being partially managed and some land set aside as clearings or old growth.

Let us see why a Hartman age is required for an even-flow steady state
solution. If the forest is harvested on a Hartman rotation, the right side of
equation (4-13) is zero. If it is harvested at any other age, the right side of
equation (4-13) is not zero, and there must be some change in the prices
over time (see left side of equation [4-13]). Changing prices over time are
not consistent with the forest being in an even-flow condition. Only the
Hartman solution is possible.

It is worth noting that setting aside some of the land area as permanent
clearings or old-growth stands is consistent with the forest being in an even-
flow condition.[29] Moreover, such solutions will often be desirable. To improve water flow, the maintenance of some clearings while managing the

remainder of the forest on a normal timber rotation can be superior to trying to provide a balanced mix of both outputs by imposing a short rotation on the whole area. (See the discussion on the convex water-flow/timber tradeoff curve in figure 4-7, in section 2 of this chapter.) Similarly, meeting the old-growth needs of wildlife will usually be better accomplished by setting aside certain areas of old growth than by selecting very long rotations. Long rotations may provide amenity value, but at a great expense of forgone timber revenue.

Cyclical Steady States. In the solution to the multiple-use problem there is a possibility of cyclical steady states, which is not possible if timber alone is valued. In the special case of valuing timber alone, it is clear that the only possible steady state solution is an even-flow forest with timber harvested at a Faustmann rotation age.[30] In the multiple-use case there may be fluctuating harvest levels over time, with the period of the harvest cycle in some sense an averaged Hartman age.

Suppose that the harvest cycle length T is not equal to the current Hartman age. We know then that the composite set of prices on the left side in equation (4-13) must be changing over time. If the rotation period were greater than the currently appropriate Hartman age, the prices would be rising. If throughout a rotation period T the currently appropriate Hartman age were always less than T, the prices would continue to rise, which would be inconsistent with the assumption that there was a steady state. However, in contrast to the timber case, the rising prices at time t are not inconsistent with the gradual lengthening of the Hartman age until it exceeds T. At that point prices will begin to decline and a cyclically repeating pattern of harvests and prices is possible.

The choice of a cyclical pattern of harvesting, with specialization over time, should be no more surprising than a decision to specialize across different land areas in the same time period. It may simply be better to receive a high value periodically than low value steadily. In some cases, even-flow schedules may force a choice between having very low levels of amenity services with any significant timber flow, and higher levels of amenity services with very low timber flows.

Cyclical steady states seem to be associated with a range of increasing marginal benefits from the holding of unharvested stocks.[31] The examples that come to mind are related to wildlife. There may be, for instance, a threshold level of clearings needed in an area before improved hunting conditions are observed, and a large improvement in value once the threshold is reached. In the early part of this century forest fires in Idaho and Montana opened up large areas of forest to young vegetation and browse; the result was a tremendous expansion in the elk population, and the forests became prime hunting areas. When extreme conditions cause a sufficiently large gain in value compared to the gain from a more moderate even-flow condition, periodic heavy harvesting may indeed be desirable.

Multiple Steady States. Several steady states can be consistent with optimality. The initial condition of the forest at the time it comes under management will determine which steady states are approached. If a forest with relatively young growth is inherited, it is unlikely that we would choose to delay harvesting long enough to allow any old-growth habitat to develop. However, if we inherited the same land with older growth we might find the amenity value sufficiently high as to justify some preservation.

The Steady State Timber Supply. The introduction of multiple-use values into the harvesting decision has an uncertain effect on the optimal long-run supply of timber. As already noted in the discussion of the Faustmann models, multiple-use values may result in a rotation age closer to the age of maximum sustained yield. Multiple-use values may justify a regeneration effort and roading entry on sites not economical for timber alone. However, the pattern of set-aside permanent clearings and older-growth stands that often seems best suited to providing multiple-use services can have a significant effect in diminishing the long-run harvest flow. Further, while we have given no attention to costs, it is clear that harvests made with multiple-use values in mind can be significantly more expensive than those made for timber alone. Higher costs would tend to depress the economically optimal supply of timber.

The Approach to the Steady State. Equation (4-11), along with equation (4-7) describing the dynamics of the production system, fully determine the harvest solution. To explicitly describe the nature of the harvest solution is a formidable task, even in the simpler case where only timber is valued.[32] There is a strong indication that with constant demands the solution does gradually converge to a steady state such as we have just described. Timber prices would generally be rising in periods in which the high stocks of older stands are being reduced (as much of the empirical evidence suggests has happened over the past 100 years). Prices would generally fall in periods during which younger growth must be harvested while allowing stocks to accumulate. Price trends and fluctuations can be expected to dampen as the forest converges to a stable pattern of harvesting.

In now noting some basic conclusions about the nature of harvesting in the period of transition to a steady state, we emphasize the difference in the nature of the current solution from the solution to Faustmann-type single-stand models. The marginal values from the services of standing stock depend upon the distribution of age classes across the forest area. As the distribution of stand ages changes over time, there will be changes in these values. And as a result of this shifting pattern of prices the harvesting rules can be very complex, not easily reflected by any rule of thumb. It is unlikely that we would choose to harvest a stand at the same age for two successive rotations. Only when the identical forest conditions exist will the same harvesting decision be made. The harvesting decision in each period balances the improvement in current value against the future benefits arising from improving the distribution of stocks. At any particular time it may be desirable to maintain some areas clear and to delay harvesting some older stands while harvesting other younger stands.

Because of the complex nature of the multiple-use harvesting solution, it is helpful to provide a few illustrations. Some solutions to a very simple multiple-use problem are described in the next section.

4. Illustrations of Multiple-Stand Harvesting Solutions

The illustrations of the multiple-use harvesting decisions presented in this section are based partly on a study of management of forested lands for water flow and timber value in Colorado (the study is described in greater detail in chapter 5). The harvests on a set of interrelated stands on the forest are considered. In using here a simple dynamic programming procedure to find harvest solutions, it should be noted that this procedure, while adequate for smaller problems, is not workable with larger problems.

The Forest Situation

Let us consider the multiple-use management of a small forested watershed capable of providing timber harvests, water flow, and recreational services. To reduce the potential complexity in our examples, a number of simplifying restrictions and assumptions about the nature of this land area and the available management options are made. (1) There are only three age classes, each of equal area and identical in productivity. (2) Management options are restricted to either fully harvesting an age class or allowing continued growth. (3) There is a thirty-year interval between harvesting decisions. (4) Amenity values depend only upon the mix of the three stand ages. (5) Land productivity and the values of goods and services are unchanging over time. (6) Stand regeneration occurs naturally, at no cost. (7) There is no stand growth after age 120. (8) There is no stand mortality within the planning horizon.

Age classes should be thought of as dispersed across the forest watershed in patches of some appropriate size. For convenience an age class will be referred to as a "stand," but we stress that this stand is not necessarily one contiguous area. Because an entire age class must be harvested, there will remain just three age classes at all times. Stand age is considered in thirty-year increments, up to age 120. The mix of existing stand ages will be referred to as the current "state" of the forest area. Timber growth and the chosen sequence of harvests result in a particular progression in the forest state over time. Water-flow and recreational values depend upon this sequence of development in the forest state, and so indirectly depend upon the harvesting decisions. Table 4-2 summarizes much of the relevant information on the forest states, on the amenity values associated with forest states, and on the options for controlling the sequence of the forest states.

Forest States and Decision Sets. By assumption, only the mix of age classes, not the order of these ages, is relevant in determining the flow of amenity values. It does not matter which particular stand is of which age. As a result, there are only 20 distinct forest states that may exist in our example. We will represent the state of the forest area by a state number. In table 4-2, all possible states (age mixes) of the forestland in our example are listed and the corresponding state numbers are given. State 6, for instance,

Table 4-2. Forest States, Per-Acre Amenity Values, and Decision Sets

State number	Age mix[a]	Water-flow value[b]	Recreation value[b]	Decision set
1	1,1,1	−40.7	−126.0	{1,2,3,4}
2	2,1,1	−19.0	−108.0	{1,2,3,5,6,7}
3	2,2,1	114.0	−81.0	{1,2,5,6,8,9}
4	2,2,2	51.6	−108.0	{1,5,8,10}
5	3,1,1	8.1	−90.0	{1,2,3,11,12,13}
6	3,2,1	114.0	−72.0	{1,2,5,6,11,12,14,15}
7	3,2,2	81.4	−81.0	{1,5,8,11,14,16}
8	3,3,1	100.4	−63.0	{1,2,11,12,17,18}
9	3,3,2	65.1	−63.0	{1,5,11,14,17,19}
10	3,3,3	32.6	−90.0	{1,11,17,20}
11	4,1,1	8.1	−90.0	{1,2,3,11,12,13}
12	4,2,1	62.4	−45.0	{1,2,5,6,11,12,14,15}
13	4,2,2	70.6	−54.0	{1,5,8,11,14,16}
14	4,3,1	46.1	−36.0	{1,2,11,12,17,18}
15	4,3,2	54.3	−18.0	{1,5,11,14,17,19}
16	4,3,3	27.1	0.0	{1,11,17,20}
17	4,4,1	46.1	−36.0	{1,2,11,12,17,18}
18	4,4,2	38.0	9.0	{1,5,11,14,17,19}
19	4,4,3	13.6	35.0	{1,11,17,20}
20	4,4,4	0.0	0.0	{1,11,17,20}

[a]1 = stand growing to age 30; 2 = stand growing to age 60; 3 = stand growing to age 90; 4 = stand growing to age 120 (or older).
[b]Per-acre values, with the value from state number 20 defined as zero.

corresponds to a forest area which has one-third of its area in stock at age 90, one-third at age 60, and the final third at age 30.

From each state there are a limited number of future states to which the forest area may progress. We may harvest no stands, any one stand, any two stands, or all three stands. We represent the decisions available when the forest is in a particular current state by a list of the state numbers of all possible subsequent states; the decision sets for each state are listed in table 4-2. There are fewer than the maximum eight possible decisions for most states because some harvesting patterns will lead to identical results.

Amenity Values.   The water-flow values in our example are based on results of a watershed simulation model for subalpine lodgepole pine forests in Colorado. (The data and the method of valuation are more fully described in chapter 5.) Studies by Leaf and Alexander and by Troendle suggest that there is a high potential for increasing water flow from these Colorado forests.[33] Water flow seems best produced from a forest area which has been partly harvested in small patch cuts. The harvests reduce evaporation and transpiration, while the size of the harvest area in relation to the surrounding tree canopy is important in controlling the collection of snow and in determining the timing of the snowmelt.

The pattern of harvesting which seems likely to be consistent with the production of high water yields may result in deteriorated aesthetic conditions. The recreational values in our example represent a combined value from hunting and aesthetic services. The values have been made up for the purposes of this example, but seem to show likely characteristics. There are higher values for a forest area which is predominantly in older stands. However, some diversity in the mix of age classes is preferred because this is assumed to be beneficial for hunting. In particular, an older forest with some small area in younger stands is highly valued.

For our problem, we need to know the present value of the flow of amenity services over each thirty-year interval between harvest decisions. The flow of services will depend upon the state of the forest area. In table 4-2, base levels for the water-flow and recreational values are associated with each of the 20 forest states. The amenity values in the table give the present value per acre of the flow of services over thirty years for a forest area that will have grown to the associated state at the time of the next harvesting decision. Since the absolute level of the amenity values has no influence on harvest solutions, the amenity values associated with state 20 have been standardized to zero. The values given for the other states should be interpreted as increments relative to the value flowing from a forest growing to state 20. A discount rate of 4 percent has been used in calculating the present values.

The highest water-flow values are seen to be from forest units which are one-third clear, as in states 3 and 6 in table 4-2. Increasing the age of the remaining stands reduces water flow. Where there are greater areas of cleared land, water flow also tends to be reduced because, in high mountain areas, winds can reduce the snowpack accumulation. Recreational values are highest from older forests with some diversity in stand age, as in a forest growing to states 18 and 19.

Timber Values and Harvest Costs. In our example, timber harvests are the only means for managing the forest. A stand may be harvested at age 60, 90, or 120. It may be cleared at age 30, but with no commercial yield of timber. Stands that are not harvested are thinned at thirty-year intervals. The stand volumes available for regeneration harvests are given in table 4-3, as are the thinning yields. The base-level price per thousand board-feet from commercial harvests is taken to be $24. This stumpage value reflects a mill price of $150 and logging costs of $126. There are sale administration and preparation costs of $16 per thousand board-feet for all commercial harvests. The precommercial thin at age 30 costs the forest $95 per acre, and the clearing of stands at age 30 is also taken to cost $95 per acre. With the prices and costs given here, a net harvest value can be associated with each decision in the decision sets listed in table 4-2.

Table 4–3. Yields per Acre, Even-Aged Stands of Lodgepole Pine

Stand age (years)	Stand volume (board-feet)	Thinning harvest (board-feet)
30	0	0
60	8,700	2,600
90	14,200	4,600
120	16,800	0

Source: C. F. Leaf and R. R. Alexander, "Simulating Timber Yield and Hydrologic Impacts Resulting from Timber Harvest on Subalpine Watersheds," USDA Forest Service Research Paper RM-133 (Fort Collins, Colo., Rocky Mountain Forest and Range Experiment Station, 1975).

The Harvest Scheduling Problem

The harvest scheduling problem for the forest area described above is remarkably easy to solve by dynamic programming methods. Representing the state of the forest area at time t by its state number, S_t, the problem to be solved can be expressed as

$$\Phi(S_t) = \text{Maximum}_{\{k\}_s} \{\pi(k,S_t) + \Phi(k)/(1+i)^{30}\} \qquad (4\text{-}14)$$

with the maximization performed for all states $S_t = 1,2, \ldots ,20$ and for all time periods $t = N,N-1, \ldots ,1$, and with N the last interval in the planning horizon. The decisions available when the forest is in state number S are represented by the decision set $\{k\}_s$, which lists the state numbers of all possible subsequent forest states. Each maximization requires the choice of a state number k, representing the selected forest condition for the next period, from the relevant decision set $\{k\}_s$.

The net flow of value from both the current harvest and the amenity services over each thirty-year interval is represented in equation (4-14) by $\pi(k,S_t)$. This value depends upon the current state S_t and the decision as to the future state k. The term Φ represents the asset value of the forest area. It is the maximum potential present value of the full future stream of goods and services from the land, given the current forest state. The recursive structure to the problem should be noted. The asset value of the forest in the current period $\Phi(S_t)$ depends upon $\Phi(k)$, the value of the forest area in the next time period. The term $(1+i)^{30}$ in the problem is the discount factor, reflecting discounting at an annual rate i over the thirty-year time interval between harvests. We discount all future values with a rate i equal to 4 percent.

The harvest scheduling problem is solved by a backwards recursion, beginning at the terminal date N. Suppose we were given values $\Phi(S_{N+1})$ for all states of the land at time $N+1$. We could then easily find the best decision for each possible state at time N simply by comparing the value resulting from each decision in the sets $\{k\}_s$. The best decisions, those resulting in the highest net present value, would determine values for $\Phi(S_N)$, as indicated by the problem in equation (4-14). With these values $\Phi(S_N)$ we can then begin the next iteration, finding the best solution at time $N-1$ and the values $\Phi(S_{N-1})$. The iterative procedure continues in this manner, back through time.

As it turns out, we can more or less arbitrarily select values for $\Phi(S_{N+1})$.[34] Because of the discounting of future values, such arbitrary selection will have no influence on the selected decisions for periods many years earlier. Eventually the selected decisions will be optimal decisions, unaffected by the arbitrary assumption. By selecting a terminal date sufficiently far into the future, the optimal harvesting decisions for the present and near future can be found. In our problem, we have assumed there is no change in the value functions over time. As a result, the optimal decision for each state will be the same in all time periods. The decisions selected in our solution procedure converge to the optimal decisions after just a few iterations.

The Optimal Harvest Solutions

To investigate the tradeoffs between timber, water flow, and recreation on our land area, various optimal harvest solutions have been generated. Each such solution corresponds to a different scaling of the price level for the two amenity output values. The base-level values for water flow and recreation (listed in table 4-2) have been variously multiplied by the factors 0, .25, .5, 1.0, 1.5, and 2.0 in many of the possible combinations. In the earlier discussion of the Faustmann-type models, a curve illustrating the possible tradeoff between present values from timber and water flow was developed. Much the same type of tradeoff analysis is presented here. The possibility for altering the distribution of age classes across several stands gives more flexibility in providing multiple-use services than was apparent when we considered only the harvest age of the single stand.

Our focus here is on the harvest solutions for a land area which is initially fully stocked with stands of age 120 (or older). Seven distinct solutions (harvest schedules) were found; they are labeled A through G in table 4-4. In that table, the progression in the forest state over the first seven time periods is detailed for each solution. For the seven solutions, the table gives the age mix of the three stands at the beginning of each successive time period. That a harvest has occurred is apparent whenever a stand aged 30 (coded 1 in table 4-4) appears in subsequent periods; in the table, to emphasize the harvests the age class to be cut has been underlined.

The price scaling levels that have resulted in the optimal selection of each of the harvest solutions are given in table 4-5. For instance, when both water flow and recreation were valued at the base levels, with price scaling (1,1), then harvest schedule E was optimal. Table 4-6 lists the present value from each of the three outputs, under the various harvest solutions. The flows of output from each harvest schedule are valued at base-level prices, rather than at the price level used to generate the solution. This standardization is needed to make the alternative harvest solutions directly comparable. The present value of the management costs incurred by the forest under each harvest solution are also listed in table 4-6.

Relying largely on tables 4-4 through 4-6, we now look more closely at the nature of the harvest solutions.

Table 4–4. Harvest Scheduling Solutions for State 20

Harvest schedule	Time 1 mix		Time 2 mix		Time 3 mix		Time 4 mix		Time 5 mix		Time 6 mix		Time 7 mix	
A	$\underline{444}$	→	111	→	222	→	$\underline{333}$	→	111	→	222	→	$\underline{333}$	→
B	$\underline{444}$	→	111	→	222	→	331	→	$44\underline{2}$	→	113	→	22\underline{4}	→
C	$\underline{444}$	→	11\underline{1}	→	221	→	332	→	$44\underline{3}$	→	114	→	221	→
D	$\underline{444}$	→	$4\underline{11}$	→	122	→	231	→	$3\underline{12}$	→	123	→	231	→
E	$44\underline{4}$	→	$4\underline{41}$	→	$4\underline{12}$	→	123	→	231	→	$3\underline{12}$	→	123	→
F	$44\underline{4}$	→	441	→	442	→	$44\underline{3}$	→	441	→	442	→	$44\underline{3}$	→
G	444	→	444	→	444	→	444	→	444	→	444	→	444	→

Note: The three numbers in the columns represent ages of the stands: 1 = age 30; 2 = age 60; 3 = age 90; 4 = age 120 (or older). Underlining indicates that a stand is harvested.

Table 4–5. Price Levels and the Corresponding Optimal Harvest Solution

Price levels (water flow, recreation)[a]	Optimal schedule
(0,0)	A
(0,.25)	B
(0,.5)	C
(.5,0); (.5,.25); (1,0)	D
(.5,.5); (1,.25); (1,1); (2,0); (2,1.5)	E
(0,1); (.5,1); (.5,1.5); (1,1.5)	F
(0,1.5)	G

[a]The numbers reflect the scaling of the base-level amenity values.

The Timber Management Solution.   The harvest solution (schedule) A is optimally selected when only the timber output is valued (see table 4-5 for results with amenity price levels [0,0]). This timber management strategy calls for harvesting at the Faustmann rotation age of 90. The initial stands are immediately harvested and then the Faustmann rotation is begun. Schedule A provides the greatest potential timber value, but at considerable sacrifice of potential water-flow and recreation values (see table 4-6). There is no economic advantage to moving toward an even-flow condition since the timber price is a constant, not influenced by the harvest level.

The Timber and Low Amenity Value Solutions.   Solution schedules B and C are optimally selected when water flow is valued at zero and the recreation values are low (see table 4-6, amenity price levels [0,.25] and [0,.5]). These solutions result in only slight modifications in the timber management strategy and maintain very high present values from timber. The management goal seems to be to introduce diversity in stand ages and to increase the average age of standing stocks, with a minimum sacrifice in timber value. Both harvest schedules result in somewhat improved distributions of age

Table 4–6. Present Value Possibilities

Harvest schedule	Timber value	Water-flow value	Recreation value	Costs
A	127.1	−6.9	−51.5	92.8
B	127.0	−4.5	−49.5	92.7
C	125.3	−0.0	−47.0	92.2
D	97.9	17.1	−37.5	71.7
E	58.8	23.9	−17.6	42.9
F	44.0	18.2	−9.3	31.7
G	0.0	0.0	0.0	0.0

Note: All values are computed using the base level.

classes in the later periods, and an improved long-run sustained flow of recreation and water-flow values. After some initial manipulation of the age distribution, the rotation period settles at 120 years.

It is informative to note how diversity in the age classes is optimally introduced. We choose not to delay the initial harvests of standing stocks; that would entail large opportunity costs from the postponed receipt of harvest revenues. Instead, we introduce diversity at a later date: harvesting one of the newly regenerated stands at age 60, with a later harvest of the other stands, results in almost no loss of present value from timber. Postponing the regeneration of one stand is a similarly good strategy.

A Timber/Water-flow Solution. Solution schedule D is optimally selected when the recreation value is zero and water-flow values are at the base level. This solution results in the initial harvesting of two-thirds of the forest area. The forest is then promptly converted to an even-flow condition with a 90-year harvest period. In the longer run the forest is identical to the one that would result under management for maximum water-flow value (schedule E). A significantly higher timber-harvest value results from the increased harvesting in the initial period. At these price levels, the prompt receipt of timber revenues more than compensates for the reduced water-flow values that result when a large fraction of the area is cleared.

The Water-flow Management Solution. Harvest schedule E provides the greatest present value from water flow (see table 4-6). In each period, a single stand is cut. The harvested stand is always that with the eldest stock. The forest area eventually settles into an even-flow steady state with a 90-year rotation period. The management goal is to maintain one-third of the area in clearings, which can be accomplished while maintaining a fairly high sustained harvest, but with a considerable reduction in the initial harvest levels. This solution is optimally selected at the base-price levels; indeed, it is selected at a wide range of prices (see table 4-5).

The Recreation/Timber Solution. Harvest schedule F is optimal when recreation values are at the base level and water flow is not valued. Under this harvest strategy, two-thirds of the area is preserved unharvested while the remaining stand is harvested on a 90-year rotation cycle. The result is a moderate sustained level of timber value and fairly high value from both recreation and water flow.

The Recreation Management Solution. Harvest schedule G is the optimal recreation management strategy. The strategy requires that the stand be perpetually maintained without harvesting. This schedule is selected when water flow is valued at zero and the recreation value is at least 50 percent above the base level. Although our recreation values indicate some preference for diversity, the immediate impact of clearing one-third of the area is too great to accept. No doubt if we had allowed for the possibility of harvesting a somewhat smaller fraction of the area occasional harvesting would have been found desirable for providing the maximum recreational value.

5. Summary and Conclusions

The general multiple-use harvesting policy is seen to be complex. No simple rule of thumb is likely to describe the harvest. Sometimes younger stands are harvested, leaving older stands uncut. Rarely is a particular stand cut at the same age twice in succession during the initial transition periods. The forest may be managed with some areas set aside as old growth or clearings. We may choose to specialize over time, with the land producing high timber yields in some periods and high amenity benefits in others. Even-flow policies are not an inherently desirable goal. Indeed, long even-flow rotations, far from being a desirable compromise policy for multiple-use management, may simply result in uneconomic timber and a poor balance of age classes for the nontimber uses.

Perhaps most important, it can be seen from these examples that the harvesting decision can be quite sensitive to multiple-use demands, about which there is little empirical knowledge. The impact that one stand may at times have upon the services available from a much wider area of the forest may be large. One very visible harvest can reduce the amenity benefits of a considerable land area for many years. An older-growth stand preserved among younger stands can be a critical element in maintaining diversity. Scarcity of older-growth stands can give the remaining stands very high value. In general, the value from improving the distribution of stand ages in a wider forest area can be sufficiently great as to justify dramatic adjustments in the optimal management strategy as evaluated for the single stand in isolation.

Notes

1. In hindsight, it is clear that the perceived problems of the private market as a supplier of timber had more to do with ill-defined ownership rights to timbered lands and the abundance of these lands.

2. M. Faustmann, "On the Determination of the Value which Forest Land and Immature Stands Possess for Forestry," in M. Gane, ed., *Martin Faustmann and the Evolution of Discounted Cash Flow*, Institute Paper no. 42, Commonwealth Forestry Institute (University of Oxford, 1968); originally published in German in *Allegemeine Forest und Jagd Zeitung* vol. 25 (1849).

3. R. Hartman, "The Harvesting Decision When a Standing Forest Has Value," *Economic Inquiry* vol. 14 (1976) pp. 52–58.

4. K. N. Johnson, *FORPLAN Version 1: An Overview* (Washington, D.C., USDA Forest Service, Land Management Planning Systems, February 1986); K. N. Johnson, T. W. Stuart, and S. A. Crim, *FORPLAN Version 2: An Overview* (Washington, D.C., USDA Forest Service, Land Management Planning Systems, August 1986).

5. A number of writers provide good descriptions of the Faustmann model. See, for example, M. Gaffney, "Concepts of Financial Maturity of Timber and Other Assets," *Agricultural Economics Information,* series no. 62 (Raleigh, N.C., North Carolina State College, 1957); P. H. Pearse, "The Optimum Forest Rotation," *Forestry Chronicle* vol. 43, no. 2 (June 1967) pp. 178–195; and C. Clark, *Mathematical Bioeconomics* (New York, Wiley, 1976). P. A. Samuelson, "Economics of Forestry in an Evolving Society," *Economic Inquiry* vol. 14 (1976) pp. 466–492, provides an excellent discussion that stresses the equivalence between competitive market results and the solution to the Faustmann model.

6. If there were a one-time cost associated with the first management cycle, as for roading or site clearing, the initial rotation should be longer than subsequent rotations.

7. The discount factor e^{-iT} corresponds closely to $1/(1+i)^T$. The number e (approximately 2.718) is the base of the system of natural logarithms. The derivative de^{-iT}/dT is equal to $-ie^{-iT}$.

8. Second-order conditions indicate that for an intersection to correspond to a maximum, the relative growth curve must cut through curve $i/(1-e^{-iT})$ from the above left, as drawn. We maintain an assumption that second-order conditions are satisfied.

9. To achieve profitability when there are fixed costs of harvesting or very low stand values, it could be necessary to delay harvests up to the age at which stand growth approaches zero.

10. We represent derivatives with respect to time by a dot over the function name, so $\dot{P}(T) = dP(t)/dt$ and $\dot{\lambda}^* = d\lambda^*(t)/dt$.

11. See D. H. Newman, C. B. Gilbert, and W. H. Hyde, "The Optimal Forest Rotation with Evolving Prices," *Land Economics* vol. 61, no. 4 (1985) pp. 347–353.

12. Samuelson, "Economics of Forestry."

13. S. Calish, R. D. Fight, and D. E. Teeguarden, "How Do Nontimber Values Affect Douglas-fir Rotations," *Journal of Forestry* vol. 76 (1978) pp. 217–221.

14. Calish, Fight, and Teeguarden, "How Do Nontimber Values Affect Douglas-fir Rotations."

15. The tradeoff curve is adapted from J. M. Conrad, "Forest Management and Nontimber Attributes: A Graphical Approach," Cornell Agricultural Economics Staff Paper (Ithaca, N.Y., Cornell University, Department of Agricultural Economics, 1981).

16. K. N. Johnson and H. Scheurman, "Techniques for Prescribing Optimal Timber Harvest and Investment Under Different Objectives: Discussion and Synthesis," Monograph 18, supplement to *Forest Science* vol. 23, no. 1 (1977). Johnson and Scheurman also provide a good discussion of the economic content of harvest timing linear programs.

17. Johnson, *FORPLAN Version 1.*

18. More correct, the harvest should meet the discrete-time equivalent of the Hartman harvest-age condition.

19. Johnson, Stuart, and Crim, *FORPLAN Version 2.*

20. Much of standard literature on the economics of harvesting of natural resources has focused on fisheries, with more attention paid to the total volume of harvest than to the age at which individual units of stock are harvested. The reader may wish to refer to the following diagrammatic presentations. D. Levhari, R. Michener, and L. J. Mirman, "Dynamic Programming Models of Fishing: Competition," *American Economic Review* vol. 71, no. 4 (1981) pp. 649–661; N. Liviatan, "A Diagrammatic Exposition of Optimal Growth," *American Economic Review* vol. 60, no. 3 (1970) pp. 302–309; C. Plourde, "Diagrammatic Representations of the Exploitation of Replenishable Natural Resources: Dynamic Iterations," *Journal of Environmental Economics and Management* vol. 6 (1979) pp. 119–126.

21. We might also consider stock-related management costs to be included in the function *A.*

22. The function *V(n)* might be viewed as giving a quality-adjusted measure of volume. In a general formulation, one might want to treat harvests from various subsets of the age classes as providing distinct products.

23. There has been considerable recent interest in theoretical aspects of the timber harvesting problem. See T. Heaps, "The Forestry Maximum Principle," *Journal of Economic Dynamics and Control* vol. 7, no. 252 (1984); M. Hellsten, "Control Theory and the Optimum Timber Rotation: A Reformulation" (paper, University of Alberta, Department of Economics); M. Hellsten, "The Optimal Management of Age-Distributed Renewable Resource Stocks" (paper, University of Alberta, Department of Economics); M. C. Kemp and N. Van Long, "On the Economics of Forests," *International Economic Review* vol. 24, no. 1 (1983) pp. 113–131; T. Mitra and H. Y. Wan, Jr., "Some Theoretical Results on the Economics of Forestry," Working Paper 270 (Ithaca, N.Y., Cornell University, Department of Economics, 1981); T. Mitra and H. Y. Wan, Jr., "On the Faustmann Solution to the Forest Management Problem," Working Paper 266 (Ithaca, N.Y., Cornell University, Department of Economics, 1981); and H. Y. Wan, Jr., "A Generalized Wicksellian Capital Model— An Application to Forestry," in Vernon Smith, ed., *Economics of Natural and Environmental Resources* (New York, Gordon Breach, 1978) pp. 141–153.

24. This problem is a linear program in the special case where functions $A(Z)$ and $\beta(X)$ are linear. With some effort, it can be approximated as a linear program, even if these functions are not linear.

25. See the solution methods described in P. Berck, "The Economics of Timber: A Renewable Resource in the Long Run," *Bell Journal of Economics* vol. 10, no. 2 (1979) pp. 447–462; and P. Berck and T. Bible, "Solving and Interpreting Large-Scale Harvest Scheduling Problems by Duality and Decomposition," *Forest Science* vol. 30, no. 1 (1984) pp. 173–182. See also the optimal control methods described in J. M. McDonough and D. E. Park, "Nonlinear Optimal Control Approach to Interregional Management of Timber Production and Distribution," in J. Meadows, B. Bare, K. Ware, and C. Row, eds., *Systems Analysis and Forest Resource Management* (Washington, D.C., Society of American Foresters, 1975); and K. S. Lyon and R. A. Sedjo, "An Optimal Control Theory Model to Estimate the Long-run Timber

Supply," *Forest Science* vol. 29 (1983) pp. 798–812. These papers focus on harvesting for timber value alone; the nontimber values may introduce nonconvexities that make solutions more difficult to find.

26. A more complete discussion can be found in M. D. Bowes and J. V. Krutilla, "Multiple Use Management of Public Forestlands," in A. V. Kneese and J. L. Sweeney, eds., *Handbook of Natural Resource and Energy Economics,* vol. 2 (Amsterdam, North-Holland, 1985) pp. 531–551.

27. Equation (4-11) could be derived directly from a continuous-time optimal control version of our multiple-use management problem.

28. These conclusions rely on the second-order conditions for optimality of the Hartman-age solution.

29. We must suppose that the amenity services from older stands approaches a constant level and that stand mortality occurs far into the future.

30. It should be stressed that we have assumed that $\beta''(H) < 0$, with increased timber supply resulting in lower timber prices. For more details, see Bowes and Krutilla, "Multiple Use Management."

31. In Clark, *Mathematical Bioeconomics,* some related discussion of "pulse" fishing can be found, with economies of scale in harvesting associated with periodic heavy fishing.

32. For comment on the related harvest condition in the timber-only case, see T. Heaps and P. A. Neher, "The Economics of Forestry When the Rate of Harvest Is Constrained," *Journal of Environmental Economics and Management* vol. 6, no. 4 (1979) pp. 297–319; K. S. Lyon, "Mining of the Forest and the Time Path of the Price of Timber," *Journal of Environmental Economics and Management* vol. 8, no. 4 (1981) pp. 330–344.

33. C. F. Leaf and R. R. Alexander, "Simulating Timber Yield and Hydrologic Impacts Resulting from Timber Harvest on Subalpine Watersheds," USDA Forest Service Research Paper RM-133 (Fort Collins., Colo., Rocky Mountain Forest and Range Experiment Station, 1975); C. A. Troendle, "The Potential for Water Augmentation from Forest Management in the Rocky Mountain Region," *Water Resources Bulletin* vol. 19, no. 3 (1983) pp. 359–372.

34. Good starting values for $\Phi(S_{N+1})$ are the values that would result from following a Faustmann timber rotation. Setting all values at zero is almost as satisfactory, and easier.

Part 2

Applications of Multiple-Use Management in Forestry Settings

Introduction to Part 2

Part 2 of this volume addresses practical problems of management by using the institutional and theoretical framework developed in part 1. In part 1 we presented some historic background on the origin and development of the forestry profession and culture in the United States, sketching briefly the tension between that European-derived school of thought which regarded forestry as the management of commercial timber plantations and that school which acknowledged the greatly different environment that characterized the New World's inherited wildlands and forests. Acknowledging the fact that the Forest Service was, and will continue to be, an important influence in the education of foresters and in the management of a very large expanse of wild land, we also addressed the theoretical background which would be appropriate to the management of publicly owned forest lands—lands expected by the public and hence by the Congress to be managed for the benefit of several different user groups, and to provide forest resource services of some kinds without charge. The theoretical material in part 1 is perhaps more intensive than is elsewhere available.

Part 2 begins by dealing with the joint production of timber and water in a very special case, an Upper Colorado River Basin subalpine setting (chapter 5). Accordingly, the results should not be generalized to other cases in which the conditions differ markedly. After considering the measurement of the economic response to management actions in a case involving sanitation cuts and wildlife (chapter 6), we go on to discuss the question of estimating the demand for recreation without going outside the National Forest System's information base (chapter 7). Having found in part 1 that joint production may have negative as well as positive connotations, we see here in the case of minerals and wilderness that these joint outputs are mutually exclusive (chapter 8).

The focus in part 2 then shifts to the question of obtaining the supplemental funding needed for nonpriced forest services (chapter 9). This is conventionally done through complicated budget and appropriations processes. With an eye to economic efficiency and the objectives of the Renewable Resources Planning Act and the National Forest Management Act, we evaluate the results of these processes to see whether they nurture or negate the intentions of the legislation. Part 2 concludes with a discussion of below-cost sales and how to perform a practical valuation of them; we do this with an application of FORPLAN, the Forest Service planning model (chapter 10).

We trust our examples of the application of analysis to practical problems of managing the multiple uses of public lands will stimulate others to go forth and do likewise.

5

Forest Management for Increased Timber and Water Yields

1. Introduction

In turning from the institutional and theoretical to the empirical context, in this chapter we address the feasibility of management activities that simultaneously affect the output of two or more forest resource services and require the coordinated management of interdependent sites.[1] The setting will be the subalpine forests in the Central Rocky Mountain region.

For a number of reasons we choose to evaluate timber and water to illustrate our joint (multiple-use) production case. First, the subalpine forests of the central Rockies are dubious prospects for commercially viable timber management. Second, this region, or more precisely its western slope, experiences a considerable amount of precipitation (25 to 30 inches of water-equivalent precipitation per year), most of it in the form of snow. The relatively ample snowfall makes possible snow management which can, under favorable conditions, increase the water available for streamflow significantly.[2] Third, this is the region for which a substantial amount of research has been done on the watershed management practice/water yield response relationship, so there are data which permit conclusive analyses. Finally, there is no consensus on the economic feasibility of manipulating vegetation for the purpose of augmenting water flows, although the practice has been extensively researched by hydrologists for a half century.[3] Our analysis of forest management activities concerns a setting, then, that would be unlikely to justify management for timber alone.

We begin (section 2) with a review of the management activities associated with timber growing and harvesting, the market value of the output, and the costs of securing it. This will provide an opportunity to evaluate the economic feasibility of timber management on the central Rockies national

149

forests. We next take a look at water augmentation practices (section 3). Then we examine the economic dimensions of water, its uses, the price it commands in different functional and geographic markets, and whether watershed water augmentation practices can be employed to increase the value of the forest outputs. Our initial conclusions are tested by observing how different assumptions alter our values (section 5). Our final conclusions are then presented (section 6).

2. Timber Considerations

Timber management may mean many different things depending upon the region, the quality and accessibility of the site, the value of the dominant species, and so forth. In the Douglas fir region of the Pacific Northwest and in the southern pine region, where yields are high and the quality good (construction-grade sawtimber), relatively intensive management regimes can be justified economically. But this is not the case in the subalpine forests of the central Rockies. Volunteer, unmanaged stands, although perhaps fairly densely stocked, have little high-quality timber so that average volumes of existing inventory are about 3,600 board-feet per acre,[4] roughly a tenth of the standing inventory on the better Douglas fir sites. An argument has been advanced, however, that with proper management these commercially indifferent stands could be replaced with viable commercial timber stands. With the removal of existing stands, the argument goes, followed by appropriate silvicultural treatment, more robust growth and better-quality timber could be expected to replace existing low-quality stands.[5]

Many different harvesting schedules could be posited, but they would tend to be variations of a pattern that would result in the gradual removal of existing inventories, followed by natural regeneration, periodic thinning, and final harvest, whereupon the cycle would begin anew with regenerated stands. This pattern would continue into the indefinite future. Given various land-management considerations, the removal of the original stands might be achieved by making patch cuts over a substantial period of time. One regime that can be proposed, and which will serve as our benchmark for evaluation, would begin with immediate road entry into a small watershed, followed by removal of timber from a third of the area in relatively small patch cuts distributed throughout the watershed. This in turn would be followed 30 years later with a similar harvest of another third, again in small dispersed patches, along with thinning of the regenerated stands on the area initially harvested. The final third would be harvested at 60 years to complete the removal of the existing inventory, and thinnings would occur on the other sites. Thinnings would continue on all of the regenerated stands at thirty-year intervals until final harvests at age 120 for each stand. The stands would regenerate and the same management cycle would continued into the future. The first 180 years of this management schedule are shown in table 5-1.

For our analysis we assume stands of lodgepole pine with a site index of 60 and thinnings to a growing stock level of 80. Yields and periodic intermediate cuts for the single-aged management units are given in table 5-2. We assume rapid natural regeneration, but with no real difficulty lags in regeneration could be considered. R. R. Alexander, the silviculturist involved in the West Slope research, has suggested that with lodgepole pine there could be prompt natural regeneration, especially with small dispersed harvests.[6] However, the significance in our conclusions is little affected by the regeneration assumption. For even with prompt regeneration the timber value would be very low relative to both the roading costs and the potential value of water-flow increments, as we shall see.

The price of timber can be calculated in different ways—at the mill or on the stump, for example. We will discuss the price in terms of the mill value. The average mill price for lodgepole pine in the Colorado subalpine forests in 1982 was $150 per thousand board-feet (mbf); that can be taken as a reasonable value for harvests from existing inventory.[7] Since both the volume and quality of managed stands are expected to improve substantially, we differentiate between prices of existing and regenerated stands. The harvest of managed regenerated stands is taken to be valued at $250 per thousand board-feet at the mill.

Direct costs associated with regeneration harvests or commercial thinning average $141 per mbf, which includes sale preparation and administration costs of $16 per mbf and logging and hauling costs of $125 per mbf. Precommercial thinnings are taken to cost $95 per acre. Access-road costs, including construction, maintenance, and reopening costs, are treated separately later in this chapter; they will depend upon the length of the access road and the type of road construction. We will not consider any overhead

Table 5-1.   A Timber Management Schedule

Year	Management unit[a] (percentage of watershed planning area)		
	1 (33%)	2 (33%)	3 (33%)
0	HVST		
30	PCT	HVST	
60	THIN	PCT	HVST
90	THIN	THIN	PCT
120	HVST	THIN	THIN
150	PCT	HVST	THIN
180	THIN	PCT	HVST

Note: HVST is regeneration harvest, PCT is precommercial thin, THIN is commercial thin.
[a]Each unit is distributed in small patches across the watershed planning area.

Table 5–2. Yields per Acre of Even-Aged Stands of Lodgepole Pine, Site Index 60

Stand age (years)	Average dbh[a]	Entire stand before and after thinning		Periodic intermediate cuts	
		Merchantable volume (ft³)[b]	Sawtimber volume (bd-ft)[c]	Merchantable volume (ft³)[b]	Sawtimber volume (bd-ft)[c]
30	4.5	170	0		
30	5.3	170	0	0	0
40	6.4	720	0		
50	7.3	1,340	0		
60	8.1	2,120	8,700		
60	8.7	1,490	6,100	630	2,600
70	9.6	2,050	8,600		
80	10.4	2,620	11,100		
90	11.2	3,280	14,200		
90	11.9	2,160	9,600	1,120	4,600
100	12.8	2,600	11,800		
110	13.7	3,090	14,300		
120	14.5	3,560	16,800		

Note: The yields reflect the assumption that dwarf mistletoe infection did not occur during the rotation of 120 years.

Source: C. F. Leaf and R. R. Alexander, "Simulating Timber Yields and Hydrologic Impacts Resulting from Timber Harvests on Subalpine Watersheds," USDA Forest Service Research Paper RM-133 (February 1975).

[a]dbh = diameter at breast height.
[b]Merchantable cubic feet = trees 6.0 inches dbh and larger to a 4-inch top.
[c]Sawtimber board-feet = trees 6.5 inches dbh and larger to a 6-inch top.

costs, as for fire protection or supervisory staff, since such costs are unlikely to be altered by the choice of a water management program.

The per-acre present value of the full sequence of timber harvests presented above will be the discounted sum of the difference between mill value and harvest costs, as in equation (5-1):

$$PV = \left[\frac{e^{-r}+e^{-31r}+e^{-61r}}{3}\right]\left[(P_E-C_h)V_E \right.$$
$$\left. + \frac{(P_R-C_h)(T_1e^{-60r}+T_2e^{-90r}+V_Re^{-120r})-C_pe^{-30r}}{(1-e^{-120r})}\right] \quad (5-1)$$

with

PV = the present value per average acre
P_E = the mill price per thousand board-feet for harvests of existing stock, equal to $150

P_R = the mill price per thousand board-feet for harvests of regenerated stock, equal to \$250

C_h = the cost per thousand board-feet of harvest or commercial thinning (including logging, hauling, sale preparation and administration), equal to \$142

C_p = the cost per acre of a precommercial thin, equal to \$95

V_E = the volume per acre on the existing unmanaged stands, equal to 3.6 mbf

V_R = the volume per acre at age 120 from the regenerated stands, equal to 16.8 mbf

T_1, T_2 = the volume per acre thinned from the regenerated stands at age 60 and 90, respectively, equal to 2.6 mbf and 4.6 mbf, respectively

e^{-rt} = the discount factor, with r the discount rate, converting revenues and costs occurring at time t to present values

The value is for an average acre in the watershed planning area, averaged over the three single-aged management units. The first bracketed term in equation (5-1) reflects the thirty-year periods between initial entry into each third of the watershed. Once initially harvested, each fraction of the area is then treated on the continuing 120-year cycle of thinning and harvesting described above. The net receipts are reflected in the second bracketed term. With the value as given and a discount rate r of 0.04, we could expect to have net present values, excluding costs of roads, totaling \$23 per average watershed acre for the harvest schedule as presented (\$12 per acre resulting from the harvest of existing stands and \$11 per acre from the subsequent harvest of regenerated stands).

These net values are presented exclusive of any associated road-construction or maintenance costs, but of course road access would be required to harvest the timber. Road costs will vary widely as a function of different factors. One important factor is the length of the road extension between the road system that currently exists and the area to be harvested. Another relates to the terrain, with the costs varying directly with the slope. Then, too, there are various types of roads that can be considered. These involve differences in construction, maintenance, and reopening costs. Temporary roads would be abandoned until the next entry required full reconstruction. Intermittent-service and seasonal roads would incur reopening expenses with each entry. At the other end of the cost spectrum are permanent all-weather roads which are kept open at all times, and which incur high construction and maintenance expenses but avoid reopening costs.

To offer a view of the interplay of these factors, the costs per acre of providing roads under the preceding harvest schedule are shown in table 5-3. We assume that the whole of a watershed planning area will be roaded for the first entry. The costs are categorized by type of road, character of terrain,

Table 5–3. Cost of Road per Acre Harvested by Type and Length of Road, Type of Terrain

(costs in dollars; discounted at 4 percent)

Miles[b]	Type of road[a]			
	A	B	C	D
Gentle slope				
2.5	$ 47	$ 71	$100	$121
5.5	104	155	220	266
8.5	161	240	339	412
Moderate slope				
2.5	64	78	108	129
5.5	141	171	237	284
8.5	216	264	366	437
Steep slope				
2.5	92	97	139	168
5.5	202	213	306	369
8.5	312	330	472	571

[a]A is temporary road, B is intermittent road, C is seasonal road, and D is surfaced road continuously open.

[b]Miles of road per square mile of land accessed.

and number of road miles required per square mile accessed. We feel that entries in table 5–3 represent the range of reasonable possibilities. Average road mileage for Forest Service lands under timber management in the West Slope region is 2 to 3 miles per square mile of land accessed. It is expected that more road mileage would have to be built to enter new, less accessible areas of the forest.[8] Leaf and Brink suggest that in the extreme as much as 12 miles of new road per square mile might be needed for the small-patch harvests desirable for water augmentation.[9]

A look at table 5–3 will suggest one reason why the treatment of access costs has been deferred until the end of our analysis. The variation in the cost per acre is so great from one situation to another that it is not meaningful to deal with "average" conditions. From the perspective of a manager, there is little one can do about slope and accessibility. It does seem, however, that a substantial difference can be realized in choosing the type of access road to be built. The cost of the permanent, continuously serviceable road will be about double that of temporary roads which are reopened for each harvest entry.

Another insight suggested by table 5–3 is that for no combination of factors will the net present value of the timber in our example be large enough to cover the harvesting costs plus the cost of road access. It is for reasons such as this that the subalpine forests of the central Rockies have not been considered prospects for prudent timber management. But before we take this as a final verdict, it will be instructive to see whether consideration of a joint product would alter the outcome.

3. Water-Yield Augmentation

It is not difficult to visualize how timber can be produced. We can imagine biological growth, with some stems selected to be favored by eliminating others through thinnings. Producing water, however, is not so straightforward. Of course, water is not actually produced. What occurs under management is the reduction of evaporation, sublimation, and transpiration losses and an alteration in the timing of the runoffs that would occur in an unmanaged state. A little elaboration will help to visualize what is involved.

Hydrologists have been investigating the relation between vegetation and various surface and subsurface hydrologic phenomena of forest and rangelands for well over a half century.[10] But the "favorable conditions of water flow" called for by the 1897 Organic Act will imply different sorts of conditions, depending upon the particular climatic, meteorologic, topographic, and vegetal characteristics of the region, and the economic characteristics of the area in which the increase in streamflows is achieved. Although the role of vegetal cover is highly complex, and not yet fully understood even by forest hydrologists and ecologists, a nontechnical grasp might be gained through a heuristic introduction.

In regard to water-yield augmentation, the influence of vegetation entails three general elements. The first concerns the fact that some of the precipitation intercepted by (that is, suspended on) the forest canopy evaporates before it reaches the ground. Interception is a function of crown density in forest communities. For conifers, which retain their leaf volume throughout the year, the interception under fully stocked conditions is as much as 15 to 20 percent of the total precipitation.[11] In cold snowy regions such as the central Rockies, the crowns of trees, intercepting some of the snow, tend to facilitate its airborne transport whenever the wind blows with much velocity. Blowing snow is subject to sublimation losses, which prevent some of it from ever reaching the ground. The second element is transpiration, involving the draft on soil moisture by growing plants; transpiration may double or triple the rate of interception loss. Reduction of transpiration is roughly proportional to the extent of cutting in the forest. Cutting reduces the coniferous forest canopy, and by reducing interception and sublimation increases the amount of precipitation reaching the forest floor by as much as 18 to 25 percent. In the central Rockies, then, removal of vegetation can be used to modify evaporation and transpiration to augment water yields. The third element, which depends on silvicultural practices, concerns the use of timber-harvest sites as repositories for blowing snow. This practice can favorably alter the period of snowmelt.

Experiments with patterns of vegetation removal in snow zones have included strip cuts having different orientations with respect to prevailing winds and aspect, as well as circular openings. Openings in the canopy change the aerodynamics. By modifying air flow from the upwind sides of a forest opening or patch cut, more snow is deposited in an opening or cut than in the forest generally. This redistribution takes place predominantly at the expense of snow deposit in the stands immediately downwind from the clearing, and approximately offsets the increased deposition within the opening.[12] A circular opening of a diameter approximately five times the height of surrounding trees is the most effective under subalpine Colorado conditions; larger openings are subject to wind scour.

Of major significance is the gain in water yield resulting from greater efficiency of the snowmelt in the small openings. These openings in the forest canopy provide direct solar radiation to the forest floor, causing an earlier snowmelt than occurs beneath the surrounding forest. The meltwater tends to be transported to, and to enter, the stream system before the period of heavy transpiration draft during the growing season. Thus a higher proportion of the infiltrated snowmelt from small openings reaches the stream system than reaches it later from the surrounding fully vegetated stands.[13]

Although the timing of water releases from snowpack in those forest openings created by timber harvesting is a major contributor to increased water yields,[14] more important for water augmentation is the reduction of evapotranspiration resulting from harvests and thinnings. This being the case, it is pertinent to examine how reforestation of timber harvest openings influences water-yield augmentation. In humid regions of the United States, the recovery of vegetation occurs quickly and the increased water yield is short-lived. In the subalpine forests of Colorado, however, the recovery period is much longer (except for aspen) and the hydrologic effectiveness of the openings persists over longer periods of time. Troendle reports that on the experimental Fool Creek drainage, twenty-five years of regrowth after clearcutting lodgepole pine and spruce-fir reduced the initial water yield by only a third. In aspen stands the recovery is much faster, eliminating water-yield increases in approximately five years.[15]

A further point ought to be made in connection with measurement of the increased yield from water-augmenting practices. The vegetation is removed from small watersheds discharging into first-order streams. Fairly accurate estimates of water yield are possible by using a paired watershed as a control, but the increments in first- and second-order stream flows from such an experimental program are not large relative to the flows of third- and higher-order streams. Failure to detect a significant increase in the yield of higher-order streams, however, will not preclude a judgment in favor of the increased headwater yields being preserved down stream. Hydrologists can say this because the magnitude of the change would not cause a significant increase in either the wetted or evaporative surface along the channel, in seepage to groundwater, or in consumptive use by vegetation.[16] But if the watersheds of a large number of first-order streams were treated to augment water yields, a significant difference could be effected, perhaps a 3 percent increase in overall flow.[17] In the discussion that follows, then, we assume as a working approximation that the increment in yield at a site (or at the outlet of a first-order watershed) is preserved throughout the stream system to which the small watershed contributes.

It is one thing to establish with reasonable assurance that water augmentation practices can increase the yield from a treated watershed; it is another to establish that the hydrologic achievement is necessarily beneficial. Two considerations must be taken into account: the first relates to the timing of the increment in yield in relation to peak flows (particularly to flood peaks), and the second concerns the valuation of an increment in yield known to occur at a time free of flood flows.

High waters in the Colorado River Basin are not storm flows, but rather are meltwater from the winter snowpack. Evidence from the experimental subalpine watersheds in Colorado suggests that the modified snowmelt occurs on the rising stem of the normal hydrograph.[18] Accordingly, while the volume of streamflow would be increased, the peak would remain relatively unchanged. Moreover, there is a great deal of storage capacity in the Colorado River Basin that usually exceeds by a large amount the storage needed for seasonal stream regulation. There is also an exceptionally large volume of over-the-year storage. Albeit 1983 was a year in which water had to be spilled because of the volume of snow and unseasonably hot weather,[19] in all but one other prior year in the history of the Colorado River Project was it unnecessary to spill water. (Perspective on the relative size of the storage in the Colorado River Basin is gained by considering that the Columbia River, which is itself highly developed and has a streamflow more than ten times that of the Colorado, has a storage capacity only about two-thirds that of the Colorado River.) Given such an incredible storage capacity on the Colorado, we would not expect the relatively small maximum potential increment (on the order of 3 percent) to the volume of flow of the river to cause any flood problems.

Although there do not appear to be any flood hazard problems associated with even a relatively large water augmentation program, there are, nonetheless, a substantial number of both beneficial and potentially harmful effects of altering the vegetation patterns in small subalpine watersheds. Some of these could occur on site where the treatment is being applied. Some may occur off site but in stream, as when falling water is used to generate electricity. Other effects would typically occur both off site and off stream— for example, where water is put to urban-industrial uses. Such effects are discussed in the next section, where their economic implications are evaluated to the extent possible.

4. The Economic Dimensions of Water

Not all of the effects of a timber harvest program can be estimated in economic terms, but this is not to say that there are no economic effects. To begin with the on-site and near-site effects, the modification of forest conditions through the described timber harvest program will have some effects on wildlife habitat. Depending on initial conditions (the presence of "doghair" lodgepole pine stands, for example) the effects could be beneficial, but it is not clear that they would always be so. Similarly, there will be changes in the riparian habitat as a result of the watershed program. It is clear that these effects may favor one species while adversely affecting another. Earth and life scientists, although willing to make such an observation, are reluctant to predict the magnitude or severity of the effects of treatment programs on various watershed or riparian habitats. But unless quantitative data can be obtained from resource managers based in the earth- and life-sciences, it is not possible to undertake an economic assessment of effects on wildlife habitat even though a conceptual framework for reckoning the costs and gains exists.[20] The impact of a timber harvest program on forest aesthetics is also difficult to assess, and in this analysis we do not attempt an economic reckoning of the net aesthetic effect of water augmentation treatments, although we expect they would not be positive.

Effects of other in-stream and off-stream uses, however, are more amenable to analysis and economic evaluation. We know the relationship between volume of water and generation of power for any given effective head.[21] We also know something about the value of water used in agriculture, and have the prices at which rights to water are exchanged. These values will depend on a number of things. Markets may be geographically different as well as functionally distinguished, so that water used for different purposes—or water used in the same way but in geographically different markets—may show a substantial difference in value. We thus need to get reasonable estimates for the functional uses and the value of water in these uses within the geographic areas relevant to our analysis.

Within the Colorado River Basin, three rivers have been chosen for study: the Fraser River, which empties into the Colorado River near the latter's headwaters; the White River, which joins the Green River and eventually empties into the Colorado River near Lake Powell in Utah; and the Gunnison River, which joins the Colorado River near Grand Junction, Colorado. All three rivers have national forests within their drainage areas, providing the possibility for management practices that would increase water flow.

The increased streamflow from a water augmentation program of any significant scale in these national forests would likely be in part diverted for use on the Front Range, primarily to cities like Denver and Fort Collins, and in part be retained for use on the West Slope and in Lower Colorado Basin states. Since these markets for water differ substantially, we evaluate water going to the Front Range separately from that part of the augmented water yield that might remain in the Colorado River Basin. These two scenarios provide our high- and low-bound estimates of the value of a water augmentation regime.

Water destined to go to the Front Range markets could be intercepted by one of the collector systems that prepares it for transmountain diversion. The Alva Adams Tunnel of the Colorado Big Thompson Project in the north, the Moffat and Robert tunnels of the Denver Water Board, and the Boustead Tunnel of the Fryingpan-Arkansas in the southern part of the state would likely be used to bring water to the various markets of the Front Range. Along the Front Range there appears to be a continuously increasing demand for water in municipal and industrial uses. Growth in nonagricultural uses on the Front Range has been very rapid from 1960 to the present, with the supply provided by transfers out of agriculture.[22] It is our assumption that any water diverted to the Front Range would be used for municipal or industrial purposes. Annualized prices at which rights to water have been exchanged in recent years have ranged from the equivalent of $100 per acre-foot (af) per year to upwards of $200 and even $300.[23] On the basis of a review of this experience, for our analysis we have taken $100/af per year as a conservative figure for in-stream water increments destined for the urban and industrial purposes on the Front Range.

To estimate the value of the water that would be retained for uses on the West Slope and in Lower Colorado Basin states, the existing uses and withdrawals of water from the three rivers were examined. The several uses, and corresponding return flow and extra-basin diversion, were tracked from the likely point of entry into the mainstem of the respective rivers downstream to the boundary of the United States and Mexico.[24] It is assumed here that the augmented flow remaining within these rivers would be used in the same way as are existing flows.

Water that remains within the Gunnison River flows through the Curecanti unit of the Colorado River Storage Project, which contains three hydroelectric plants and two large diversion points for irrigation use. After a few other small diversions, the Gunnison River empties into the Colorado River near Grand Junction. The White River joins the Green River after several small diversions for irrigation. Some additional water is then diverted from the Green River, primarily for irrigation, before it empties into the Colorado River above the Glenn Canyon Dam.

In the Fraser River Basin, any additional water yields produced would probably not follow the course of the Colorado River. In this watershed, the Fraser River transmountain diversion taps the lower-order streams and transports the water through the Moffat Tunnel across the Continental Divide. The water is eventually used for municipal and industrial purposes by the

City of Denver. For comparative purposes, an alternative scenario in which the water increment remains in the Fraser River and flows into the Colorado River is also considered in our analysis.

After the Green and Gunnison rivers empty into the Colorado, the remainder of the water from each of the three rivers flows through the same diversion points and dams. Facilities at the Glenn Canyon, Hoover, Davis, and Parker dams produce hydropower. In the Lower Basin, a series of canals transports water to farms and cities in Nevada, central Arizona, and California. It should be noted that we distinguish among hydropower use, in which 100 percent of the water is retained in channel, available for further uses; irrigation agriculture, which has a roughly 50 percent return flow; and those uses, primarily municipal or industrial, that result in the transport of water out of the Colorado River drainage.

While it is relatively easy to track the existing uses of water in the three rivers under study, the more appropriate task is to determine how increased water yields might flow to each functional market. Because of the complicated legal and institutional factors that influence the distribution and sale of water rights, it is very difficult to predict the net change in water use in each sector that might result from a program of forest management for increased water yield. We assume that, for water remaining within the Upper Basin rivers, each existing water withdrawal or use would increase in proportion to the increment in flow remaining at the point of diversion or use.[25] An exception to this assumption relates to our treatment of the Central Arizona Project (CAP), which came on line in 1985: we assume that water delivered by the CAP would be increasingly used for urban-industrial purposes, and that there would be a compensating decrease in the fraction delivered to agricultural use over time. Expected decreases in California's use of Colorado River water after the CAP is fully on line are also included in our analysis.[26]

Market conditions on the western slope on the Upper Colorado suggest that rights to incremental water for municipal and industrial purposes would bring about $60 per acre-foot each year, while the marginal value product of water withdrawals for agriculture there is estimated to be about $10/af per year.[27] In the Lower Basin states—Arizona and California—the marginal value product of water withdrawals for agriculture is known to be higher, falling within a range of $25 to $35/af.[28] On that basis, we have taken $30/af as our estimate for the yearly value of water going to Lower Basin agricultural uses. For water going to urban-industrial uses in the Lower Basin states, we use a value of $100/af, the same value estimated for Front Range urban and industrial uses.

One word concerning valuation of the use of falling water to make hydroelectric power is in order. The value of water used in the production of power depends on a number of factors. Quite apart from the issue as to whether excess power and transmission facilities exist (as they do in this situation) or need to be built, there is the question of when the water (or power) is available. If it is available only at times of low (nonpeak) demand, the electricity produced by water is valued at only the fuel costs that its use displaces, since there is excess generating capacity during off-peak periods. If, on the other hand, water is available for release during the season of peak demands, the annual credit earned by hydropower is equal to the annual cost of an equivalent amount of dependable capacity, in addition to the fuel replacement value. If, however, the hydroelectricity enters a system in which all power is subsidized and sells for less than real marginal costs, the marginal value of water in hydroelectric production is probably no greater than the marginal value product in the use to which it is put, and thus may not be greater than the price of the subsidized power.

We have not been able to undertake a comprehensive audit of the price-cost relationship in the region in which the Southwest Power Administration markets Colorado Storage Project power. This should be done before any estimate of the marginal value of water in power production there is taken to be reliable. We have thus taken a relatively conservative view and credited the power at each dam according to fuel replacement values which were separately calculated for each state. Fuel replacement value averaged 17.03 mills per kilowatt hour (kWh) (1 mill = $0.001). Under our assumptions on the usage of water, an acre-foot of increased water flow in one of the three rivers would result in hydroelectric power production of between 1,000 and 1,600 kWh a year.[29]

At this point we should say something about traditional views on the valuation of water. It has been customary to impute as the value of "new" water the value of the marginal product of water in agriculture. The rationale for this seems to be as follows. In perfectly competitive markets for factors and products, the factors that are first diverted to a more productive competitive use will be the units currently in the least productive applications. Accordingly, if there is a more productive use and no new water is available, the water that will be diverted will be the marginal units from agriculture. On this basis it is suggested that new water merely permits "old" water to remain in marginal applications in the agricultural sector. Increments in water provided to municipal and industrial users simply substitute for transfers of water rights from the agricultural sector that would have otherwise occurred, and hence should be valued at the marginal value product in agriculture.

Now, the market for water rights in western United States is an institution of fairly recent origin. Moreover, it operates in an environment quite different from that required for perfectly competitive outcomes. One troublesome fact is that the marginal value of water in residential and industrial uses, as measured by exchange prices, continues to greatly exceed the marginal value product of water in agricultural use. Carried to its logical conclusion, the market transfer argument, far from suggesting that increments in water should be valued at the marginal value in agriculture, suggests that water would be transferred between sectors until in-stream marginal values are equal for each use. Increments of water would continue to be transferred to the high-valued urban sector until such equality of value occurs. Only under unusual conditions, with demand in municipal and industrial sectors being satiated without a corresponding decline in the marginal value, could real differences in marginal value be maintained under the competitive transfer assumption.

To maintain the assumption of competitive markets in the face of such differences in values, one might suggest that the measured marginal value of water as used in agriculture understates the true marginal value of water to this sector. Indeed, if farmers are passing up the chance to sell their water rights, it seems fair to suggest that they consider these rights to be worth more than the current exchange price. The extra value might perhaps reflect the importance of water to their continuing not just a commercial activity, but a preferred way of life. A second factor might also be postulated. Howe and his associates suggest a case could be made that the difference between the marginal value product in agriculture ($30/af in their study) and the annualized price at which water rights are exchanged ($160–$174/af by their estimate) could reflect a speculative motive.[30] Water rights are seen as an asset that has been fairly steadily increasing in potential sale value, and under such conditions there has been little reason to sell. This motivation, in addition to the essential role of water in providing for a style of living that is strongly preferred, could perhaps account for the divergence in price between the exchange price for water and the value of the agricultural product attributable to water. If true, these arguments would call for the valuation of water increments at a price at least as high as that at which rights exchange in transfer to the urban and industrial sectors. In fact, however, it seems likely that farmers do not sell their water rights largely because of institutional difficulties in arranging the sales, and because of a lack of incentive to go through with such transactions. It should be noted in particular that receipts from sales of water rights often go to quasi-governmental water authorities, and not directly to farmers.

It is possible, of course, that a program of forest management for increased water yields might, even in the absence of competitive markets in water rights, trigger a reduction in the transfer of water out of agriculture. Were the water provided to municipal and industrial users to substitute for the transfer of water from agriculture, then indeed the net value of a water management program should be measured as the value of the continued availability of water to the agricultural sector. It is our feeling, though, that a forest water management program would not produce a guaranteed change in the Colorado River flow sufficient to trigger any change in water rights or in the potential sales of water rights. Further, with the continued relatively much higher value of water in municipal and industrial uses, it is hard to understand why any small allocation of new water would cause potential transfers of water to be forgone. History suggests that water rights would continue to be transferred to the urban and industrial sectors to the extent they are made available.

Estimating the Value of an Increment of Water

It is in acknowledgement of the difficulty in predicting how changes in river flow might in net be allocated, given the imperfect market in water, that we choose to evaluate two alternative scenarios for water use. As the high-valued scenario, we consider a case where all increments in water go in net to Front Range municipal and industrial uses. As our conservative alternative, we consider a case where no water increments go to the Front Range, with the water going instead to other existing uses in proportion to the increased river flow. As it turns out, our choice of this conservative alternative, rather than valuing all water as if it went to marginal agriculture, may have little effect on our conclusions as to the economics of managing the forests for water. Because irrigation use results in approximately a 50 percent return flow of water, agricultural users could withdraw twice the amount of the actual water increment and still preserve in-steam flow below them. This, and the possibility of maintaining the production of hydropower at all facilities upstream of the irrigation withdrawals, tends to offset the higher value of those few withdrawals for consumptive water use in municipal and industrial sectors under our conservative alternative. In the discussion that follows, we will occasionally note how our estimated value for water increments might differ if the water rights were to go fully to either Upper Basin or Lower Basin agriculture.

Given the preceding background, and with our estimates of unit value for water in the different geographic and functional markets, we now turn to a determination of the increased water yield that would result from water augmentation treatments. Doubtless the most exact water-yield measurement to have been made is that by the Fool Creek experiment facilities on the USDA Forest Service Fraser Experimental Forest. An estimate of almost a third of an acre-foot per watershed acre, in response to an initial harvest of 40 percent of the land area, was obtained. This estimate was based on the water flows from the treated watershed in comparison to those from an untreated control area, the East St. Louis Creek. Although these data were obtained from monitoring conducted with excellent equipment and instruments and based on scientific research, they are nonetheless suspected of being unrepresentative of results that would be obtained from a large number of watersheds.

We will present results from what would be considered a broadly representative subalpine watershed under a likely timber and water augmentation management program. We believe that the results shown in table 5–4 offer a

Table 5-4. Projected Change in Yearly Water Yield from an Average Acre Under the Water Augmentation Program

Years	Yearly water increment (acre-feet)
0–10	0.142
11–20	0.142
21–30	0.108
31–40	0.200
41–50	0.177
51–60	0.125
61–70	0.258
71–80	0.242
81–90	0.175
91–100	0.233
101–110	0.167
111–120	0.117

Source: C. F. Leaf, unpublished data, 1983.

typical response to a water augmentation program for the more representative watersheds.[31] The increments in water yield given in table 5–4 result from the application of the harvest schedule previously described (see table 5–1). The flows are presented in terms of the increase in yearly water flow from the average acre of a watershed planning area following the implementation of the water/timber management program. Under the schedule, the planning area is only partly harvested at any one time. The increased water flow results from the relatively large fraction of the area which is either harvested or commercially thinned in each thirty-year interval. Cover density, which plays an important role in determining interception and evapotranspiration, is reduced by timber management.[32] Cover density on one-third of the area is reduced to zero with each regeneration harvest. Thinnings then maintain the cover density at about one-half the level existing on the stand before management. Other harvesting schedules could be followed, and we will look at some subsequently.

The gross present values (per acre) of increased water flow that might result from application of the described water-augmentation harvest schedule to timber lands in each of the three Upper Colorado tributary areas are given in table 5–5. A 4 percent rate of discount has been used in the calculation of these present values. Since the several headwater streams have somewhat different volumes of flow and the water in each is somewhat differently used before its confluence with the Colorado, separate estimates are made for each of the three. Even so, the credit to each acre does not differ appreciably among the several drainages presented in table 5–5. The marginal value of water in each sector is also summarized in table 5–5.

Table 5–5. Gross Present Value per Acre of Augmented Water Flow, by Geographic and Functional Disposition
(120-year rotation; discounted at 4 percent)

Function		Marginal value	Present value
	Fraser–Colorado		
Hydroelectricity	@	$0.01791/kWh	$ 83
Irrigation (Upper Basin)	@	$10/af	10
Municipal and industrial (Upper Basin)	@	$60/af	—[a]
Irrigation (Lower Basin)	@	$30/af	93
Municipal and industrial (Lower Basin)	@	$100/af	52
Total			$238
	Gunnison–Colorado		
Hydroelectricity	@	$0.01508/kWh	$ 93
Irrigation (Upper Basin)	@	$10/af	24
Municipal and industrial (Upper Basin)	@	$60/af	2
Irrigation (Lower Basin)	@	$30/af	72
Municipal and industrial (Lower Basin)	@	$100/af	41
Total			$232
	White–Colorado		
Hydroelectricity	@	$0.01900/kWh	$ 77
Irrigation (Upper Basin)	@	$10/af	8
Municipal and industrial (Upper Basin)	@	$60/af	2
Irrigation (Lower Basin)	@	$30/af	96
Municipal and industrial (Lower Basin)	@	$100/af	54
Total			$237
	Front Range		
Municipal and industrial	@	$100/af	$388
Total			$388

Note: All values are rounded to the nearest dollar, and totals shown may not add up to those that would be found by adding the given present values.
[a]No data.

Our analysis indicates that there is a substantial difference in the value of incremental water in Front Range markets as compared with the value from the West Slope and the Colorado Lower Basin. If the water were to go to the Front Range, at a value of $100/af per year, the present value of the augmented yield would be about $388 per acre of managed watershed. This is more than half again the value found for water that might remain in the Colorado drainage, even when the return flow and further uses in the latter are taken into account. This result is also presented in table 5–5.

The more important comparisons to be made, however, are with the costs of access to the watersheds given in table 5–3. The lengths of access roads are likely to be quite variable. It can be seen that where timber values alone could not cover the cost of access, the value attributable to the water augmentation program could cover the costs under most of the conditions shown in table 5–3 (see the Timber Considerations section earlier in this chapter). Thus the value of water would exceed costs of access on terrain of gentle slope beyond the 8.5-mile-length per square mile of access for roads of temporary and intermittent types. The timber harvest revenues would provide an additional $23 in present value. The combined value of the water and timber that would result from the watershed treatment program would cover the cost of either an extension of the road or higher-grade roads. Indeed, if the water increased by the treatment were destined for the Front Range, almost all of the cost of the most expensive roads for any terrain, out to roughly 7 to 8 miles, could be covered.[33]

An essential caveat concerning a neglected element that may arise in the context of an extensive water-yield augmentation program should be noted. This has to do with aesthetics and amenities. Our early efforts to involve wildlife biologists on subalpine Colorado forests met with no success. As a current tentative rule of thumb, we posit that improving wildlife habitat for some (say game) species alters the environment unfavorably for other species. We do not know whether this rule is also applicable to dog-hair stands of lodgepole pine; in a natural state they tend to provide quite sterile habitats.[34] However, landscape disfigurement from more or less uniformly shaped specialized patch cuts, if visible, is likely to introduce some diminution of utility otherwise available from scenic overlooks. This is not by itself reason to dismiss a program evaluated on a watershed-by-watershed basis. A multiple-use management regime accommodates timber harvests wherever they can be economically undertaken. The possibility that the adverse aesthetic consequences of such a program would affect the economics of particular (marginal) watersheds, however, must be entertained. Performing the necessary analysis would require detailed analyses of numerous watersheds to determine their visibility from recreational observation points. In addition, it would require an extensive and difficult survey research and econometric effort to estimate the locational distribution of the recreating population relative to the drainage in question, and the change in utility experienced by those recreating individuals to whom the patch cuts would be visible. The magnitude of such a study, and the required research resources, fall quite outside the scope of the present study.[35] In short, it seems to us that only in response to a specific forest watershed program proposal would such a study be feasible. If it were done for locationally specific areas, the effects of the proposed patch cuts would be, in principle, susceptible of evaluation. Such an approach is potentially feasible, but it would be very costly.

Returning to the main theme of our analysis, we should point out that in the absence of the value of the water that is provided through management for both timber and water, virtually no management for timber would be economic. With the cost of watershed access and treatment attributable jointly to both forest outputs, the amount of land that could come under management for joint production is greatly increased.

5. Testing Initial Conditions

In section 3 we took a rather conservative stance on the amount of watershed modification that should be done to augment water flows. We assumed that a third of the area would be harvested in scattered patches at the time of initial entry. Then half of the remainder, similarly, would be harvested in 30 years, with the last third patch-cut in the sixtieth year. Thinnings to a growing stock level of 80 were scheduled in the regenerated stands at thirty-year intervals. In the analysis that follows we refer to that harvest schedule as A_1. Another watershed treatment schedule could be designed to remove a half of the timber on initial entry, in an effort to promptly increase the water yield; the remainder of the area would be harvested in equal amounts at thirty-year intervals. Final harvest of each unit would be at age 120, with periodic thinnings at thirty-year intervals, as before. This we will call harvest schedule A_2.

We also consider a schedule B_1, a variant on A_1, for which we harvest a third of the area in the initial, thirtieth, and sixtieth years, but in addition thin the residual stands at the time of initial entry to half their density. If we thin the residual stands in schedule A_2 at the time of initial entry to half their density, we have harvest schedule B_2. Such thinning of the existing stock can be expected to provide greater increases in water flow in the initial years. Taking the associated water and timber values of each variant, we show the results in table 5–6, where the values are for an average watershed acre.

It is clear that as the area harvested on initial entry increases (up to a half of the total watershed), the value of the combined timber and water yields rises. If the fraction of the area cut at initial entry remains constant but thinning of the existing stands is undertaken at the same time, the value of the yield will also increase. Given the various fractions of the area that can be considered for harvest at initial entry, and their associated schedule of harvests, the A_1 variant analyzed in section 3 of this chapter would appear to be relatively conservative. That being the case, and taking into account the fact that the benefits of A_1 exceed the costs of access for moderate slopes, moderate extension of roads, and construction for all but the highest-standard roads, we can feel secure about the prospects of a water augmentation program having a net value remaining after deducting moderate access costs.

Table 5–6. Present Value per Acre, with Timber and Water Values Discounted at 4 Percent

	Harvest schedule			
	A_1	A_2	B_1	B_2
Water values				
Fraser–Colorado	$238	$337	$295	$363
Gunnison–Colorado	232	330	289	355
White–Green–Colorado	237	336	294	362
Timber values	23	30	30	35

Sources: The estimates of water yield on which the values above are based have been provided by C. F. Leaf; the associated timber yields have been provided by R. R. Alexander. See also Leaf and Alexander, "Simulating Timber Yields."

Costs of roads in difficult terrain and of high standard would doubtless exceed the value of program outputs in all cases except the one in which roads for access would be quite short.

If, as we have seen, the West Slope uses of incremental water provide benefits that exceed costs, it is clear that the Front Range uses do so as well. This follows because the value of the marginal product of Front Range uses will be higher than those on the West Slope, and their benefits will thus exceed costs by a larger margin, other factors remaining equal.

Another issue that must be addressed has to do with the discount factor used for our analyses. We have assumed that a real rate of 4 percent is appropriate to our analysis. The Office of Management and Budget (OMB) has authorized the use of such a rate for the Forest Service, so in our analysis we have discounted future benefits and costs at this rate. At the same time, OMB has requested that analyses of investments in forest management on the public lands also indicate the results that would follow from the use of a 7 percent discount rate. Accordingly, we need to check what our results would be if a 7 percent discount rate were used. Our new estimates of water and timber benefits, reflecting 7 percent discounting, are given in table 5–7; access-road costs calculated at 7 percent are shown in table 5–8.

Combined water and timber values are considerably reduced when the present value is computed with discounting at 7 percent. This is also true of access costs, although to a lesser extent, since the largest element of these costs occurs in the first year. The number of miles of road that can be covered by the combined timber and water values net of all costs other than access are thus fewer for the costs and values calculated at 7 percent than for those calculated at 4 percent. With discounting at 7 percent, roughly two miles of road represent the tradeoff—or, put another way, there are roughly two-thirds of a mile of access road reduced per one percentage point increase in the discount rate.

Table 5-7. Present Value per Acre, with Timber and Water Values Discounted at 7 Percent

	Harvest schedule			
	A_1	A_2	B_1	B_2
Water values				
Fraser–Colorado	$149	$232	$197	$255
Gunnison–Colorado	142	222	189	245
White–Green–Colorado	149	233	198	257
Timber values	7	10	15	16

Another potential tradeoff also requires examination. The harvest schedules for existing stands reflect the amount of the watershed that is treated (harvested) at initial entry. The entries for thinning and harvesting of regeneration stands, however, assume a 120-year rotation as a silvicultural imperative. Yet such a harvest age may turn out to be less desirable than some other age. Although it is Forest Service policy to harvest at the culmination of mean annual increment (that is, at the age which maximizes average annual harvest volume), section 6(m) of the National Forest Management Act of 1976 provides latitude in this policy in the interest of multiple-use management. It is our view that the decision on the economic age for

Table 5-8. Cost of Road per Acre Harvested by Type and Length of Road, Type of Terrain

(cost in dollars; discounted at 7 percent)

	Type of road[a]			
Miles[b]	A	B	C	D
	Gentle slope			
2.5	$ 38	$ 58	$ 81	$103
5.5	83	128	179	227
8.5	128	198	267	350
	Moderate slope			
2.5	51	65	89	111
5.5	112	144	196	244
8.5	173	216	303	377
	Steep slope			
2.5	73	85	120	150
5.5	160	186	265	330
8.5	247	288	409	510

[a]A is temporary road, B is intermittent road, C is seasonal road, and D is surfaced road continuously open.

[b]Miles of road per square mile of land accessed.

harvesting is within the control of management in a way that the discount rate is not.

The management of a forest for water flow is not a simple function of rotation age. The size of clearings and the age of surrounding stands has been shown to affect water flow. Beyond that, the harvesting of a very large proportion of an area over a short time is not likely to be accepted on either silvicultural or aesthetic grounds. We will look at two management variants which combine shorter rotations with a fairly well-balanced mix of recently harvested sites and older stands—variants C_1 and D_1.

Variant C_1 is much the same as the earlier schedule B_1, although with a 90-year harvest age. Under C_1 one-third of the area would be harvested in small patch cuts in the initial, thirtieth, and sixtieth years. Subsequent thinnings would follow at thirty-year intervals, with regeneration harvests at age 90. The two-thirds of the existing stock not harvested in the initial year would be thinned at that time to one-half their existing density.

Variant D_1 calls for a 60-year harvest rotation period. To accomplish this, the first harvests would be staggered at twenty-year intervals. One-third of the forest would be harvested in the initial year, one-third in year 20, and the remainder in year 40. Again, the stands not harvested in the initial year would be thinned at that time. One precommercial thin would precede the regeneration harvest, which would occur at age 60 on each unit. Road costs for this alternative increase because the greater frequency of harvest entries results in higher maintenance and road-opening expenses for the same road network.

In schedules C_1 and D_1, as before, the value of the existing stands is taken to be $150 per thousand board-feet at the mill, as reflected in the Forest Service experience with Colorado subalpine forests in the early 1980s. In considering the improved management of the regenerated stands, we have used $250 per thousand board-feet as the value at the mill of the regenerated stands, and stocking rates consistent with managed stands on the Fraser Experimental Forest. For illustrative purposes we have evaluated only a West Slope–Lower Basin scenario, namely the Fraser-Colorado.

Playing out these assumptions and data in our scenario, we get results as shown in table 5–9. The benefits of shorter rotations and entry intervals are illustrated. Variant D_1, with its short rotation, is superior in producing both timber value and water yield, although the timber value may be exaggerated because of the smaller diameter of the logs resulting from a short rotation.

In concluding this analysis, we consider the option of maintaining some of the older stands largely unmanaged while managing the remaining fraction intensively. We present a schedule—E_1—of the timber and water yields that would be associated with a 50 percent area harvest at initial entry. This fraction of the planning area would then be periodically reharvested. The other 50 percent of the area would be initially thinned to half its volume, but

Table 5–9. Present Values per Acre: Effect of Rotation Age
(discount rate = 4 percent)

Schedule	Rotation age/ entry interval	Timber value	Water value	Roading costs	Total value
B_1	120/30	$30	$295	$237	$ 88
C_1	90/30	36	298	237	97
D_1	60/20	53	332	263	122

Note: Data are for the Fraser–Colorado, assuming moderate slope, seasonal roads (permanent construction, open except for seasonally bad conditions), and 5.5 miles of access road per square mile.

would not be entered in later years. The costs and benefits have been computed for harvest rotation periods ranging from 10 to 120 years, in increments of 10 years. Road costs would depend upon the cycle of harvests and thinnings. Thinnings of the managed area are assumed to occur at thirty-year intervals (for rotations which are sufficiently long).

The results of schedule E_1, presented in table 5–10, are for the Fraser-Colorado scenario with 5.5 miles of seasonal road in moderate terrain. The values, as always, are per-acre values for the full area, not simply for the fraction under timber management. When timber and water values for each harvest age are summed and the joint cost of roads is deducted, we find that the maximum net present value per (watershed) acre is achieved with a thirty-year rotation. That is, to maximize the net present value of the two outputs

Table 5–10. Present Values per Acre by Rotation Ages, under Partial Area Management
(discount rate = 4 percent)

Rotation age	Timber value	Water value	Roading costs	Total value
10	$−74	$415	$281	$ 60
20	−18	415	248	149
30	−1	390	237	152
40	4	362	242	124
50	6	343	239	111
60	49	331	237	143
70	46	322	238	130
80	42	313	237	118
90	38	311	237	113
100	36	308	237	107
110	33	306	237	103
120	31	304	237	98

Note: Data are for the Fraser-Colorado, assuming moderate slope, seasonal roads (permanent construction, open except for seasonally bad conditions), and 5.5 miles of access road per square mile.

combined, the harvested areas would continue to be cleared periodically in the interest of water yield. For rotations longer than thirty years, each ten-year delay would result in a diminished net present value of the two outputs combined. Of course, no commercial timber would be yielded by the thirty-year clearing cycle. It can be seen in this schedule that short rotations which maintain one-half the forest area as clearings provide an effective means of water augmentation. In effect, this is management for water flow alone. Silviculturists might argue, however, that inattention to the residual stands would eventually result in a forest with a broken canopy, disrupting conditions that have led to high experimental yields.

Although the results based on the Fool Creek experiments and management regime tend to be more dramatic than is the case in the more typical watershed, the results are nonetheless generally similar to the typical case. In general, in the absence of the value of water, timber values alone would not be able to cover the access as well as all management and harvesting costs. Indeed, in the scenarios run with a 7 percent discount—exclusive of the Front Range case—only those regimes involving initial harvest from half the total watershed area, or heavier thinnings of the residual stands at time of initial entry, could cover the costs of even a 5.5-mile intermittent road in moderately sloped terrain. In all of our analyses, among the most critical elements are the length and type of access road and the terrain it traverses. Accordingly, the economic feasibility of a water augmentation program for any particular watershed would doubtless turn on the distance and condition of the access.

6. Summary and Conclusions

It appears quite clear that subalpine forests in Colorado, and perhaps in the Central Rocky Mountains, are not economic for timber production, or are at best of dubious economic value when viewed from the perspective of an owner who depends exclusively on the market for his revenues. If, however, timber is managed in conjunction with water augmentation programs, the joint product, even though not captured by the market for the owner, may have a value that will exceed joint resource management costs, exclusive of access to the candidate sites. Whether the net benefits from timber and water combined will exceed the cost of access in addition to other management costs depends ultimately on a combination of factors having to do with the construction and maintenance costs of the road access. Using a 4 percent discount, where slopes are gentle and roads are temporary the distance from the road system can be considerable, yet both access and management costs can be covered. At 7 percent, the latitude even for joint production is much more limited.

If we consider West Slope and Lower Basin uses in particular, the feasibility of joint production will depend also upon the water-augmentation management regime. Our most conservative treatment—that is, harvesting one-third of the acreage in each of the thirty-year intervals—can be undertaken with temporary or intermittent road extensions of roughly 5 miles per square mile of land treated. Greater distances or roads of higher standard would require a more extensive removal of vegetation and periodic thinnings. At a discount rate of 7 percent, many potential sites would be precluded, and a careful selection of sites within the constraints ruling would have to be made.

For reasons given earlier, we have not said anything substantive about one element of the costs that may arise from an extensive water augmentation program: the aesthetic and amenity issues. All we can do is draw attention to this aspect of the problem and hope for a more complete analysis when an actual locationally specific program is considered.

A general concluding comment is in order. This case study of a small multipurpose land-management proposal illustrates one of the findings of chapter 4—namely, the site interdependencies in the production of multiple forest outputs. This phenomenon is repeated in every forestland management case study we review in the chapters ahead, and comes forcefully to attention in chapter 10.

Notes

1. The original version of this chapter appeared as M. D. Bowes, J. V. Krutilla, and P. B. Sherman, "Forest Management for Increased Timber and Water Yields," *Water Resources Research* vol. 20, no. 6 (June 1986) pp. 655–663; copyright by the American Geophysical Union.

2. See C. A. Troendle, "The Potential for Water Augmentation from Forest Management in the Rocky Mountain Region," *Water Resources Bulletin* vol. 19, no. 3 (June 1983).

3. R. C. Kattelmann, "Water Yield Improvement in the Sierra Nevada Snow Zone: 1912–1982," paper presented at the Western Snow Conference, Reno, Nevada, April 20–23, 1982 (Berkeley, Calif., USDA Forest Service, Pacific Southwest Forest and Range Experiment Station, 1982).

4. James R. Beavers to the authors, January 5, 1983. Beavers is director, Timber, Forest, and Pests, and Cooperative Forestry Management, USDA Forest Service, Rocky Mountain Region.

5. C. F. Leaf and R. R. Alexander, "Simulating Timber Yields and Hydrologic Impacts Resulting from Timber Harvests on Subalpine Watersheds," USDA Forest Service Research Paper RM-133 (Fort Collins, Colo., Rocky Mountain Forest and Range Experiment Station, 1975).

6. Robert R. Alexander and Carleton B. Edminster, "Management of Lodgepole Pine in Even-Aged Stands in the Central Rocky Mountains," USDA Forest Service

Research Paper RM-229 (Fort Collins, Colo., Rocky Mountain Forest and Range Experiment Station, 1981).

7. C. W. Rupp, unpublished data, 1983.

8. R. J. Lowe, USDA Forest Service Region 2, unpublished data, 1983.

9. C. F. Leaf and C. E. Brink, "Land Use Simulation Model of the Subalpine Coniferous Forest Zone," USDA Forest Service Research Paper RM-135 (Fort Collins, Colo., Rocky Mountain Forest and Range Experiment Station, 1975).

10. J. D. Hewlett, *Principles of Forest Hydrology* (Athens, Ga., University of Georgia Press, 1982); Kattelmann, "Water Yield Improvement."

11. C. A. Troendle and C. F. Leaf, "Hydrology," in USDA Forest Service, *An Approach to Water Resources Evaluation of Non-Point Silvicultural Sources,* EPA-600/8-80-012 (Springfield, Va., National Technical Information Service, 1980).

12. C. A. Troendle and C. F. Leaf, "Effects of Timber Harvest in the Snow Zone on Volume and Timing of Water Yield," in D. Baumgartner, ed., *Interior West Watershed Management* (Pullman, Wash., Washington State University, Cooperative Extension, 1981) pp. 231–243.

13. Troendle and Leaf, "Effects of Timber Harvest."

14. Ibid.

15. C. A. Troendle, "The Effects of Small Clearcuts on Water Yield from the Deadhorse Watershed, Fraser, Colorado," paper presented at the Western Snow Conference, Reno, Nevada, April 20–23, 1982 (Fort Collins, Colo., USDA Forest Service, Rocky Mountain Forest and Range Experiment Station, 1982).

16. Troendle, "The Effects of Small Clearcuts."

17. A. R. Hibbert, "Vegetation Management for Water Yield Improvement in the Colorado River Basin" (Fort Collins, Colo., USDA Forest Service, Rocky Mountain Forest and Range Experiment Station, 1979).

18. Troendle and Leaf, "Effects of Timber Harvest."

19. Reservoir operating procedures failed to accommodate the objective of stream regulation, which should have been given top priority under the congressional mandate for the Colorado River Storage Project.

20. J. V. Krutilla, M. D. Bowes, and P. B. Sherman, "Watershed Management for Joint Production of Water and Timber: A Provisional Assessment," *Water Resources Bulletin* vol. 19, no. 3 (1983) pp. 403–414.

21. W. P. Creager and J. D. Justin, *The Hydroelectric Handbook* (2d ed., New York, Wiley, 1950).

22. C. W. Howe, D. R. Schurmier, and W. D. Shaw, Jr., "Innovation in Water Management: Lessons from the Colorado–Big Thompson Project and the Northern Colorado Water Conservancy District," in K. D. Frederick, ed., *Scarce Water and Institutional Change* (Washington, D.C., Resources for the Future, 1986) pp. 171–200.

23. R. L. Anderson, "Transfer Mechanisms Used to Acquire Water for Growing Municipalities in Colorado," paper presented at the Western Farm Economics Association Meeting, Blackburg, Va., July 14, 1978; R. L. Anderson and R. Milton, "Trends in Price and Delivery of Water in the Northern Colorado Front Range,"

report issued by the Natural Resource Economics Division, USDA Economics, Statistics, and Cooperatives Service, in cooperation with Colorado State University, Cooperative Extension Service (Fort Collins, Colo., Colorado State University, Cooperative Extension Service, 1979). Discussions by the present authors with the Denver Water Board similarly fixed yearly rates of raw water at $130/af to $169/af, depending on the security of the supply.

24. P. B. Sherman, "Methodology for Tracing Flow and Use from Augmented Water Yields: An Exercise in Valuation," research memorandum, Resources for the Future, 1982.

25. This implies that users below points of additional inflow (from joining rivers) get proportionately less new water than do those above.

26. For details, see Sherman, "Methodology for Tracing Flow."

27. M. D. Frank and B. R. Beattie, "The Economic Value of Irrigation Water in the Western United States: An Application of Ridge Regression," Technical Report no. 99 (College Station, Tex., Texas Water Resources Institute, Texas A&M University).

28. R. E. Howitt, W. D. Watson, and R. M. Adams, "A Reevaluation of Price Elasticities for Irrigation Water," *Water Resources Research* vol. 16, no. 4 (1980).

29. See Sherman, "Methodology for Tracing Flow"; and Krutilla, Bowes, and Sherman, "Watershed Management for Joint Production," p. 406.

30. Howe, Schurmier, and Shaw, "Innovation in Water Management," p. 194.

31. The simulation model which generated these estimated water yields is described in Leaf and Alexander, "Simulating Timber Yields," pp. 1–17.

32. See Leaf and Brink, "Land Use Simulation Model."

33. If all water rights were to go to Lower Basin agriculture, the present value from water use would be $223, plus hydropower value (almost all of which could still be provided). The program would still be economic under most conditions. On the other hand, if water rights went fully to Upper Basin irrigators, the water augmentation program would be only marginally profitable, at best. The present value from water use would be $74 per acre plus any hydropower value. However, much of the hydropower might be forgone. In the Gunnison, perhaps up to 49 percent of the power production could be maintained, depending upon the point of irrigation withdrawal. For the Fraser River, a corresponding figure would be 13 percent, while for the White River there would be no hydropower production.

34. We owe this point to William McKillop.

35. See chapter 6 for an example of the magnitude of effort encountered in an attempt to define the proper population and select a sample for a questionnaire or interview, when the relevant data must be obtained by primary source acquisition modalities.

6

Valuing Recreational Quality: Hedonic Pricing

1. Introduction

While the production of timber and water are basic objectives stipulated in the Organic Act of 1897, in more recent times, as we have seen, recreation has emerged as a dominant use of the national forests. The public forests may provide some recreational opportunities on every land area within a forest. While typically just a few areas of a forest will be managed predominantly for recreational purposes, a much larger area is likely to provide some recreational and hunting services as a more or less incidental result of the vegetational and other natural characteristics of the land. Management decisions as to the location and intensity of timber treatments on such lands are of primary importance in determining the flow of amenity services over time. With the growing demands for both timber and the other resource services of the forest, there has been an increased need for methods of assessing the impact of timber harvesting and other management actions on the recreational services of the public forestlands.

Unfortunately, the multiple-use management effort has been hampered by the limited attention given in the past to empirical methods for estimating an economic value of changes in the conditions of forest lands. Much of the early recreation demand analysis tended to focus on the total value of sites in their existing condition. The travel-cost methods of Clawson, of Burt and Brewer, or of Cicchetti, Fisher, and Smith illustrate this focus on total site value.[1] In contrast, the forest manager is most often concerned with management actions that effect changes in physical attributes of forest sites. This concern presents the recreation demand modeler with a very difficult problem. A national forest offers numerous potential sites for recreational use. It must be determined how the value of each of these sites would change in response to adjustments in the site attributes. In order to make such a determination, the individual's choice among the many recreation sites must be modeled.

There is now increasing attention being given to this type of recreation demand problem. Two general techniques are employed, both of which use the observable expenditures and site choice of the individual as the basis for estimating the value of changes in recreational site attributes.[2] In one category are those travel-cost procedures which estimate the demand for an individual's visits to sites as a function of both the cost of trips and the attributes of sites. The travel-cost procedures focus on the number of visits taken by otherwise similar individuals at different locations relative to the forest. These individuals face different travel costs to use sites. The wide variation in these travel costs, along with some variation in attributes across sites, allows the estimation of demand equations for visits to sites as dependent upon the site attributes. From these demand equations the value of changes in site attributes can be measured by evaluating the resulting shifts in demand. Several versions of the travel-cost procedures exist, the essential difference among them being in the assumptions made about the role of substitute recreation sites.[3] Travel-cost techniques have the advantage of being reasonably easy to understand. With appropriate data, they can be straightforwardly estimated and easily used to value changes in the attributes of particular sites.

In the second category of valuation procedures are the so-called hedonic pricing methods, or hedonic travel-cost methods. The hedonic techniques seek to directly estimate the marginal value of site attributes by observing choices among sites. The individual's decision to visit one particular site, selected from among the many sites of various attributes, reveals his willingness to pay for greater site quality. Suppose, for example, we had two otherwise similar recreation sites, one with somewhat better facilities. A choice made to travel an extra distance to the better site reveals that its facilities are worth more than those at the other site by at least the cost of the extra travel. We might then suppose that an improvement in facilities at the poorer site would provide this same increment in benefits to each visitor. The application of hedonic pricing techniques to the valuation of recreational site attributes is a recent development, associated with Robert Mendelson and Elizabeth Wilman.[4] The techniques can seem complex, especially in the estimation procedures.[5] Further, the results are not always easily applied to the valuation of discrete changes in the condition of a particular site.

The consumer-choice theory underlying the hedonic methods, as applied to recreation sites, exactly parallels the theory underlying a particularly simple version of the travel-cost demand procedures. Despite these underlying parallels, it is important to realize that there will be situations for which the hedonic methods are well suited while the travel-cost methods prove inapplicable, and vice versa. The methods tend to complement each other.

In this chapter we will develop the consumer-choice theory behind the hedonic methods. We first describe the corresponding travel-cost model in some detail, then take advantage of the parallels to describe the hedonic methods. It is hoped that some of the difficulty in understanding the hedonic methods can be avoided by drawing on the analogous features in the travel-cost procedure. Two variations on the hedonic method will be presented. The second of these most closely corresponds to the hedonic travel-cost methods of Mendelson and Wilman. We present the other variation partly because it seems in many ways to be the more desirable method, and we want to show that the distinction between these two variations can be rather subtle. They are estimated in essentially the same manner and it can be hard to distinguish which version an author has in mind. However, the distinction is important because the procedures for evaluating benefits with these two variations of the hedonic model are quite different.

In the third section of this chapter we will describe an application of a hedonic pricing approach to the valuation of changes in hunting conditions on the Black Hills National Forest. That application, by Elizabeth Wilman, sought to estimate the value to deer hunters of certain changes in the character of the forestlands which might result from timber management.[6] The study was undertaken as part of a larger effort to investigate the economies of multiple-use forestland management; it illustrates the dependence of recreational values on the dispersion of forage and timber cover.

2. Measuring Benefits of Changes in Site Attributes

The forest environment can be thought of as providing an array of recreation sites of various physical attributes. The costs of using a particular site will depend upon the location of the individual relative to the forest. Each individual, in effect, faces a somewhat different set of recreational opportunities. After considering the travel costs and the attributes of the sites, the individual selects the sites to visit and the number of visits to be made. These observable choices, or rather the variations in choice across individuals, provide the basis for valuing site attributes. The methods to be described rely on a somewhat restrictive assumption about the choice among sites: it is implicitly assumed that each individual will choose to visit just one site. The same theory of choice among sites is the basis for both a simple version of the travel-cost method and the hedonic travel-cost methods.

The Utility Theory Framework

It will be assumed that individuals select among recreation sites so as to maximize their utility, subject to a constraint on their expenditures. Let V_j be the number of visits the individual makes to a site j; let z_j represent the measured physical attributes of site j; let P_j be the travel cost (both time and money costs, including any site entry fee) of visiting site j; and let Y be the total amount of income the individual can allocate to these recreation travel expenditures. The utility function is taken to depend upon site attributes weighted by the number of trips taken to the site, summed over each site visited. For simplicity, the role of other goods purchased is ignored.

Utility Maximization—The Travel-Cost View. Utility maximization requires a choice of the number of visits to make to each of n sites, solving the problem

$$\underset{V}{\text{Max }} U\left[V_1\phi(z_1) + V_2\phi(z_2) + \ldots + V_n\phi(z_n) \right]$$
$$\text{subject to: } P_1V_1 + P_2V_2 + \ldots + P_nV_n = Y$$

The first-order conditions describing the solution to this problem are as follows:

$$\alpha\phi(z_i) = P_i \quad \text{for the optimal site } i \text{ (for which } V_i > 0\text{)}$$
$$\alpha\phi(z_j) < P_j \quad \text{for all other sites } j \text{ (for which } V_j = 0\text{)} \quad (6\text{-}1)$$

where α represents U'/λ the marginal utility divided by the marginal value of income.

Put simply, these first-order conditions tell us that the individual will choose to visit only one site. Specifically, if we interpret $P_i/\phi(z_i)$ as the price per unit of attribute service, the site with the lowest price per attributes will be visited. The number of trips taken to this site is then determined, so that $\alpha\phi(z_i)$ the monetary value of the marginal benefit from the last trip taken is just equal to P_i the travel cost of the trip. Variation in the price of visiting the chosen site across similar individuals allows us to estimate a demand function for visits, relating the marginal benefits from a trip to the number of visits.

Under the assumptions of our utility model, the demand function for visits to the selected site will depend only upon the price and attributes of this one site. The price and attributes of other sites will not be explicitly reflected in the demand function. However, the other prices and attributes are relevant in that the lowest value $P_j/\phi(z_j)$ among the other sites gives us a reservation price—the price at which the individual would switch all use to the other site. If there were an entry fee charged at the selected site, at some sufficiently high entry fee, the individual would switch his use entirely to the next-best alternative site. These conclusions are the basis for the simple travel-cost demand method described below.

Let us suppose now that there is a site j with slightly poorer attributes than those at the chosen site i, and further suppose that this site is a little closer than the chosen site, so that the individual is almost indifferent as to which is selected for use. From the first-order conditions we have the following approximate equality:

$$\alpha[\phi(z_i) - \phi(z_j)] \simeq P_i - P_j \qquad (6-2)$$

That is, the monetary value of the increment in marginal benefits received from a site with attributes z_i over a site with attributes z_j is approximately measurable by this difference in travel cost $P_i - P_j$. More generally, the travel-cost differences will give upper and lower bounds on the marginal value of site attributes. These results are the theoretical basis for the hedonic methods.

Utility Maximization—The Hedonic View. It is usual to describe the utility theory framework for the hedonic methods in a slightly different but equivalent form. In effect, we begin by imposing the condition that only one site will be chosen, and represent the utility function by $U(V\phi(z))$. The travel costs of using sites of various qualities will be represented by $p(z)$. The function $p(z)$ indicates the least travel cost required to use a site with attributes z; it is simply a convenient summary of known travel-cost information. The function $p(z)$ will vary across individuals, depending upon their location.[7] The choice of a particular level of attributes z (and the corresponding price P) implicitly determines the site that is chosen for use. Utility maximization requires the choice of visits V and attributes z, solving the problem

$$\operatorname*{Max}_{V,z} \ U(V\phi(z))$$

$$\text{subject to: } p(z)V = Y$$

In practice, it is usual to assume that $p(z)$ can be approximated by a continuous and differentiable function.[8] The first-order conditions describing the solution to this problem are then as follows:

$$\alpha\phi(z_i) = p(z_i) \qquad (6\text{-}1')$$

$$\alpha\phi'(z_i) = p'(z_i) \qquad (6\text{-}2')$$

where α again represents U'/λ and z_i is the chosen level of attributes; that is, z_i gives the attributes of the chosen site and $p(z_i)$ equals P_i the travel cost of using that site.

The first of these conditions (6-1') describes the choice of the number of visits. It is identical to the condition (6-1) derived above from the more explicit version of the utility model. The second condition (6-2') describing the optimal choice of attributes (and, implicitly, the choice of site) appears to be something new. In fact, it is nothing more than a restatement of condition (6-2), derived earlier. The derivative $P'(z_i)$ corresponds to the price difference $P_i - P_j$, for a site j with attributes z_j only marginally different from z_i. Similarly, the derivative $\phi'(z_i)$ corresponds to the benefit difference $\phi(z_i) - \phi(z_j)$. Again, this condition indicates that the extra cost spent on visiting a site of somewhat higher quality can be used as an estimate of the marginal benefits of an improvement in the quality of site attributes. The variation in these marginal benefits across individuals choosing sites with different attributes allows us to estimate what might be called a demand curve for attributes, relating marginal benefits to the level of attributes chosen.

Comment. The equivalence of first-order conditions demonstrates the essential equivalence of our two utility maximization models. Both the simple travel-cost model and the hedonic model rely upon this same utility framework. The travel-cost model is largely based on the marginal utility condition describing the choice of the number of visits. On the other hand, the hedonic method is based primarily on the marginal utility condition describing the choice of site quality (and site). This difference is the basis for our conclusion that the two methods tend to complement each other.

The travel-cost method relies on estimation of the demand curves for visits. Travel-cost methods are most suitable when there are relatively few recreation sites to choose among and there is a wide dispersion of distances traveled by users. A wide range of distances traveled is required in order to fully identify the demand curves for trips. However, we might often be satisfied with having demand curves well-estimated over only some limited range of site attributes. The hedonic methods rely on the estimation of demand curves for site attributes. They are well suited to situations where a substantial number of sites is used, and those sites have a great variety of attributes. The hedonic techniques do not require a wide range in the levels of observed travel costs; rather, they can be estimated from a sample of users from locations spread out around the forest area (perhaps all close to the forest) so that they face varied costs of using any particular site.

Now, despite a common underlying basis for these models, in practice the two may turn out to be quite different. The compromises required by the specific empirical procedures tend to introduce different sources of error into the two models. For example, in the hedonic method any difficulty in accurately measuring the exogenous marginal costs $p'(z)$ introduces a corresponding error into the estimation of marginal benefits. Further, the assumption that only one site is chosen for use often will prove unrealistic. With the travel-cost model, we would most likely abandon our simplistic utility maximization framework and follow another that allows for the use of several sites; that is, we would probably use a travel-cost model for which the demand curve depends upon the prices and attributes of both the selected site and other alternative sites. It is somewhat harder to fix the hedonic methods, which rely strongly on the assumption that a single alternative is selected. In practice, we can stretch the definition of the alternative so that it no longer corresponds to a single site. The individual can be thought of as choosing among possible "package tours" of several sites. That leaves open the question as to how we should best measure the attributes of this package of sites. Such are the trials of empirical economists.

Let us now look at the travel-cost model implied by our utility maximization model.

A Simple Travel-Cost Model

We wish to measure the influence of site attributes on the demand function for visits to a forest. In general, the demand for a particular recreation site will depend not just upon its attributes and the cost of using the site, but also upon the travel costs and attributes of substitute sites. The travel-cost model to be described here reflects an extreme view of these substitution effects. The individual's demand function for visits will depend only upon the travel cost and attributes of the site selected for use. Such a demand function alone would leave unclear how the choice of sites might be altered in response to changing forest conditions. Here it is assumed that individuals choose to visit only the one best site—that site providing the greatest net benefits.[9] Changes in the attributes of another site will have no effect on site use until that other site is sufficiently improved to become the preferred site. The individual then shifts all his use to this alternative site.

The Demand Function. The individual's trip demand equation is of the general form

$$V = V(P,z) \tag{6-3}$$

where V is the number of visits to the selected site, P is the travel cost per visit to the selected site, and z is a vector of attributes describing the selected site. The travel cost is the least cost required to make a trip, including all travel expenses, the time costs of travel, and any site entry fee. Measurement of travel cost should be straightforward except for the component of this cost which reflects the opportunity cost of travel time. The site attributes measured are those physical features of the site that are related to the recreational value of the site. The focus would typically be on those attributes which might be influenced by management treatments.

The demand curve is estimated by using data on visitor use of all sites on the forest. It is usual to divide the area around the forest into a number of geographical zones; the per-capita visitation rate for each zone would be used as the data for visits. In effect, this procedure assumes that all potential visitors have similar tastes and so have the same demand function for visits. We could introduce other variables to explain some differences in individual taste. Estimation of the demand curve requires wide variation in travel costs across the zones of visitor origin and some variation in the attributes of the sites selected for use.

Care must be taken in interpreting the demand curve. It is not in this case the demand curve for a specific site, and in itself it tells nothing about which site is used. The function might be best interpreted as showing what the demand function for a site would be if there were no alternative sites. Really, it simply tells us that if an individual uses a site with attributes z at a travel cost P, we expect him to make V visits.

Forest Benefits and Choice of Site. The consumer's surplus for the forest as a whole can be measured as the area above travel costs and below the trip demand curve evaluated at z, the attributes of the chosen site. That is, the net benefits from the forest to an individual visiting a site with attributes z at travel cost P can be represented by the integral

$$S(z,P) = \int_P^\infty V(p,z)dp \tag{6-4}$$

The chosen site of each individual will be that which provides the greatest value of these net benefits $S(z,P)$.

We might note that the net benefits from the chosen site itself should be found as the area under the demand curve (for given attributes z) between P_r and P, where P_r is the price at which the individual would switch to the next-best alternative site. The benefits from sites not selected for use are zero.

The demand function and the choice among sites is illustrated in figure 6-1. Suppose individuals at one origin have a choice between two sites. Site 1 is available at travel cost P_1 and has attributes z_1. Site 2 has attributes z_2 and is available at cost P_2. In figure 6-1, the lower demand curve indicates the demand for visits as a function of travel cost, given that the attributes of the chosen site are z_1. The higher curve illustrates the demand for visits when site attributes are at the higher level z_2. Improving the attributes generally shifts out the demand for visits in this manner.[10]

If site 1 were selected for use, the consumer's surplus from the forest would be given by area *ABC*. If site 2 were selected, the consumer's surplus from the forest would be area *DEF*. In this case, site 1 would be selected, since it provides the greatest value of these net benefits.

Benefits of Changes in Site Attributes. Of primary interest to us here is the valuation of changes in the attributes of sites. Benefits from changes in

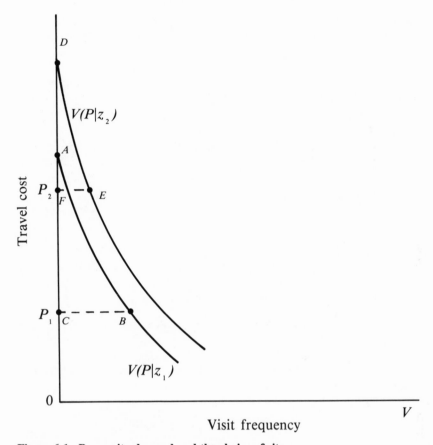

Figure 6-1. Per-capita demand and the choice of site

site attributes can be measured by evaluating the resulting increase in the overall net benefit from the forest. The individual's net benefits from attribute changes across the forest can be measured for each origin as

$$S(z',P') - S(z^0,P^0) = \int_{P'}^{\infty} V(p,z')dp - \int_{P^0}^{\infty} V(p,z^0)dp \qquad (6\text{-}5)$$

where P^0 and z^0 are the price and attributes of the initially chosen site and P' and z' are the price and attributes of the site selected after any modifications in sites.[11] It is assumed that individuals do not value the attributes of the sites they do not visit.[12] If they do, we will be understating benefits of attribute improvements to some extent.

In figure 6–2, the two visit demand curves from figure 6–1 have been redrawn, and a third demand curve has been added. These curves represent individual demand for visits as a function of travel cost, given three different

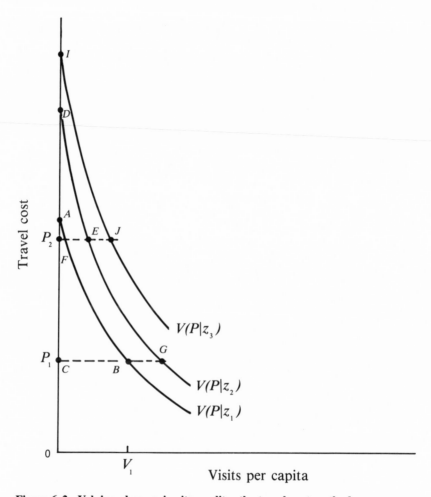

Figure 6-2. Valuing changes in site quality: the travel-cost method

attribute levels z_1, z_2, and z_3. As before, we assume there are two alternative sites available to the individuals at a given origin. Site 1 is available at travel cost P_1 and has attributes z_1. Site 2 is available at travel cost P_2 and has attributes z_2. As in figure 6–1, site 1 is selected for use because it provides greater net benefits ($ABC > DEF$).

Now suppose site 1, the currently selected site, is modified so that its attributes are improved to z_2. The net benefits to the individual using site 1 would then be DGC. The increase in net benefits resulting from the site improvement is the area $DGBA$ (DGC minus ABC). The net benefits of the quality improvement at a site already selected for use by the individual are measured as the net increase in area above travel cost and between the two travel demand curves.

Suppose, on the other hand, that attributes of site 2 were improved to z_3, with site 1 left unmodified. If site 2 were then selected for use the net benefits would be *IJF*. If this net benefit exceeds *ABC* (the net benefits attained by selecting site 1), individuals would switch their use to site 2. If they switch, the resulting increase in per-capita net benefits would be found as the area *IJF* − *ABC*. However, if the improvement in attributes were not sufficient to induce individuals to switch to site 2, there would be no increase in benefits from the site improvement.

Comments. This travel-cost model provides a relatively easy means for estimating the value of changes in site attributes. The method can be viewed as taking direct observations on the slope of the individual's net benefit function with respect to price P (z held constant), with the negative of this slope $-S_p(z,P)$ equal to the observable units $V(P,z)$. Integrating under the slope or marginal net benefit function gives us the overall measure of benefits from the forest for individuals using a site of attributes z at price P. The measure of benefits is the area under the visit demand curve above travel cost P for given attributes z. With that measure of overall benefits, we can then evaluate the change in the overall benefits from the forest that result from site attribute improvements.

The First Hedonic Method: The Net Benefit Approach

The hedonic methods are similar in principle to the travel-cost methods. Hedonic methods focus initially on the slope of the benefit function with respect to changes in site attributes, rather than on prices. Integrating under this marginal value function will also give us the individuals' benefits from the forest. Changes in the forest condition can then be valued directly. Two variations of the hedonic method will be described. The first focuses on the direct estimation of the net benefit function $S(z,P)$, and is most directly analogous to the travel-cost method. The second approach focuses on esti-mation of the gross benefit function, from which travel-cost expenditures must be subtracted to find net benefits.

The first hedonic approach relies on an alternative method of estimating the individual's net benefit function $S(z,P)$. The choice of site reveals the individual's willingness to incur extra travel expense for improved site qual-ity. It is assumed that the extra cost of going to the chosen site, over one of somewhat lesser quality, is just equal the extra benefits gained from the higher-quality attributes. We will first describe the measurement of benefits and then comment on the procedures used for estimation.

The Net Benefit Function.   The per-capita net benefit function $S(z,P)$ represents the net value to the individual from the forest as a whole when a site of attributes z is visited at a per-trip travel cost of P. This is the same benefit function previously represented in equation (6-4), although we will now be looking for an alternative means to measure these benefits, not using the travel-cost demand curves. In figure 6–3 the net benefit function corresponding to the demand curves in figure 6–2 is illustrated. For convenience, we assume that a single attribute is of interest. The curves $S(z|P_1)$ and $S(z|P_2)$ represent the net benefits as a function of attributes when the cost per visit is held constant at either P_1 or P_2. The point labeled X on the $S(z|P_1)$ curve corresponds to the consumer surplus benefits ABC in figure 6–2; point Y shows the consumer surplus benefits DGC. A few obvious features of the net benefit curves should be noted. Improving the attribute at a site, given a constant price, increases the net benefits. An increased cost of using a site, given a constant attribute, lowers the net benefit.

The Marginal Value of Quality.   The slope of the net benefit function with respect to attribute changes $S_z(z,P)$ gives a measure of the marginal value of the attribute. That is, this slope measures the increase in individual net benefits that would result from a small improvement in quality at a site currently with attributes z available at travel cost P.[13] It would perhaps be better to refer to this as a potential increase in benefits, since the benefit is dependent upon the site being selected for use. The slope is easily visualized as the increased area under the travel demand curve that would result from a very small attribute improvement (much as was measured by area $DGBA$ in figure 6–2 for the larger change in attributes); that is,

$$S_z(z,P) = \int_P^\infty V_z(p,z)dp \qquad (6\text{-}6)$$

where V_z is the partial derivative of the trip demand curve with respect to the attribute z (price P held constant). However, the estimation procedure to be described is not based on the travel-cost demand curves; rather, the hedonic method relies on the direct estimation of these marginal value curves.

The net benefits can be evaluated by integrating under the estimated marginal value curve; that is, the consumer's surplus benefits $S(z,P)$ are found by computing the area under the attribute demand curve, with

$$S(z,P) = \int_0^z S_z(z,P)dz$$

The marginal value curves labeled $S_z(z|P_1)$ and $S_z(z|P_2)$ in figure 6–4 graph the slopes of the benefit curves drawn in figure 6–3. The hedonic technique seeks to directly estimate marginal value curves for attributes,

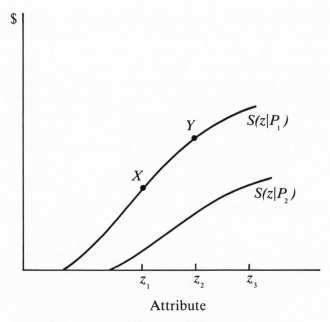

Figure 6-3. The net benefit function

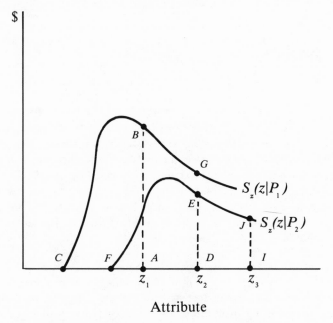

Figure 6-4. The marginal benefit of attribute quality

such as those illustrated in figure 6–4. The marginal value curves are sometimes referred to as demand curves for attributes, or hedonic demand curves.
Certain features of the marginal value curves should be noted. If there is no
use of the forest, the value of its attributes will be zero (by assumption we
have associated value with use). The marginal values of improvements in
quality may themselves be zero, until there is sufficient improvement to
generate visitor use. The marginal value curve may be increasing over some
lower range of quality, but in general it is expected to decline with improved
attributes, much as does the usual demand curve; that is, we expect there
would eventually be diminishing marginal value to further improvements in
site quality.

Evaluating Attribute Changes.   If the marginal value curves for site attributes can be estimated, then the benefits of changes in site quality can be
measured by computing the area under these marginal value curves. With
one attribute, the individual's net benefits from attribute changes can be
measured for each user origin as

$$S(z',P') - S(z^0,P^0) = \int_0^{z'} S_z(z,P')dz - \int_0^{z^0} S_z(z,P^0)dz \qquad (6\text{-}7)$$

where P^0 and z^0 are the price and attributes of the initially chosen site and
P' and z' are the price and attributes of the site selected after any modifications to sites. Note that if the same site is still selected for use, then P'
equals P^0.[14]

To illustrate the use of this hedonic approach we again consider the individual making a choice between two sites. Site 1 is available at price P_1 and has
current attributes z_1. Site 2 is available at P_2 and has current attributes z_2.
Area *ABC* in figure 6–4 gives the net benefits if site 1 is selected. Area *DEF*
would be the net benefits from the forest if site 2 were selected. As illustrated,
site 1 currently provides the greater benefits and would be preferred. Attribute
changes are valued as the change in the overall benefits from the forest. If the
quality of the attributes at site 1 were improved to the level z_2, the individual's
net benefits would increase by the amount represented by area *DGBA*. If, on
the other hand, site 2 attributes were improved to z_3 (site 1 unchanged), the
overall benefits from choosing site 2 would increase to *IJF*. If this value
exceeds the benefits from site 1, individuals would switch sites and their
benefits would increase by the amount *IJF* − *ABC*.

Principles of Estimation.   The estimation of the marginal value curves relies
on the observable cost differences in reaching sites of different attributes. Let
us suppose that for each origin the least possible travel costs for reaching sites
with attributes z can be approximated by a smooth continuous function $p(z)$

(we comment on this assumption shortly). Because of differences in location, this travel-cost function will not be the same for each origin. The individual's choice among sites may now be conveniently represented.

Each individual chooses the site which provides the greatest net benefits. In so doing, he can be viewed as choosing a level of attributes z to solve the maximization problem

$$\text{Max}_z \{S[z,p(z)]\}$$

which (we might note) is identical to the maximization

$$\text{Max}_z \left\{ \int_{p(z)}^{\infty} V(p,z)dp \right\}$$

The solution must meet the first-order conditions

$$S_z(z,P) = p_z(z)V(z,P) \tag{6-8}$$

where z and P are the attributes and travel cost of the selected site;[15] that is, the individual chooses among sites so that the marginal value (over all trips) from choosing a better-quality site just equals the marginal cost of doing so. The term $p_z(z)$ is the slope of the travel-cost function at attribute level z. It represents the extra cost per trip of shifting to a slightly better site. The marginal cost of taking all trips at the higher-quality site $p_z(z)V(z,P)$ is, in principle, observable data, so the marginal value of an attribute $S_z(z,P)$ is revealed. Estimation of the marginal value of attributes as functions of z and P should be possible if there is sufficient independent variation (across origins) in the attributes of chosen sites and in the per-trip travel costs to the chosen sites.

In fact, the marginal cost of an attribute is observable only in a limited sense. The trouble arises because the assumption that $p(z)$ is a smooth and continuous function is not completely accurate. In practice, $p(z)$ might often be a discontinuous set of points. The actual travel-cost function $p(z)$ for an origin is unlikely to be differentiable, since individuals face a choice among a discrete set of sites which may not be systematically located across space in terms of their attributes and cost. This is not necessarily a disastrous conclusion.

Suppose that for each origin we can identify sites that are roughly equal in value to the preferred site. With z^* and P^* the attributes and trip cost for the selected site and z and P the attributes and trip cost for the other site, we could attempt to estimate the marginal benefits for each attribute by using the observed differences in travel costs. To a first approximation, the marginal value $S_z(Z^*,P^*)$ can be estimated by

$$\frac{S(z^*,P^*)-S(z,P^*)}{(z^*-z)} = \left(\frac{P^*-P}{z^*-z}\right)V(P^*,z) \qquad (6\text{-}9)$$

This condition is analogous to equation (6-8), but with the discrete spatial location of sites recognized.

In practice, regression methods have been used to estimate travel-cost functions $p(z)$ for each origin, based on the observed travel costs to various sites visited. From this estimated function, the marginal costs of quality $p_z(z)$ could be computed. Of course, in approximating the costs with a regression model some error is introduced. The chosen functional form and the artificially introduced continuity of costs can significantly influence the estimation of marginal benefits.[16]

Comments on the Net Value Hedonic Approach. The advantage of this first hedonic approach is the relative ease of valuing changing conditions at specific sites once one has estimated the hedonic or quality demand curves. Further, it has the desirable feature of estimating the attribute demand as functions of exogenous variables, travel cost, and site attributes. Despite some potential for difficulty in estimating the marginal value function, the technique might well be chosen over travel-cost methods in certain situations. For example, suppose it were possible to survey forest recreational use in a sample of households from towns within or near the forest. There would then be insufficient data on visitation and too little variation in the observed distances traveled to allow estimation of travel demand curves. Nevertheless, if a sufficient variety of sites were chosen and if the marginal cost of attributes varied over origins, one could estimate the hedonic demand curves for site attributes. The value of changes in site attributes could then be found for those individuals living within the observed range of travel costs.

A Second Hedonic Method: The Gross Benefit Approach

The second hedonic procedure can be viewed as providing estimates of the gross consumer benefits from the forest. Travel expenditures must be subtracted from the estimate of gross benefits in order to find the relevant net per-capita benefits. This type of hedonic method corresponds most closely to the approaches used by Mendelson and Wilman, and seems to be a natural adaptation of Rosen's classic description of hedonic pricing theory (perhaps for this reason it is used).[17] The data requirements and the techniques of estimation for the gross benefit approach correspond closely to those of the net benefit approach just described. Despite the great underlying similarity, the gross benefit method generally proves to be less easily used for the valuation of discrete changes in attributes at a particular site. The Wilman study illustrates an exception to this rule. One practical distinction between the two models is that in the gross benefit formulation, V the endogenously determined number of visits to the forest site is a variable in the regression equation to be estimated.

The Gross Benefit Function. Let us define the gross benefit function $G(z,V)$ as representing the value from the forest to an individual taking V trips to a site of attributes z. In terms of the travel demand curves, the gross per-capita benefits from the forest are the consumer surplus benefits plus the expenditure on travel cost. To be more precise, we will have to look at our travel demand curve in a slightly different manner. Define the inverse demand function for visits by $P = R(V,z)$. This function tells us the per-trip cost P that would lead an individual to choose to make V trips, when a site of attributes z is visited. With this inverse demand function we may now express the benefit function as

$$G(z,V) = \int_0^V R(V,z)dp \qquad (6\text{-}10)$$

The measure of gross benefits $G(z,V)$ includes the travel-cost expenditure along with the consumer-surplus type net benefits measured by the first hedonic method. The net benefits from the forest are of ultimate interest. The net benefits can be represented by $B(z,V)$, and are found by subtracting travel-cost expenses from the estimated gross benefits, with:[18]

$$B(z,V) = G(z,V) - PV \qquad (6\text{-}11)$$

where P is the per-trip cost to the site actually selected for use.

In figure 6–5, which reproduces the travel demand curves from figure 6-2, area $0ABV_1$ represents the gross benefits from V_1 visits to a site of attributes z_1. Similarly, area $0DKV_1$ represents the gross benefits that would result if the same number of visits V_1 were made to a site with attributes z_2. Area ABC gives the net benefits that should be measured if the individual does choose to make V_1 visits to the site of attributes z_1.

In figure 6–6 the gross benefit function corresponding to the demand curves in figure 6–5 is illustrated. The curves $G(z|V_1)$, $G(z|V_1')$ and $G(z|V_2)$ give the gross per-capita benefits as functions of attributes, with visits held constant variously at the levels V_1, V_1', or V_2 marked in figure 6–5. As illustrated, the benefits increase with higher-quality attributes, given the number of visits. Benefits also increase with the number of visits, given a constant attribute level.

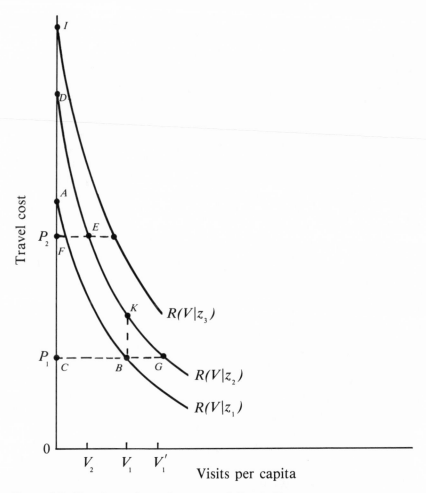

Figure 6-5. Travel-cost demand curves and site quality

The Marginal Value of Attributes. The marginal value of a change in site attributes is given by $G_z(z,V)$, the slope of the gross benefit function with respect to attribute z (holding V constant). This slope can be expressed in terms of the inverse travel demand curves as

$$G_z(z,V) = \int_0^V R_z(V,z)dp \qquad (6\text{-}12)$$

where R_z represents the shift in inverse demand curve that results from a change in the attribute z. The value G_z can be viewed as a measure of the area between the shifted travel demand curves as attributes improve; that is,

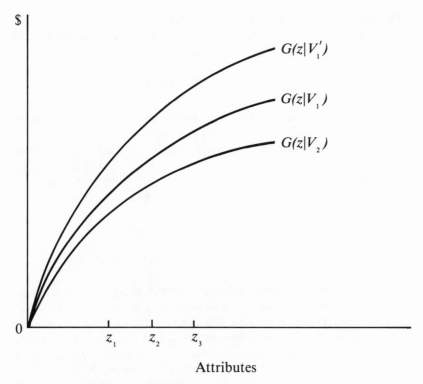

Attributes

Figure 6-6. The gross benefit function

we are concerned with areas like *ADKB* in figure 6–5, the improvement in gross benefits that results from the change in attributes from z_1 to z_2.

The marginal value curves $G_z(z|V_1)$, $G_z(z|V_1')$, and $G_z(z|V_2)$, in figure 6–7, depict the slopes of the three benefit curves that were drawn in figure 6–6. The curves plot the marginal value of attribute improvement as a function of attributes for given visit levels. For comparison, the corresponding marginal value curves $S_z(z|P)$ from the net benefit approach are indicated in figure 6–7 by the dashed curves. The marginal value curves derived by the two techniques are quite different. In particular, these current marginal value curves are reasonably assumed to be downward-sloping over the full range of attributes.

Evaluating Attribute Changes. The second hedonic method seeks to estimate the marginal value function $G_z(z,V)$. The per-capita gross value of the forest site can be computed by evaluating the appropriate area under the marginal value curve. Subtracting the travel expenditures will then give us the estimate of net value of interest.

For an individual making V trips to a site with attributes z, the gross benefits $G(z,V)$ are found as the area under the marginal value curve G_z up to chosen attribute level z; that is, the gross benefits are

$$G(z,V) = \int_0^z G_z(q,V)dq \qquad (6\text{-}13)$$

where attribute level zero is defined as that level at which the demand for visits drops to zero and so value $G(z,V)$ becomes zero.

The value of changes in site conditions can be measured in the following manner. Suppose an individual is currently using a site of attributes z^0 at trip cost P^0, but after a change in site conditions will use a site of attributes z' at trip cost P'. The change in site conditions should be valued by measuring

$$G(z',V') - G(z^0,V^0) = [G(z',V') - G(z^0,V^0)] - [P'V' - P^0V^0] \quad (6\text{-}14)$$

where $V' = V(P',z')$ and $V^0 = V(P^0,z^0)$ are the visit levels that would be selected at two sites.

In practice, the determination of the value from changes in site attributes will prove difficult when using the gross benefit hedonic technique. The difficulty arises from having visits V as the argument of the marginal value function, rather than P the exogenous travel cost. Implementation of the valuation equation (6-14) would in general require supplemental information on the choice of visit level under changed forest conditions. Indeed, without the trip demand function it might prove impossible to predict which site would be visited following the change in forest conditions. The appropriate values for V', P', and z' to be used in equation (6-14) would then be unknown. This is often a serious shortcoming of the method, at least with respect to its potential use in valuing discrete change in the condition of forest sites.

To illustrate the difficulties in valuing changes in site conditions by using the gross benefit hedonic method we will again consider the example of an individual making a choice between two sites. Site 1 is available at travel cost P_1 and has attributes z_1. Site 2 is available at travel cost P_2 and has attributes z_2. Suppose the individual has chosen to visit site 1, making V_1 visits. In figure 6–7, the gross benefits from the forest in its current state are measured in the hedonic approach by evaluating the area $0abz_1$. The net benefit is then found by subtracting the travel-cost expenditure P_1V_1. The net benefit corresponds to area ABC under the travel-cost demand curves of figure 6–5.

Suppose now that the attributes of site 1 were raised to the level z_2. From the travel-cost demand curves of figure 6–5 we know that the individual gross benefits would increase by the area $ADKB$ and visits would increase to the level V_1'. Using the hedonic value curves of figure 6–7, the net benefit of

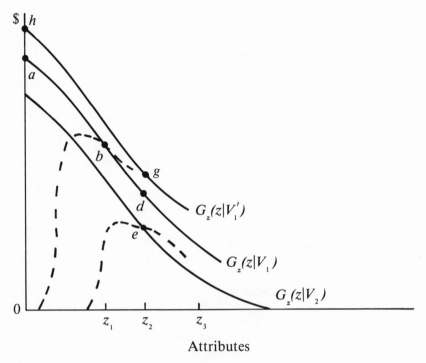

Figure 6-7. The marginal value function

this attribute change should be measured as the combination of areas *ahgd* and z_1bdz_2 minus the increased travel expense $[P_1V_1' - P_1V_1]$. The area z_1bdz_2 in figure 6-7 corresponds identically to the area *ADKB* from figure 6-5.This is the benefit from the attribute change under the assumption that the number of visits remains constant. The benefit can be easily measured by using this second hedonic method. The remaining benefits, *aghd* from figure 6-7 minus $[P_1V_1' - P_1V_1]$, are associated with the increased visits and correspond to area *BKG* in figure 6-5. These will prove impossible to measure by the hedonic method alone because the level of visits V_1' will be unknown, except under unusual circumstances. However, ignoring this hard-to-measure component of benefits would be justified if little change in the number of visits is anticipated.

The difficulty with the gross benefit hedonic procedure is actually more serious than the preceding example might suggest. The problem arises when, as is quite likely, changes in the forest condition might lead to a switch in the site selected for use. Suppose site 2 were improved in quality, but that unfortunately the potential benefits from choosing site 2 are unknown to us. Although the site attributes and travel cost could be determined, the potential level of visitation $V(P_2,z)$ to site 2 would not generally be known. Without this visit level, the potential benefits simply could not be evaluated. If potential benefits from site 2 are not measurable, then it cannot be determined whether use would switch to the improved site 2. As a result, it will be impossible to measure the value of the attribute improvements. The values for V', P', and z' for use in equation (6-14) would be unknown.

The discussion above suggests little hope for use of the gross benefit hedonic method. There are exceptional situations, however. The prime exception arises when it can be determined that, within the range of travel costs observed, there is very little variation in visits in response to attribute changes or to travel cost. Further, there must be some basis for predicting the site that individuals will use. These conditions were both met in the Wilman study of hunting in the Black Hills, described below.

Estimation of the Marginal Value Function. The principles behind the estimation of the marginal value functions G_z are much the same as those described in the discussion of the net benefits approach. Let $p(z)$ again represent the least per-trip travel cost to use a site having attributes z. There will be a different function $p(z)$ for each origin of travel, reflecting different locations relative to the forest site.

The individual's choice among sites can be represented as the choice of an attribute level z which solves the problem

$$\text{Max}_{z} \ \{G(z,V) - p(z)V\}$$

Each individual chooses the site which provides the greatest net benefits.

At the optimally chosen site, the marginal benefits from shifting use to a higher-quality site must have declined to just equal the extra travel cost from doing so; that is, the solution meets the first-order condition

$$G_z(z,V) = p_z(z)V \tag{6-15}$$

So, if the marginal cost of attributes $p_z(q)V$ is observable, it may be used as the estimate of the marginal value of attributes $G_z(z,V)$. The estimation of the marginal value functions can be based on a regression of the marginal value against the number of visits V (rather than P as in the net benefit method) and the attribute levels z for the chosen site.

The procedure used by both Mendelson and Wilman calls for the initial approximation of the travel-cost functions $p(z)$ for each origin by a regression of travel costs to selected sites against the attributes of these sites. From these regressions, marginal costs of attributes are computed. The estimates of the marginal costs are then used as data in the estimation of the marginal value of attributes as a function of visits V and site attributes z. However, as previously noted, the marginal costs of attributes often cannot be precisely observed; at best upper and lower bounds on each individual's marginal value of an attribute might be observable.

Comments on the Gross Benefits Hedonic Approach. The gross benefits technique is seen to be generally not easy to use for the valuation of changes in the attributes of specific sites. As noted, the gross benefits method is best used when there is little variation in visits in response to the observed differences in travel cost and site attributes. This is the case in the Wilman study, in which the number of hunting trips taken by individuals living close to a forest seems largely a function of the length of the hunting season and the available time. Little variation in the per-capita visitation rate was observed across the origins close to the forest. (It should be noted that Mendelson's use of the hedonic technique has a somewhat different justification. His concern was primarily with the value of marginal changes in overall attributes, rather than with the valuation of nonmarginal changes in attributes at specific sites.)

3. Valuing Changes in Hunting Quality: The Black Hills National Forest

The structure of the gross benefits method proves to be well suited to the study of the effect of timber management on hunting values in the Black Hills National Forest. The Black Hills National Forest is located in the southwestern corner of South Dakota. The Black Hills are an anomalous geological feature, an attractive mountainous area in striking contrast to the surrounding plains. The land is mostly forested, with ponderosa pine the dominant species. Recreational use of the forest is high. Mount Rushmore is within the forest boundaries. Rapid City, a town with a population of about 50,000, or 7 percent of the state's population, is immediately to the east of the forest. Other smaller towns such as Lead, Deadwood, Spearfish, Sturgis, Custer, and Hot Springs are located within or adjacent to the forest. Thirty percent of the hunting use comes from these smaller towns; much of the remaining hunting use is from or comes through Rapid City.

Elizabeth Wilman's study is an attempt to assess the benefits of management actions that may affect hunting quality on the Black Hills National Forest. The basis for the study is the expectation that changes in the forest's vegetative characteristics can affect the value of the hunter's recreational experience. Timber harvest practices may alter the suitability of the land as wildlife habitat, altering as well the probability of hunter success. It is likely that hunters are also influenced by the visual attractiveness of the forest landscape. Data for the study were drawn primarily from the towns within or adjacent to the forest. A 1981 survey of deer hunters provided observations on the choice of hunting sites within the Black Hills forest. Data on the cost of access to these sites were computed. A variety of measures of the attributes of the forest lands and vegetation were available through the national forest staff. Because there was little observed variation in travel cost or the number of days of hunting use, travel-cost demand methods did not seem likely to be useful; instead, a hedonic pricing technique was selected as the method of valuation.

The Hunter Survey

The survey of Black Hills deer hunters carried out in the summer of 1981 reflects use of the forest during the 1980 hunting season (November 1 to November 30). Roughly 15,000 individuals had licenses and tags to hunt Black Hills deer in this period. A stratified random sample of hunters was drawn, primarily from the set of hunters who returned hunter report cards to the South Dakota Department of Game, Fish, and Parks. To supplement the primary sample an additional 300 individuals who obtained either "antlerless" or "bucks only" permits were also sent questionnaires. A total of 1,500 people were surveyed, 637 of whom returned questionnaires—a response rate of 42 percent. The 637 responses provided information on the hunters and detailed information on where they hunted.[19]

Ideally, the population from which the sample was drawn would have included all Black Hills hunters. However, for all practical purposes such a sampling frame was not available. The primary set of 1,200 hunters who returned report cards was unfortunately only a selection of the relevant population of all hunters, and from all indications a biased one. Studies by the Department of Game, Fish, and Parks indicated that successful hunters were twice as likely to return report cards as were unsuccessful hunters.[20] This meant that successful hunters would be overrepresented if random sampling were used: instead of being 32 percent of the sample, which would reflect their frequency among Black Hills deer hunters, they would account for 48 percent of the sample. To correct for this, in the Wilman study samples from each county were drawn so that 32 percent of the questionnaires were sent to successful hunters.

Stratified sampling was necessary for another reason. It was of primary importance that the data exhibit variation with respect to accessibility to different parts of the Black Hills National Forest. Simple random sampling would have included predominantly hunters entering the forest from the direction of Rapid City. So the sample was stratified by county group, with a lower rate of sampling among hunters who would enter the forest from the direction of Rapid City than among those who would enter from other directions. The nearby counties of Butte, Custer, Fall River, Lawrence, and Meade were sampled much more heavily than Pennington County (Rapid City) and other counties to the east of Rapid City.

Site Characteristics

Information on the characteristics of the forest areas where individuals hunted was available from the Black Hills National Forest staff.[21] It proved rather troublesome to determine the land area that best represented a hunting site. The stand, which might be only a few acres in size, was clearly too small. The subcompartment—a grouping of stands—was believed to be the unit which hunters would most naturally consider as a hunting site. A subcompartment might be a valley between two ridges, for example. Acreage-weighted aggregations over stands were used to obtain average values of the descriptive variables for each subcompartment.

The next possible level of aggregation was the compartment, a grouping of several adjacent subcompartments. The compartment was perhaps too large an area to be meaningfully viewed as a hunting site. However, it became clear as the data were analyzed that it would be impossible to identify with certainty which subcompartment within the compartment had been used by a hunter. As a result, the data actually used were compartment average values for the vegetative and land-form variables. The subcompartment values were not acreage-weighted in forming the compartment average. In effect, each subcompartment was assigned an equal probability of being the selected hunting site.

A large number of variables were available from which to choose the characteristics to be considered in the valuation model. After preliminary investigation, two site-quality attributes were selected for use in the Wilman study. One variable, MGDEAD (defined below), was intended to represent the effect of management action on deer habitat quality and the probability of hunting success. The second variable, HEIGHT, was intended to be associated with the aesthetic quality of the forest area.

The variable HEIGHT is actually a measure of elevation. In using this variable as a proxy for aesthetic quality, the supposition was that the areas of the Black Hills that are at higher elevation are wilder, more rugged, and more natural than the lower elevations. The variable HEIGHT was of course not subject to management action, but it was expected to be correlated with characteristics that do attract hunters.

Forage-cover ratios and opening-size distributions seem to be the key vegetative characteristics related to deer habitat suitability.[22] Unfortunately, the variables available as descriptors of the Black Hills National Forest did not provide a precise way to measure these two characteristics; hence proxies of some sort were necessary. MGDEAD, representing the combined effects of management on deer habitat, was formed by multiplying the two proxy variables.

First, a measure of available forage per acre was computed. Using the results of a study by Pase and Hurd, with modifications to reflect local conditions across the Black Hills, forage per acre was predicted from the available measures of basal area of ponderosa pine per acre.[23] Second, a search was made for those measured variables which were highly correlated with hunter success within the sample. The highest correlation was found to be with the variable AVEDEAD, the acreage of standing dead trees per acre. The variable AVEDEAD seems to be strongly related to pine beetle infestation. (Other variables found to be highly correlated with hunter success seemed to also reflect characteristics of the general area of the pine beetle infestation.) It is the treatment of pine beetle infestations in the northwestern area of the Black Hills that appears to have contributed to desirable deer habitat. Treatment has reduced the basal area of standing timber per acre, thus contributing to forage production, and has resulted in the distribution of this forage in a desirable manner by creating a series of fairly small openings where forage and cover are in close proximity. The descriptive variable MGDEAD that is used in the Wilman study was formed as the product of the available-forage measure and AVEDEAD. It was intended that MGDEAD be a proxy for the degree to which forage is both available and well distributed in close proximity to cover.

Estimating the Marginal Costs of Quality

The first step in the estimation of the marginal value of the MGDEAD quality variable was the determination of marginal costs. Estimates of the marginal costs of reaching a high-quality site were based on the measurable costs of travel to sites from the five main towns of origin through which hunters were thought to enter the forest—Rapid City, Sturgis, Custer, Hot Springs, and Lead–Deadwood. A separate travel-cost curve was estimated for each town.

Because of differences in tastes, hunters originating from the same town selected a variety of sites for use. The total costs for a season of use were regressed on the levels of the site variables MGDEAD and HEIGHT and on a variable (DISTANCE) measuring the individual's distance in miles from home to the closest of the five primary towns. Total costs were evaluated at $.08 per mile (the American Automobile Association variable cost estimate for 1980). A calculation of time costs was also made.[24] Weekday time was estimated at a cost of $26 per day, while weekend or holiday time was estimated to have no opportunity cost. A variety of forms for the total cost functions $p(z)V$ were considered. Some of the better-fitting cost equations for each town are presented in table 6–1.[25]

The marginal cost of the quality attribute MGDEAD was calculated for each town by taking the partial derivative of the cost equation with respect to MGDEAD. The marginal cost estimates derived from table 6–1 are presented in table 6–2, where they are listed in order of decreasing value. It should be noted that MGDEAD is a component of several of the variables in the cost equations. Only those variables with significant coefficients were considered in computing the marginal costs. These costs were greatest for the towns of Hot Springs and Custer in the southern part of the forest. The marginal costs were the least—not statistically different from zero—for the Lead–Deadwood area, which is near the higher-quality hunting sites in the northern part of the forest. Because the variable MGDEAD entered the cost equations linearly, the marginal cost is a constant independent of the level of MGDEAD. This linearity simplifies the estimation of the marginal value function in the next stage of the valuation procedure.

A few comments on the cost estimation procedure seem necessary at this point. First, in the Wilman study preliminary investigation indicated that there was little variation in the number of days of hunting selected by individuals. Nevertheless, it might have been preferable to estimate the costs per trip rather than the overall cost per season. It might have been better to estimate cost functions $p(z)$, which are based only on observable and exogenous cost data. Estimation of the total costs $p(z)V$ confounds what should be a simple estimation procedure by including the individual's level of visits V, which generally depend in a complex manner on the prices and qualities of all sites and on the consumer's tastes. Second, the method of estimation relies on the costs of travel to selected sites only. If there had not been a wide range of sites selected by different individuals at each origin, it would have been necessary to independently determine the available alternative sites and compute the costs of using them. Marginal costs would have to have been estimated from the generated data on this expanded set of sites. Such a procedure might have been desirable in any case. The hedonic method relies on accurate estimation of the slope of the cost function $p(z)$, and all relevant information should be used.

Estimating the Hedonic Demand for Quality

The marginal cost estimates of table 6–2 were used in the Wilman study in the estimation of a demand curve for the quality variable MGDEAD. It was assumed that the demand curve was the same across all hunters, except for the effect of certain demand shifters. The demand shift variables included the number of years of hunting experience and the income of the hunter, which were essentially reflected in differences in hunter preferences for site quality. Hunters living in different towns had different accessibility to various parts of the forest. The difference in access resulted in variations in the

Table 6–1. Estimates of the Total Cost Relationship

Dependent Variable	Rapid City total cost	Sturgis total cost	Custer total cost	Hot Springs total cost	Lead–Deadwood total cost
INTERCEPT	25.26 (8.01)**	42.44 (23.41)	–13.17 (56.15)	225.35 (117.39)	34.13 (25.98)
HEIGHT	0.12 (0.02)**		–0.12 (0.26)		0.11 (0.05)**
MGDEAD	0.041 (0.018)**	0.005 (0.05)	0.17 (0.08)**	–0.77 (0.44)	–0.0053 (0.030)
DISTANCE	0.16 (0.03)**	0.08 (0.14)			
DHEIGHT	0.84×10^{-4} (0.11×10^{-3})				
DMGDEAD		0.00048 (0.00039)			
OPEN		0.0050 (0.00019)**		0.0019 (0.00052)**	
DOPEN		-1.31×10^{-6} (7.16×10^{-7})			
STAY				–238.16 (153.32)	83.70 81.1

	(1)	(2)	(3)	(4)	(5)
	0.54×10^{-4} (0.18×10^{-4})**				
POPEN				−0.0014 (0.00083)	
PMGDEAD				0.80 (0.54)	0.019 (0.14)
SQUARE			0.00032 (0.00026)		
Regression statistics					
R^2	0.53	0.56	0.43	0.39	0.23
Adjusted R^2	0.52	0.51	0.36	0.25	0.09
F	60.11	9.57	6.16	2.91	1.63
L	−1,320.1	−186.8	−127.9	−119.9	−127.6
N	276	43	28	29	27

where:
HEIGHT = the average elevation of hunting sites chosen minus 4,500 ft. (1,371.6 m)
MGDEAD = the forage generated by the average basal area per acre of ponderosa pine at the hunting sites visited, multiplied by the average number of dead trees per acre. The latter is a proxy for the probability of forage being in small openings (less than 10 acres [4.0 ha])
OPEN = HEIGHT × MGDEAD
DISTANCE = distance in miles from the origin town to the closest point in the Black Hills National Forest
DHEIGHT = DISTANCE × HEIGHT
DMGDEAD = DISTANCE × MGDEAD
DOPEN = DISTANCE × OPEN
STAY = whether any trips were overnight trips
POPEN = STAY × OPEN
PMGDEAD = STAY × MGDEAD
SQUARE = HEIGHT × HEIGHT

Note: Numbers in parentheses are standard errors. **Indicates significance at the 0.05 level.
Source: Adapted from Elizabeth A. Wilman, "Valuation of Public Forest and Rangeland Resources," Discussion Paper D-109, Resources for the Future (Washington, D.C., August 1984).

Table 6–2. Marginal Cost Estimates for MGDEAD

Town	Marginal cost
Hot Springs	$0.0019 \times$ HEIGHT
Custer	0.17
Sturgis	$[0.0005 - (0.13 \times$ DISTANCE $\times 10^{-5})] \times$ HEIGHT
Rapid City	0.041
Lead–Deadwood	0.0

Note: The variable HEIGHT has a mean value of about 200 over all towns.

Source: Adapted from Wilman, "Valuation of Public Forest."

marginal costs of quality, and that variation allowed estimation of the demand curve for quality.

The demand curve was estimated by regressing the chosen level of MGDEAD on the variables PRICEJ (the estimated marginal cost for MGDEAD) and HEIGHT (the elevation of the selected site), and also on the variables reflecting hunter preferences. Since the number of days spent in the forest was roughly constant across observations, visits or days of hunting were not explicitly included in the regression equation. Both linear and semi-log versions of the demand functions were estimated. The results of the demand curve estimation procedure are given in table 6–3. The coefficients for PRICEJ, the marginal cost of MGDEAD, are significant in the regression results in that table. Both the semi-log and linear forms of the demand curves were used in the calculation of the benefits from a change in the forest condition, providing some measure of sensitivity of the benefit estimation to the form of the demand equation. The low R^2 of these demand regressions should be noted. Even these provide only a partial measure of the adequacy of the estimation because of the possibility of measurement error in the estimated values for the independent variable itself.

Evaluating a Management Action: The Norbeck Timber Sale

With the demand curve for hunting quality in the Black Hills having been estimated, enough information was generated to measure the benefits to hunters from a specific timber sale. It has been noted that the marginal cost for hunting quality for individuals from the Lead–Deadwood area was zero; this is because of the easy accessibility to a forest area having high levels of the variable MGDEAD. Were a site of equal quality made as easily available to the individuals of other towns, there would be a significant gain in consumer benefits. A proposed timber sale at Norbeck was anticipated to make an improved hunting area available to residents of Custer, and provided an opportunity for assessing benefits.

Table 6-3. Demand Curves for MGDEAD

	Independent variable	
Dependent variable	MGDEAD	LOG(MGDEAD)
INTERCEPT	316.43	5.69
	(29.23)**	(0.06)**
HEIGHT	0.35	0.00064
	(0.0039)**	(0.00009)**
PRICEJ	−522.63	−0.92
	(211.82)**	(0.41)**
ANTLESS	10.38	0.09
	(17.74)	(0.03)
INCOME	−0.94	−0.00003
	(0.10)	(0.00020)
YRSHUNT	2.22	0.004
	(0.77)**	(0.0014)**
Regression statistics		
R^2	0.17	0.16
F	20.66	19.09
N	510	510

where: MGDEAD = the forage generated by the average basal area per acre of ponderosa pine at the hunting sites visited, multiplied by the average number of dead trees per acre

HEIGHT = the average elevation of hunting sites chosen minus 4,500 feet (1,371.6 m)

PRICEJ = 0.041 for Rapid City, [0.00050 − (0.0000013 × DISTANCE)] × HEIGHT for Sturgis, 0.17 for Custer, 0.0019 × HEIGHT for Hot Springs, and zero for Lead-Deadwood

ANTLESS = 1 if the hunter applied for antlerless licenses, 0 if he did not

INCOME = the hunter's income level in hundreds of dollars

YRSHUNT = the number of years the hunter has hunted

Note: Numbers in parentheses are standard errors. **Indicates significance at the 0.05 level.
Source: Adapted from Wilman, "Valuation of Public Forest."

The Norbeck timber sale was scheduled to take place on forest compartment 302. This compartment is roughly the same distance from Custer as compartment 703 (currently having a high MGDEAD value) is from Lead-Deadwood. A primary purpose of the timber sale was the improvement of deer habitat. This would be accomplished by reducing the average basal area of the timber stands and by distributing the cutting to create a pattern of small openings. The average basal area and forage per acre for compartment 703 are compared in table 6-4 with the current situation and the anticipated postharvest situation in compartment 302. From the table it can be seen that as a result of the Norbeck sale the marginal cost of MGDEAD faced by a Custer hunter would be reduced to the same level as that faced by a Lead–Deadwood hunter.

The net consumer's benefit to a Custer hunter resulting from the Norbeck timber sale is presented in table 6-5. Consumer's surplus changes are derived for the two alternative demand equations of table 6-3. The consumer's surplus gain for a Custer hunter is in the $63 to $71 range. These amounts reflect the increased benefit from the higher quality of the site and the cost saved from using a closer site. In 1980 there were an estimated 844 Black Hills hunters from Custer.[26] This would have meant aggregate benefits from the timber sale for Custer hunters in the neighborhood of $58,000, or $9 per member of the population of Custer County. The consumer surplus gains have been calculated for existing hunters only, assuming the number of hunters does not change. The current hedonic technique does not allow us to estimate whether there would be any increased participation in hunting activity; nor do we know how to value such an increase in use.[27]

Table 6-5 also provides estimates of the value of a similar management action for each of the other towns, under conditions like those in the vicinity of Lead–Deadwood. It should be that much smaller benefits would accrue to hunters in most of the towns. The greatest value would be to residents of Custer and Hot Springs, primarily because of the relatively large travel-cost

Table 6-4. Comparison of Vegetation Characteristics by Subcompartment

Town	Subcompartment no.	Basal area (ft²/acre)		Forage (lbs/acre)	
		Existing	Post-Norbeck	Existing	Post-Norbeck
Custer	30204	161	57	56	494
	30206	116	64	143	427
	30207	146	101	76	196
	30208	146	39	76	721
	30209	147	80	75	305
	30210	124	186	121	269
Lead–Deadwood	70301	46		620	
	70302	85		275	
	70303	82		292	
	70304	76		332	
	70305	98		208	
	70307	66		409	

Source: From Wilman, "Valuation of Public Forest."

Table 6–5. Individual Consumer's Surplus Changes ($/year)

Town	Linear	Semi-log
Rapid City	16	14
Sturgis	11	9
Custer	71	63
Hot Springs	50	49
Lead–Deadwood	0	0

Source: Adapted from Wilman, "Valuation of Public Forest."

savings that would be experienced by hunters from these two towns. Hunters from Hot Springs and Custer currently face the greatest marginal costs for MGDEAD. Substantial reductions in travel costs could be expected to yield substantial benefits.

The benefits discussed so far are for a single hunting season. But suppose a management policy were instituted to maintain the postharvest conditions that provided the increased hunting benefits. If, for example, the new vegetation pattern resulting from harvesting in compartment 302 were maintained for twenty years and annual benefits of $69 per hunter were to accrue at a 6 percent discount rate, the present value of discounted benefits would be $800 per hunter. If the number of hunters did not change, this would amount to $675,000 in the aggregate. Allowing the participation rate to increase by 1 percent a year would bring this amount to around $811,000.

4. Summary and Conclusions

In this chapter we have directed attention to the importance of valuing changes in forest conditions that result from timber management activity. Although this chapter has presented an innovative way to investigate the change in recreational hunting benefits resulting from habitat-improving management activities, it has also pointed up the importance of the linkage between forage and cover sites—a key to understanding the essentials of multiple-use forestland management.

The hedonic methods for the valuation of changes in recreational benefits are neither well known nor well understood. In this chapter the basic principles of the hedonic techniques have been presented. The travel-cost method has been shown to focus on the consumer's choice of the number of visits to recreation sites as a response to differences in costs of using these sites. The hedonic methods have been shown to focus on the observed responses of individuals to differences in attributes across the available sites. The decision to visit a particular site reveals an individual's willingness to sacrifice extra travel expense for increased site quality.

It is fair to say that the hedonic methods are complex and leave considerable room for error. Nevertheless, there are situations where the hedonic methods could be used while travel-cost methods would be of no use. Elizabeth Wilman's study of the impact of timber harvesting on hunting values in the Black Hills National Forest illustrates an innovative use of the hedonic methods. With most of the observed hunting use coming from towns within or close to the forest, there was little chance of successfully using a travel-cost procedure in that study. However, the hedonic methods proved to be applicable because there was significant variation across towns with respect to their access to areas of quality hunting.

Notes

1. M. Clawson, "Methods of Measuring the Demand for and Value of Outdoor Recreation," Reprint no. 10, Resources for the Future (Washington, D.C., 1959); O. D. Burt and D. Brewer, "Evaluation of Net Social Benefits from Outdoor Recreation," *Econometrica* vol. 39 (September 1971) pp. 813–827; C. J. Cicchetti, A. C. Fisher, and V. K. Smith, "An Econometric Evaluation of a Generalized Consumer Surplus Measure: The Mineral King Controversy," *Econometrica* vol. 44 (November 1976) pp. 1259–1276.

2. Among techniques not discussed in the text are the contingent valuation (CV) methods and the household production function approaches. With CV methods, questioning of individuals is used to reveal a direct estimate of value. The CV methods do not seem well suited to the recreational demand problem discussed here, for which we must estimate the values of many related forest sites. An application of the household production function approach is described in N. E. Bockstael and K. E. McConnell, "Theory and Estimation of the Household Production Function for Wildlife Recreation," *Journal of Environmental Economics and Management* vol. 8, no. 3 (1981) pp. 199–214. That work is best viewed as an attempt to decompose the total value of a site among various subjective aspects of the recreational experience, rather than to estimate the value of physical changes in sites.

3. In A. M. Freeman, *The Benefits of Environmental Improvement: Theory and Practice* (Baltimore, The Johns Hopkins University Press for Resources for the Future, 1979), various travel-cost methods used for the valuation of attribute changes are discussed. All such travel-cost techniques require an assumption that different recreation sites are providing an essentially similar service. If the services of each site were unique, and there was a distinct demand curve for each site (with data from only one time period), we would have no variation in site attributes upon which to base estimation. In practice, this means structural similarities must be imposed on the demand functions for each site.

4. R. Mendelson, *An Application of the Hedonic Travel Cost Framework for Recreation Modeling to the Valuation of Deer* (Seattle, University of Washington, Department of Economics, 1983); Elizabeth A. Wilman, "Valuation of Public Forest and Rangeland Resources," Discussion Paper D–109, Resources for the Future (Washington, D.C., August 1984).

5. Some critical comments on the empirical procedures used in the hedonic travel-cost methods can be found in V. K. Smith and Y. Kaoru, "The Hedonic Travel Cost Model: A View from the Trenches," *Land Economics* vol. 63, no. 2 (1987) pp. 179–192.

6. Wilman, "Valuation of Public Forest"; portions of Wilman's study previously published in Discussion Paper D-109 (Resources for the Future, 1984) are used with permission of the author.

7. In many ways the hedonic travel-cost methods are similar to the standard hedonic pricing methods described in S. Rosen, "Hedonic Prices and Implicit Markets: Product Differentiation in Pure Competition," *Journal of Political Economy* vol. 82 (1974) pp. 34–55. The standard hedonic pricing methods are often used to study the demand for attributes of housing. One important distinction should be observed: the function $p(z)$ here summarizes exogenous travel-cost data, while the corresponding price function in housing studies represents endogenous equilibrium prices. As a result, we will not have the simultaneity and identification problems that plague housing studies. On the other hand, our prices $p(z)$ will not be as adequately represented by a smooth continuous function.

8. The approximation of $p(z)$ becomes the source of much difficulty in applications of the hedonic technique. Indeed, despite the notational elegance of this second formulation of the utility framework, there is much about it that tends to obscure understanding of the empirical problem.

9. The sites are perfect substitutes; that is, the indifference curves of $U(\Sigma V_i \phi(z_i))$ are linear, with constant slope $dV_i/dV_j = -\phi(z_j)/\phi(z_i)$. The site with the lowest price per attributes $p/\phi(z)$ is selected.

10. Although not necessarily at the lowest prices. If that seems puzzling, try to draw demand curves for bottles of milk as a function of the price per bottle; do it for gallon bottles, then for half-gallon bottles.

11. Because of the special role of substitute sites, simultaneous changes in the attributes of several sites add no complications. It also should be noted that the measure of the change in benefits from the forest as a whole will correspond exactly to the usual measurement of benefits from the demand curves for the individual sites visited.

12. This assumption is referred to as weak complementarity. See K. Mäler, *Environmental Economics: A Theoretical Inquiry* (Baltimore, The Johns Hopkins University Press for Resources for the Future, 1974) pp. 183–189; or D. F. Bradford and G. G. Hildebrandt, "Observable Preferences for Public Goods," *Journal of Public Economics* vol. 8, no. 2 (1977) pp. 111–131.

13. $S_z(z,P)$ is the partial derivative of function S with respect to z; that is, it is the derivative with P held constant.

14. If several attributes are changed, valuation requires a sequential determination of the areas under the marginal value curves for each attribute.

15. We have made a substitution of $V(z,p)$ for $-S_p(z,p)$. In our earlier utility maximization framework, the corresponding condition would have been $\alpha V \phi'(z) = p'(z)V$.

16. Function $p(z)$ is often assumed, with hazy justification, to be linear. It would seem best to fit the known costs as well as possible.

17. Mendelson, *Application of the Hedonic Travel Cost Framework;* Wilman, "Valuation of Public Forest"; Rosen, "Hedonic Prices."

18. The previous measure of net benefits $S(z,P)$ is related to the current measure according to $S(z,P) = B[z,V(P,z)]$.

19. More detailed descriptions of the survey and the variables obtained from it can be found in Wilman, "Valuation of Public Forest," chapter 3 and appendix B.

20. J. J. Kranz and L. E. Petersen, *Reliability of Deer Hunter Report Card Data, 1971, South Dakota: A Progress Report* (Pierre, S.Dak., Department of Game, Fish, and Parks, 1972).

21. A list of available descriptive variables can be found in Wilman, "Valuation of Public Forest," appendix C.

22. J. W. Thomas, *Wildlife Habitats in Managed Forests: The Blue Mountains of Oregon and Washington,* Agricultural Handbook no. 533 (Washington, D.C., USDA Forest Service, in cooperation with the Wildlife Management Institute and the USDI Bureau of Land Management, 1979); S. G. Boyce, *Management of Eastern Hardwood Forests for Multiple Benefits (DYNAST-MB),* USDA Forest Service Research Paper SE-168 (Asheville, N.C., Southeastern Forest Experiment Station, July 1977).

23. C. P. Pase and R. M. Hurd, "Understory Vegetation as Related to Basal Area, Crown Cover and Litter Produced by Immature Ponderosa Pine Stands in the Black Hills," in *Proceedings of the Society of American Foresters* (Syracuse, N.Y., 1957) pp. 156–158.

24. See Wilman, "Valuation of Public Forest," chapter 3.

25. A variety of other results are reported in Wilman, "Valuation of Public Forest."

25. There are 270 hunter report cards from Custer County. The average return rate of 32 percent, which seems to be fairly constant across the counties, would have meant there were 844 hunters from the county.

27. A rough estimate of the increased value from new participants is provided in the Wilman study (see Wilman, "Valuation of Public Forest," pp. 4-33, 4-34). The procedure relies on the new participants valuing attributes according to the same hedonic demand function estimated for current participants.

7

Recreation Valuation for Forest Planning

1. Introduction

In our analysis (in the preceding chapter) of the relation between forest practices and the value of deer-hunting sites, some extensive measures were undertaken to obtain data relevant to the problem, and econometric expertise was required to carry out the study. But persons in management positions on a national forest can justifiably point out that the working complement on a forest is composed of persons who have the kinds of skills needed to manage the land under their charge and are not likely to have the capability for carrying out sophisticated research. It is practical to ask whether accumulated research may suggest ways of estimating the value of forest recreation for planning purposes that do not require sophisticated research techniques. The issue is not whether such estimates meet the criteria of scientific discovery, but rather whether they provide advances over ad hoc techniques that are widely applied in forest planning.

The purpose of this chapter is to attempt to value non-priced recreation resource services for inclusion, along with timber values, in an evaluation of management alternatives on an eastern forest, applying simple techniques to data that are available on national forests. The forest studied will be the White Mountain National Forest in New Hampshire and Maine, one of the first national forests to be established under the Weeks Act of 1911. It contains about 750,000 acres, located mostly in New Hampshire, with a small portion in Maine. The forest is the largest public land tract in New England, and represents about 12 percent of the area of New Hampshire. The White Mountains include one of the largest alpine zones in the eastern United States, with eight square miles above timberline. This area is a major regional recreation resource. It accounts for upward of two-and-a-half million recreational visitor days annually.[1] Some 65 million people live within a day's drive of the forest.

The White Mountain National Forest is well consolidated, with 87 percent of the land within its boundaries being national forest land. Franconia Notch, Crawford Notch, and the summit of Mount Washington are New Hampshire state parks located within the forest boundaries. Substantial areas are under special-use permits, including seven alpine ski areas and a number of sites operated or supervised by members of the Appalachian Mountain Club (AMC). The AMC was active on the present forest even before a national forest had been established there. Members built the first hut, or hostel, in 1888, based on the Austrian model. The club now operates eight huts on the Appalachian Trail, accessible only by foot. The AMC also manages a number of more primitive shelters during the summer season.

Other similar recreational resources in the region are Baxter State Park in Maine, the Green Mountain National Forest in Vermont, and the Adirondack and Catskill mountains in New York. In spite of the number of outdoor recreation grounds of this kind in the region, the ratio of acres of such public lands to, say, a thousand persons in the population is minuscule in comparison with the ratio in a western state like Montana.

The White Mountains were among the last areas to be logged in New England. This occurred largely in the 1890s. Much of the logging done on difficult terrain was accomplished with the aid of cograil. Because of the high cost of entry with such methods, the harvesting was intense, removing virtually all of the fir and spruce. The pioneer species colonizing the logged-off land were predominantly northern Appalachian hardwoods. These have now grown to even-aged stands in the 70- to 90-year age classes.

With this brief introduction to the White Mountains, we are ready to consider a forest-level recreation valuation effort that should be consistent with the quality and precision of information used in forest plans. We begin by reviewing briefly (section 2) the travel-cost recreational demand model to extract information relevant to the problem of outdoor recreation valuation (taking the model off the shelf after it was put there by past research on outdoor recreation). Section 2 also addresses in a limited way the change in value of a recreational site in response to management actions which either enhance or degrade the site. In section 3 we look at the kinds of data one can expect to have available in greater or lesser degree on all national forests, and other data that will be available from the files of the regional offices of the Forest Service. Estimates of recreation demand and an analysis of the estimated demands are given in sections 4 and 5. Since jointness in production is the essence of multiple-use forest management,[2] in sections 6 through 8 we review briefly timber management and the relation between recreation and timber on the White Mountain National Forest.

2. An Approach to Estimating the Demand for Forest Recreation

For those who ponder the lag between the results of research and their application, the case of forest recreation must be sobering to contemplate. Early conceptual breakthroughs in the valuation of outdoor recreation occurred over a quarter of a century ago, yet estimates of economic demand for, and value of, forest recreation are conspicuous by their absence.[3] One can empathize with a forest supervisor's lack of enthusiasm for an additional data-generation and assembly task during his forest's work on its land-use and management plan. But it is not necessary to duplicate the work of a sophisticated econometrician to get estimates equal to FORPLAN-level precision. We present here a simplified approach which we believe will be generally applicable to the National Forest System to help obtain data that are conceptually consistent with standard allocative criteria.

For planning purposes, the travel-cost model appears to be the most practical for estimating the demand function for outdoor recreation.[4] Through our review of the data available in the National Forest System we have concluded that there are sufficient data to implement the travel-cost model to obtain acceptable results. Estimating the demand for forest recreation, employing the travel-cost method, will produce results conceptually consistent with the results of demand studies for other forest outputs (such as wood products) that are processed and consumed off the national forests.

Review of the Travel-Cost Model

The travel-cost method focuses on travel as an input in producing visits to a site for recreational purposes. Because of the spatial dispersion of households about any recreation site, the costs of access will vary. Per-capita use of a site, consequently, will tend to decrease directly as the distance from the site increases. This observed response to higher travel costs is used to estimate a demand curve for site use.

Estimating Per-Capita Demand. The individual's demand for a site is estimated with an equation that relates the per-capita use of a site to travel costs, characteristics of the site, alternative recreation opportunities, and socioeconomic factors. The estimating equations are based on data gathered at an existing site or a cross section of existing sites, with the resulting statistical coefficients employed to estimate use under changed conditions. The general form of the per-capita demand equation is

$$V_{ij}/P_i = f(C_{ij}, Q_j, E_i, S_{ij})$$

where V_{ij} is the total visits to site j from origin i, P_i is the population of origin i, C_{ij} is the site entry fee plus the round-trip travel cost from origin i to site j, Q_j is a vector of variables reflecting physical characteristics of site j, E_i is a vector of socioeconomic characteristics of origin i, and S_{ij} is an index of substitute recreation opportunities.

The interpretation of the estimated equation as the representative individual's demand function for site use relies on an assumption that increased costs owing to increased distance have the same effect on consumers' behavior as equivalent increases in admission fees. Once estimated, the per-capita demand equation can be used to predict the number of visits from each origin that would result from either current or changed levels of the explanatory variables.

Deriving Aggregate Site Demand. An aggregate site demand curve relating entry price to total use can be derived using the per-capita demand curves. For each origin, the predicted per-capita use under current conditions is computed and multiplied by the population of the origin. This provides the total estimated use from each origin. Summing over all origins then yields the total visits anticipated at the current price. This is one point on the site demand curve.

To find other points on the aggregate site demand curve it is necessary to calculate total predicted use at incrementally higher entry prices. This exercise is assumed to correspond to requiring individuals from each place of origin to incur higher travel costs to reach their destination. The resulting relationship between all possible prices and the calculated total use yields the desired demand curve for the site.

Computing Site Benefits. The benefits from the recreational site can be computed from either the per-capita or aggregate demand curves. It is usually more practical to use the per-capita curves directly. For each origin, find the area under the per-capita demand curve above travel cost. Multiply these amounts by the corresponding population of the origin and sum over all origins. This procedure is equivalent to finding the area under the aggregate site demand curve. It will provide an estimate of the consumer surplus, which is the value of recreational services received by the consumer above his actual expenditure. If user charges or entrance fees are collected, these receipts are added to the consumer surplus to establish the total resource value of a recreational facility or site.

Of most relevance to the forest situation is the change in benefits resulting from a change in the physical character of a recreation site. Such site-quality changes are likely to shift the demand, with greater demand for an improved site. The increased area under the per-capita demand curves, or under the aggregate site demand curve, provides an estimate of consumer surplus due to the quality change.

A Specification for a Travel-Cost Model

Theory provides little guidance in the selection of the functional form for the per-capita demand curve. We have selected a semilog specification, consistent with the better statistical fit this specification has shown in a large number of other travel-cost demand studies.

Our specification has as the dependent variable the natural logarithm of visits per capita. The independent variables, including travel cost C, the vector of site characteristics Q, and the socioeconomic variables E, are untransformed and enter the equation linearly. Limited time and funding did not permit inclusion of an index of substitute sites. The regression equations were of the form

$$ln(V/P) = b_0 + b_1C + b_2Q + b_3E$$

where b_0, b_1, and the vectors b_2 and b_3 are coefficients to be estimated. The demand equation expressing individual visitation as a function of C the travel cost, Q the site attributes, and E the various socioeconomic variables is given by the exponential form of the regression equation

$$V/P = \exp\{b_0 + b_1(C) + b_2Q + b_3E\}$$

We now look at some features of our specification.

Substitute Sites.   Because of the large number of sites within a forest some simplification of the demand structure is essential. A simplified view of the substitution possibilities between sites is reflected in our approach.

Our view is that the recreation sites in a forest can be divided into a few classes. Those sites within a class can be considered perfect substitutes in the following sense. Individuals, after considering the relative travel cost and site qualities, will choose the best site in a class and visit only that site. The chosen site is that providing the greatest consumer surplus. Because of differences in the location of origins relative to the forest, the chosen site need not be the same for each origin.

The practical implication of this view is that it greatly reduces both the number of demand functions that need to be estimated and the number of variables to be considered. A demand equation is estimated for each class of sites rather than for each individual site.

Ideally, one should still include in each demand function an index of availability of the alternative classes of recreation. This proves to be difficult because recreation areas throughout a region must be considered, not just those areas on the forest. Our data did not permit the inclusion of an index of substitutes. The omission could bias our estimation and affect our ability to predict changes in the distribution of visitors across the forest under changed forest conditions.

Benefit Evaluation.   The calculation of consumer surplus when demand is semilog is particularly easy. The consumer surplus per visit can be shown to equal $-1/b_1$ where b_1 is the estimated coefficient of travel cost in the demand equation. An unusual and convenient feature of the semilog specification is that the consumer surplus per visit is a constant, not dependent upon either the number of trips taken or the site quality.

One point requires clarification. The consumer surplus, or willingness to pay, for a particular site is not in general to be found as the predicted use multiplied by the estimated consumer surplus per visit; this would give the consumer surplus attributable to the set of sites in the class. It must be realized that our demand functions are not technically demands for a site, but rather for a class of sites. The consumer surplus due to any particular site may be a smaller amount. This amount would reflect the possible shift in use to substitute sites within the class that would occur if any attempt were made to charge for the use of one site alone.

The consumer surplus that arises from the change in attributes at one or several sites is found by evaluating the resulting change in the overall consumer surplus from the forest. Consumer surplus for the forest as a whole is found by multiplying projected use of each site by the estimated consumer surplus per visit for the site class and summing over all classes and sites. The value of a change in quality at any one site, or set of sites, can be found by multiplying the change in projected visits to each site in the forest by the appropriate consumer surplus per visit, and summing over all sites. For example, suppose the next-best alternative site within the class were initially of the same quality as the altered site. The benefit from the changed site would then be, approximately, the savings in travel cost over the cost of visiting the alternative site for all trips. This method of valuing site-quality changes is consistent with the conclusion that the willingness to pay for any one site among many might be small.

Site Demand.   Our method of estimation is rather cumbersome to be used for estimating site use. The procedure does not directly yield the demand for a site. Nevertheless, site demand can be derived.

For each origin, we can predict the chosen site as that which would provide the greatest consumer surplus. The estimated per-capita visitation curve then indicates the expected individual use of the chosen site for the origin. With a higher entry price, the per-capita site demand will decline along the estimated visitation curve, although demand will drop to zero once the site entry price is so high as to make an alternative site the preferred choice for the origin.

Problems of Application

There are various caveats that a user of the travel-cost method might heed. In this connection one point needs to be emphasized. In the social sciences, where knowledge is obtained from analysis of data that most often were not obtained in the research environment of carefully controlled experiments, strong assumptions are sometimes made that will attract scrutiny and reexamination. To one who is unfamiliar with the research environment it may appear that research produces only caveats and contention. This should not suggest absolution from criticism for not using those positive contributions that thoughtful analysis can make. Even tentative results may be preferred over dismissal of incomplete but relevant information.

The travel-cost method does pose several empirical problems. For a variety of reasons it may be difficult to accurately determine the true costs of travel. For example, the use of out-of-pocket travel expense alone to estimate travel costs would ignore the importance of travel time. Travel time may be a major consideration in the recreational decision. Although time costs are not readily measured, their exclusion will result in an underestimate of benefits. In this exercise we have proceeded without attributing costs to time in travel.

Similarly, there will be difficulties if there are unidentified differences in individuals across origins. For example, suppose nearby residents have chosen their location because of a strong preference for access to outdoor recreation; or suppose that more distant areas have a better available choice of alternative recreation. An observed decline in visitation at greater distances will reflect the effects of both higher cost and lower interest. Attributing the full decline in visits to travel cost alone would result in an underestimation of benefits.

The assumption that travel costs are an accurate proxy for site entry fees also fails to hold if visitors stop at multiple sites on the same trip. This situation raises the familiar problem of joint cost allocation, for which there is no satisfactory solution.

We should also note that there are several alternative models for demand or benefit estimation. Our selection of the travel-cost model was made specifically because our intent was to avoid any method for which special surveys or extensive new data collection would be required. The probability that data collected on the forest would permit implementation of the travel-cost method, with useful results, seemed much higher than any other approach. We now turn to a review of the available data.

3. Sources of Forest Data

As we have stressed, the ground rule of this study has been to limit the data to that readily available to most national forests. In general, the basic data necessary to implement the travel-cost method is accessible throughout the Forest Service. Detailed information on site characteristics is maintained for all forests in the Recreation Information Management System. Census statistics provide socioeconomic data once counties of origin are identified. The difficulty is in finding data identifying the origin and destination of recreationists. While considerable data on visitors are collected, there is currently no routine policy of recording and reporting many of these data. We will review below the generally available sources of information that deal in one way or another with forest recreation and that have been used in this study.

Site Characteristics

Recreation Information Management System (RIM). The RIM system was established in 1974 as a comprehensive storage and retrieval system for Forest Service recreation management. The system is fully automated, and data are updated annually. Data are provided by individual forests, and consist of three types: site characteristics, facilities, and use.

Site characteristics, reported in the Basic Address Record, include an assigned site code, location information, natural setting, and various descriptors indicating whether the site is developed or undeveloped (as for dispersed recreation areas). Facilities data are reported in the Facility-Condition Record, and include detailed information on facility types, condition, and maintenance, as well as estimated costs of operation, maintenance, rehabilitation, and the like. Estimated use is reported annually by site and activity category. Methods of estimating use vary widely, and the accuracy of this information is doubtful in many instances. These data do not include visitor origin data.

Recreation Opportunities Spectrum (ROS). The Recreation Opportunity Spectrum provides a broad classification of land areas according to their suitability for various types of recreation.[5] Land areas are classified in categories ranging from primitive to urban. Although ROS categories are becoming widely used for planning purposes in the Forest Service, the usefulness of the ROS data for our study is limited by the small variation among recreation sites on the White Mountain National Forest.

Socioeconomic Data, Travel Distances, and Costs

In this study we have employed the Census Bureau's PICADAD tape to match zip codes of origin with county of origin and county population centroid. This tape is available to the public, and the accompanying documentation includes a FORTRAN program for calculating highway distances. An alternative would be to use the Forest Service's own RECZIP program to calculate distances. Unfortunately, there is some doubt as to whether this program will be maintained.

Once counties of origin are identified, census statistics provide various socioeconomic data. We have used median age, median income, and various other characteristics of the counties of origin.

Travel cost is based on a National Highway Administration report on 1981 costs. For our purposes a cost of $0.20 per mile for automobile operating costs has been used. We have estimated that there are three individuals per vehicle, sharing costs.

Visitation Data

Although considerable data on visitors to the National Forest System are collected, some containing the crucial information (zip codes) on visitor origins, there is currently no policy of routinely recording and reporting these data. Law enforcement information is the one important exception.

Fee Site Registration. On the White Mountain National Forest, all fee sites use an envelope deposit system for collecting fees. Visitors are required to fill in a small questionnaire on the envelope, place the correct payment inside, and deposit the envelope in a marked box. The questionnaire requests zip code information that would be useful in identifying the place of residence (hence origin) of the visitor. The data on the questionnaire, however, are not tabulated routinely on the White Mountain National Forest, except for one campsite. To assist with this study, forest recreation personnel did tabulate that information for two representative campsites during a typical high-use period in the 1983 season.

Wilderness Permits. On the White Mountain National Forest wilderness permits are currently required only in the Great Gulf Wilderness Area. Access is limited to seventy-five groups per night. The permit form requests information on entry and exit points, the number of people in the party, and length of visit, as well as origin data. This information is not currently being tabulated. The permit requirement is enforced and citations are issued for noncompliance, but enforcement activity in backcountry areas has varied considerably in the recent past. The fiscal 1982 backcountry enforcement was limited because of budget constraints.

Law Enforcement Management and Reporting System (LEMARS). The White Mountain National Forest employs a number of officers who oversee enforcement of the law and regulations on the forest. Patrols tend to be more heavily represented in locations experiencing the highest incidence of infractions. They are also more evident during high-use periods such as weekends and holidays.

The form on which law enforcement data are recorded is the same whether a warning is given or a citation issued for a violation. All of the data are routinely tabulated, and while the names and street addresses are not retained, the origin (as given by the zip code) and site of the citation are stored. Annual summary reports are issued by the Washington Office of the Forest Service, and data tapes are accessible through the service's information processing system. (Our use of the LEMARS data is examined in detail below.)

Special Surveys. Many national forests have access to special surveys that would provide useful data. A number of special surveys have been undertaken on the White Mountain National Forest by the forest itself, by the Appalachian Mountain Club, and by independent academics. We reviewed all of these studies; the information provided would have been quite useful in some cases, but as noted we have chosen to limit our study to data that would be generally available.

The LEMARS Data

Since special surveys are not universally available, we have elected to focus on the LEMARS information that is more or less routinely available throughout the National Forest System. Law enforcement data provide a sample of visitor data for a large number of sites for which information is not otherwise routinely reported. Although the sites visited on the White Mountain National Forest tended to be developed campsites or other high-use areas, there was fairly wide variation in site characteristics. Such variation in site attributes makes it possible to evaluate the effects of those attributes on visitation.

The LEMARS sample provided 1,030 usable zip code observations from among the 1,918 total observations in the White Mountain National Forest. By aggregating visits from a given county of origin or county grouping, and dividing by origin population, per-capita visitation rates were formed. Visits from outside New England were aggregated by state because of the large number of counties for which no visits were observed in these more distant states. Upon aggregation by origin, the data provided 428 origin-destination pairs involving 89 sites where violations had been recorded and 171 origin-destination pairs involving 51 sites where warnings had been issued.

It is unknown what fraction of total visitation is represented by the LEMARS sample. This does not affect our estimation of consumer surplus per visit, nor does it affect the estimated coefficients (other than the constant term) in the semilog specification of per-capita demand. We were not able to directly estimate total use, however, and used the RIM data on visitation to supplement our analysis.

To avoid problems resulting from multiple destinations, we somewhat arbitrarily further restricted our use of data to those visitors from origins within the New England and central Atlantic states listed in table 7–1. This left us with a total of 446 observations on per-capita visits.

Bias in the Sample.   The potential bias of the LEMARS sample relative to a true random sample is unknown. Such a random sample could have been obtained with a carefully designed survey instrument and a data collection effort, but this option was ruled out because of our intention to limit data to that readily available from National Forest System sources. In order to provide some check on the suitability of the LEMARS data, we made comparisons with data and regression results from three special surveys.

White Mountain National Forest personnel supplied us with samples from two campsites, consisting of all registrations for two weeks during July and August 1983. The Dolly Copp site, located at the northern end of the forest, and the Russell Pond site, at the southern end, were chosen as "typical" developed summer recreation sites. The Appalachian Mountain Club provided us with the registrations for the year 1981 for their Guyot shelter, a supervised semiprivate site on the Appalachian Trail.

The distribution of visits from the two sets of data for New England and central Atlantic states are compared in table 7–1. Comparing LEMARS developed sites with the Dolly Copp and Russell Pond sites, and LEMARS backcountry sites with the Guyot shelter, Massachusetts in particular seems overrepresented in the law enforcement data. There is a significant difference between Massachusetts and other locations in the distribution of visitor origins. Discussions with law enforcement officers confirmed expectations

Table 7–1. Sample Comparison: Percentage of Visits by State Origin

State	Copp and Pond sites	LEMARS developed sites	Guyot shelter	LEMARS backcountry
Mass.	40.9	50.9	44.5	55.6
R.I.	5.1	4.6	2.7	5.7
N.H.	18.4	25.5	24.4	20.0
Me.	8.0	4.0	3.0	2.9
Vt.	1.6	0.9	3.6	0.3
Conn.	9.5	5.0	10.4	5.4
N.J.	5.3	3.5	3.3	3.2
N.Y.	7.3	4.2	6.9	6.0
Pa.	3.9	1.5	3.3	1.0

that young white males were overrepresented in the LEMARS sample. These gross comparisons are suggestive of bias.

A more meaningful comparison of the data sets focuses on the estimated demand equations and benefits, presented in section 4 of this chapter. Although differences in results are apparent, they do not appear to be of a sufficient magnitude to cause great concern given the present rough precision in forest planning. Nevertheless, results based on LEMARS data must be evaluated in light of potential bias.

A Comparison of Samples. Table 7–2 compares regression results among the special site surveys and the LEMARS samples. The combined Dolly Copp and Russell Pond sample is compared to the LEMARS data from developed sites; the Guyot shelter sample is compared to the LEMARS data collected at backcountry sites. Since there is no variation in site characteristics for a single site, only travel cost (TRCOST) and socioeconomic independent variables are included in this comparison. The dependent variable for all models is the natural log of the ratio of visits from each origin to the population of the state or county of origin.

In addition to travel cost, participation in outdoor recreation activities is related to age, income, and the socioeconomic status of individuals. Tests of alternative statistical measures of these variables in various models have indicated that median county age, median county household income, and percentage of white population performed well in most regressions. These data are readily available from published state reports in the 1980 census.

Our statistical results indicate, as is commonly found to be true, that the counties having a high proportion of white households participate more in forest recreation activities. The coefficient on percentage of whites

Table 7–2. Sample Comparison: Regression Results

Variable[a]	Copp and Pond sites ($N = 158$)	LEMARS developed sites ($N = 285$)	Guyot shelter ($N = 62$)	LEMARS backcountry ($N = 96$)
TRCOST	-0.0795^b	-0.0638^b	-0.1118^b	-0.0578^b
%WHITE	0.1726^b	0.9804^b	0.9940^b	0.1521^b
AGE	-0.0321	-0.0653^b	-0.1513	-0.1505
INCOME	0.0262	-0.2237^b	0.0594	-0.2134^b
Adjusted R^2	.84	.72	.85	.73

[a]TRCOST = travel cost from county of origin to site within forest
  %WHITE = percentage of white population in total population of origin
  AGE = median age in county of origin
  INCOME = median household income in county of origin
[b]Indicates coefficient significant at the 10 percent level or better.

(%WHITE) is positive and is consistently significant, with very high t-ratios. Since participation rates in forest recreation are higher for young persons, we would expect counties having lower median age populations to have higher visitation rates. This is also confirmed by our data, although the AGE coefficients are not statistically significant in the site samples.

The INCOME variable is more problematic. Although we would expect higher visitation rates from more wealthy counties, the signs of the coefficients are not consistent. Signs are positive but not significant for the site samples, and significant but not positive for the LEMARS samples. There is some indication that the effect of income is nonlinear. Although not shown in table 7–2, other regressions indicate the effect is positive over some income ranges and negative over others.

Although socioeconomic coefficients are useful for predicting future changes in visitation for assumed changes in population characteristics, we defer discussion of this point until later. The most important variable for estimating willingness to pay for access to recreation sites or opportunities is travel cost. This variable is a proxy for site entry price in the demand equation.

We expected, of course, that the visitation rate would be inversely related to distance to the site. The sign for the price proxy is universally negative and highly significant in all model specifications tested. This result is required in order to estimate the economic demand function or willingness-to-pay schedule. Possible selection bias in the LEMARS sample may have led to an oversample of visitors from Massachusetts. Given the location of the White Mountain National Forest relative to population centers, this apparent bias tends to result in larger distance coefficients for the LEMARS data than for the site samples. Consumer surplus estimates from the LEMARS sample may therefore be overestimated.

These relatively simple models provide consumer surplus estimates of $12.58 per visit when using the Copp and Pond data, which compares to $15.69 per visit when LEMARS developed-site data are used. The Guyot shelter sample provides an estimate of $8.95 in consumer surplus per visit, which compares to an estimate of $17.32 from the LEMARS data. The difference in the latter case may reflect limited comparability between the sites involved.

4. Demand Estimation

As noted earlier, the primary source of data on visitor origins and destinations used for this study was the fiscal year 1981 LEMARS data for violations and warnings issued on the White Mountain National Forest. Census data on county characteristics were used to control for taste factors. The

1981 data on violations at 89 sites and warnings at 51 sites provided sufficient variation among site characteristics to incorporate these descriptors in
estimating demand.

The RIM data and ROS classifications were used to reflect quality differences among sites. In addition, after consultation with the forest's recreation
staff, a scenic index was assigned to each site to reflect observable scenic
values not captured by other reported site descriptors. Table 7–3 defines the
variables that were considered for use in demand estimation.

The natural log of the county (or state) visitation rate was regressed on a
set of explanatory variables for various subsets drawn from the LEMARS
data. The first three subsamples involved particular kinds of sites: vistas and

Table 7–3. Definitions of Variables Considered for Estimating Demand

ADMINR	Ratio of administrative costs to total management costs
AGE	Median age in county of origin
CAMP	Activity dummy equal to one if camping is available
CAR	Activity dummy equal to one if driving for pleasure is available
CLEANR	Ratio of cleaning costs to total management costs
COST2R	Ratio of estimated vegetation treatment costs to total management costs
COST3R	Ratio of estimated surface treatment costs to total resource costs
COSTT$	Estimated total resource treatment costs
DEVACRE	Number of developed acres at the site
DVSCALE	Development scale index
DIST	One-way distance from county of origin to site within the forest
ELEV	Elevation
FEESR	Ratio of fee collection costs to total management costs
FSHHNT	Activity dummy equal to one if fishing and hunting are available
HIKE	Activity dummy equal to one if hiking is available
INCOME	Median household income in county of origin
INTERPR	Ratio of interpretive services costs to total management costs
MAINTR	Ratio of maintenance costs to total facilities treatment costs
MANAG$	Total management costs
NATCHAR	Combined index for topography and vegetation on the site
NETDSP	Net dispersed (undeveloped) acreage at the site
%WHITE	Percentage of white population in total population of origin
PICNIC	Activity dummy equal to one if picnicking is available
REHABR	Ratio of estimated rehabilitation costs to facilities treatment costs
REPLAR	Ratio of estimated replacement costs to facilities treatment costs
ROS	Recreation Opportunity Spectrum Index
SCALE	Forest Service development scale
SCENIC	Index of scenic value surrounding site: $1 = $ low, $2 = $ moderate, $3 = $ high
TOPOG	Topography index, ranging from flat to mountainous
TOTAL$	Estimated total facilities treatment costs
TRCOST	Travel cost from county of origin to site within forest
TRVD	Total recreation visitor days reported for the site
VDUM	Dummy variable equal to one if observation is a violation
VEGET	Vegetation index, ranging from grass to dense conifers
WATER	Distance to lake, pond, or stream

roads, campgrounds and picnic areas, and general undeveloped areas. Alternative subsamples were drawn on sites classified as semiprimitive (ROS classes 2 and 3) and roaded natural (ROS 4). Finally, subsamples were also formed for sites where opportunities for a particular activity were available: camping, picnicking, fishing and hunting, and hiking. It must be emphasized that visitors were not observed engaging in such activities; rather, they were reported as being at locations where such activities were available.

Regression Results

The results of our regressions are presented in table 7–4. In the table we have omitted site variables whose t-statistic fell below 1.0 from the regression. The socioeconomic variables %WHITE, AGE, and INCOME have the same signs as in table 7–2 and are generally statistically significant at the 1 percent or 5 percent level for all models.

NATCHAR is a natural characteristics index combining topography and vegetation. Flat terrain with grass or bushes has a lower value than mountainous terrain with trees. The variable is significant in all models in which it is included, and has the expected positive effects on visitation rates. SCENIC, the assigned scenic index, also has the expected positive sign.

The next group of variables measures development and management characteristics of each site. SCALE of development has ambiguous signs, possibly because it confounds attractiveness and congestion. It was not possible to get a good site congestion measure. RIM contains estimated recreation visitor days (RVDs) for each site, but this variable was not significant in any of the models tested.

The ROS index does not vary much on the White Mountain National Forest. About a quarter of the sites are classified as nonmotorized semiprimitive (ROS 2), the remainder being roaded natural (ROS 4) except for a small number of motorized semiprimitive sites (ROS 3). The forest does not have any sites classified as primitive. The coefficient has a positive sign and is significant only for the camping subsample. DEVACRE is significant for only the semiprimitive ROS subsample, probably because it is picking up sites with heavier use.

The RIM files contain substantial data on expenditures and estimated costs of repair and rehabilitation. As with the SCALE variable, these cost variables seem to confound elements of both quality of service and deterioration of facilities, and were not altogether satisfactory variables. ADMINR is significant for fishing, hunting, and hiking sites. CLEANR is negative for picnicking, and is probably a proxy for congestion or physical unattractiveness. Interpretive expenditures have an inexplicable ambiguous sign, although the positive sign for camping seems reasonable. MANAG$ is also ambiguous, being positive for the complete sample but negative for areas more heavily used, implying that it may be picking up congestion effects. A similar

Table 7-4. Regression Results for LEMARS Subsamples

Variable	Vistas and roads (N=56)	Campground and picnic areas (N=256)	General undeveloped sites (N=123)	ROS 2 & 3 sites (N=136)	ROS 4 sites (N=310)	Camping (N=300)	Picnicking (N=191)	Fishing and hunting (N=274)	Hiking (N=290)
TRCOST	-0.0608	-0.0718	-0.0683	-0.0690	-0.0713	-0.0755	-0.0754	-0.0763	-0.0761
%WHITE	0.079	0.913	0.119	0.119	0.847	0.990	0.097	0.094	0.101
AGE		-0.061*	-0.118	-0.114	-0.057	-0.075	-0.073	-0.081	-0.050*
INCOME	-0.207	-0.231	-0.223	-0.226	-0.216	-0.223	-0.236	-0.215	-0.185
NATCHAR						0.303	0.171	0.185	
SCENIC	0.35			0.59	0.17		0.32		0.31
SCALE		0.28*				-0.59		0.27*	-0.25
ROS						1.01			
VDUM	0.25*	0.22	0.39	0.40*	0.33	0.28	0.36	0.34	0.27
DEVACRE				0.26*		-0.104*			
ADMINR				0.004				1.04	1.24
FEESR					-1.69	6.0	-12.2		2.1*
CLEANR							-0.67		
INTERPR									
MANAG$		-0.00003*	0.00001*		-0.00005	-0.00004		0.00002	0.00001
COST2R		-0.60			-0.81	-0.88			
COST3R					-0.17*		0.58*	0.46*	
COSTT$		0.00013		0.00025	0.00019				0.00005
MAINTR		0.63							
REHABR		0.28*			-0.60	1.27	-0.54*		0.26*
REPLAR					-1.25	-0.81		0.28*	
TOTAL$			0.00104	-0.00003*	0.000004*	0.00001	0.00001*		
CAMP		0.17*			0.19*				
FSHHNT					-0.23				
Adjusted R^2	.76	.76	.78	.78	.78	.77	.79	.78	.78

*Coefficient not statistically significant at the 10 percent level or above.

problem occurs with MAINTR, REHABR, REPLAR, and TOTAL$. The COST2R is consistently negative and reflects relative resource quality. The activity variables CAMP and FSHHNT have the expected signs, but CAMP is not statistically significant.

5. Application of the Analysis to Selected Issues

In most instances forest managers are concerned primarily with how changes in management prescriptions or demographics are likely to affect demands on the forest. For the White Mountain National Forest, the regression equations shown in table 7–4 can be used to predict changes in visitation and consumer benefits associated with specified changes in the independent variables. The primary caveats are that changes must be limited to the ranges encompassed by the original data, and that predictions are subject to statistical error depending on the statistical variance of the estimated demand coefficients.

The elasticities of demand are presented in table 7–5. Each entry in the table indicates the percentage change in predicted visitation for a 1 percent change in an independent variable. These elasticities were computed by using the per-capita demand curve to evaluate the change in visitation from each origin resulting from a 1 percent increase in the dependent variable. After summing the changes over all origins, the overall percentage change in visitation was found. For example, if median county age increases by 1 percent in all origins, predicted total visitation to the White Mountain National Forest would decline by 3.6 percent. The decline at campgrounds and picnic areas would be only 1.9 percent.

Management options that alter the quality of sites can be evaluated in a similar fashion. For example, if resource treatment expenditures (COSTT$) in all roaded natural areas were to increase by 1 percent, predicted visitation in these areas would increase by 0.84 of a percent. The same percentage increase in expenditures in all semiprimitive areas would increase predicted visitation by only 0.14 percent. Other scenarios not limited to equal percentage changes in variables at all origins or sites could be evaluated. Such projections would require somewhat more complicated calculations. One difficulty in the more general scenario is the increased likelihood of users altering their choice of sites.

Calculation of Consumer Surplus

Consumer surplus is defined as the amount over and above the "price" that a consumer of a good or service would be willing to pay rather than do without. In the case of forest recreation, the price is the entrance fee, if any,

Table 7-5. Demand and Benefit Elasticities for LEMARS Subsamples

Variable	Vistas and roads (N=56)	Campground and picnic areas (N=256)	General undeveloped sites (N=123)	ROS 2 & 3 sites (N=136)	ROS 4 sites (N=310)	Camping (N=300)	Picnicking (N=191)	Fishing and hunting (N=274)	Hiking (N=290)
%WHITE	0.76	0.88	0.12	0.11	0.82	0.95	0.93	0.90	0.98
AGE		−1.86	−3.57	−3.46	−1.74	−2.28	−2.22	−2.46	−1.53
INCOME	−3.42	−3.83	−3.70	−3.73	−3.60	−3.71	−3.88	−3.57	−3.09
NATCHAR						1.08	0.67	0.73	0.73
SCENIC	0.77			1.50			0.78	0.60	−0.32
SCALE		0.95			0.47				
ROS				0.83		−1.16			
DEVACRE						3.18			
ADMINR						−0.08		0.27	0.33
FEESR							−0.21		0.08
CLEANR							−0.29		
INTERPR					−0.10				
MANAG$		−0.59	0.04		−0.79	0.71		0.28	0.14
COST2R		−0.32			−0.35	−0.57			
COST3R					−0.02	−0.29		0.02	
COSTT$		0.71		0.14	0.84		0.03		0.16
MAINTR		0.24							
REHABR		0.11			−0.18	0.29	−0.08		
REPLAR				−0.03	−0.19	−0.35		0.12	0.04
TOTAL$			0.46		0.16	0.32	0.19		
CAMP	0.15	0.15			0.12				
FSHHNT					−0.13				

plus travel cost, and the procedure used to calculate consumer surplus relies on the estimation of the per-capita demand curves for use of the forest. The estimation of demand curves is possible because visitors residing at different origins face different access costs to given recreation sites. The individual's consumer surplus, or willingness to pay, for access to forest sites can be found as the area under the per-capita demand curve, above travel cost. For our specification of demand, the consumer surplus per visit is found as $-1/b$, where b is the estimated coefficient of the travel-cost variable. The accuracy and reliability of the calculation of consumer surplus depend principally on accurate estimation of this distance coefficient in the per-capita demand curve.

The estimates of consumer surplus per visit to the White Mountain National Forest for various samples are shown in table 7–6. Consumer surplus was calculated under the assumption that there would be three passengers per vehicle, and is the willingness to pay per visitor for a trip to the forest. As noted earlier, the consumer surplus values should be thought of as the value of access to the forest rather than the consumer surplus attributable to a particular site. As also noted, travel-cost estimates were based on a per-mile automobile operating cost of $0.20, derived from a National Highway Administration report on 1981 costs. Because there were no data on individual socioeconomic characteristics, we were not able to include the opportunity cost of time. This omission results in consumer surplus estimates which are biased downward.

Also of interest is the change in consumer surplus that results from a change in the quality of a site. The demand elasticities in table 7–5 can be interpreted to be estimates of the elasticities of consumer surplus benefits.

Table 7–6. Estimated Willingness to Pay for a Visit to White Mountain National Forest

Sample	Travel-cost coefficient	Consumer surplus per visit (in dollars)
Vistas and roads	−0.0608	16.46
Campground and picnic areas	−0.0712	14.05
General undeveloped sites	−0.0683	14.65
ROS 2 & 3 sites	−0.0690	14.49
ROS 4 sites	−0.0713	14.02
Camping	−0.0755	13.24
Picnicking	−0.9754	13.27
Fishing and hunting	−0.0763	13.11
Hiking	−0.0761	13.14

Each entry in the table indicates the percentage change in aggregate consumer surplus that would result from a 1 percent change in an independent variable at all sites or origins.

Visitation for recreation on the White Mountain National Forest is typically reported in recreation visitor day units (RVDs) rather than in visits (trips), as we have done. Accordingly, calculation of total estimated recreation benefits requires an appropriate conversion from visits to RVDs. The White Mountain National Forest employs a rule of thumb which equates a visit to 1.5 RVDs. Based on an estimated 2.4 million RVDs in 1983,[6] and an average willingness to pay of about $14 per visit (or $9.33 per RVD), a lower-bound estimate of the total fiscal year 1983 value of recreation on the forest would be about $22.2 million, plus about $0.4 million in receipts from fee sites.

The Net Value from Recreation

The estimated value of recreation on the White Mountain National Forest needs to be adjusted for the costs incurred to provide service. The costs can be estimated using data available from two different sources. First there are the actual expenditures charged to recreation in the 1983 season, which total about $1.2 million. While these are actual expenditures, there are grounds to suspect that this level of expenditure is inadequate to service the volume of visits without degradation of facilities. An alternative level of cost may be inferred from the data provided for the RIM Facility-Condition Record to assist in budget preparation. It is probable that these data err on the high side. The $4.5 million proposed budget for combined operating and maintenance expenditures exceeds by a significant amount the funds actually available for recreation administration. These budget requests were only partially met, so they may have reflected unrealistic aspirations. It is also possible that following some years of restricted budgets, the requests reflected a backlog of deferred expenditures necessary for preventing degradation of recreational facilities.

We think it highly probable that the actual expenditures on behalf of recreation on the forest were too low at $1.2 million, and that, on the other hand, the RIM data may overstate the annual costs since previously deferred expenditures have been carried over into the current period. Something between the actual expenditure of $1.2 million and the expenditures required to rehabilitate facilities suffering from insufficient maintenance would probably be a better estimate of actual annual cost. For our purpose, since the precise level is not critical, we arbitrarily pick a value midway between these—say, $2.8 million for 1983. This gives us an amount in excess of $19.8 million as the net value of recreation for 1983, or about $26 per acre. This value is large relative to the value of timber on the White Mountain

National Forest, to which we attend in the next section. Of course, not all of the recreation value is attributable to the recreation expenditures, and further, not all of this value would be forgone as a result of intensification of timber management.

6. White Mountain Timber Resources and Management

Because recreation is a product of joint production under multiple-use management, we need to address management issues affecting timber as well as recreation. The White Mountain National Forest has been managed for a number of land-management objectives, with one result being a harvest schedule involving longer rotations than where timber management is the only objective. The forest now is one of the very few sources of large-size, high-quality hardwood usable for specialty turned products and other high-quality wood products.[7] The forest harvests about 38 million board-feet a year, almost half of which is of sawtimber quality; the remainder is resold by the sawmills to the pulp mills.[8]

The predominant species on the forest are northern hardwoods, including ash, beech, birch, and maple. The demand for hardwoods has not been very strong over the years in comparison to the demand for softwood sawtimber. Even some of the traditional demand for good-quality hardwoods has been eroded by the advent of particle board veneer core, among other technical developments. The commercial outlook for hardwoods thus has not provided motivation to manage woodlots—or, indeed, private industrial forests—to produce high-quality hardwood sawtimber. On the contrary, the bulk of the forestland outside the White Mountain National Forest (excepting also the Green Mountain National Forest) has been selectively harvested for the most valuable stems, or clear-cut for pulpwood. The result has been that over time many residual stands have come to contain a high proportion of poor-quality stems, largely usable only for pulpwood, firewood, and perhaps the fiber content of particle board. Many stands are too young to have the large trees needed for high-grade sawtimber.

The question we need to consider here is whether timber management on the White Mountain National Forest is ultimately a useful undertaking—a question that arises not only because there may be conflicting demands for the land which timber occupies, such as scenic vistas, but also because timber management might not be economic even if there were no competition for the land. That is, we have to ask whether the costs of road access, sale preparation and harvest, and timber stand improvement would be justified by the estimated value of the timber.

A fairly extensive road network provides access for about half of the commercial forestland on the forest. The remainder, if it is to be harvested and thereafter managed for timber, would require about a mile of road per 800 acres of commercial forestland. We consider next that part of the commercial forestland which has not yet been accessed to see if this less economic part of the forest could yield a return that would cover harvest and management costs, including the cost of access.

Characterizing the Land Base

For transport purposes, the White Mountain National Forest is segregated into four different land classes.[9] Classes I and II are in the lower-elevation bottomlands and gently sloping areas which have for the most part already been roaded. These represent about two-thirds of the land base. Class III lands are not well drained and require frozen conditions to permit logging activities. The condition of this land results in significantly greater roadbuilding costs than do the more favorable conditions of class I and II lands. In class IV are higher-elevation lands having steeper slopes; these are the most costly to access.

Among the important cost factors, in addition to drainage conditions and terrain, are the type of road and construction standards. In order of increasing initial costs, there are intermittent winter roads, intermittent all-weather roads, and permanently open all-weather roads. If we were to consider the cost of managing and harvesting forestland that is already accessible, road construction, but not maintenance, costs could be ignored.

Aside from depending on access and other costs that will differ among different sites, the ultimate feasibility of economic timber management is also determined by the quality of the site for growing timber. On the White Mountain National Forest three classes of land are defined by their productivity characteristics at age 50. The forestland class with the lowest productivity will foster the growth of trees of about 46 to 55 feet in height at age 50. The intermediate class will produce trees of roughly 56 to 65 feet in height, while the most productive land will grow trees upwards of 66 feet. For convenience we shall refer to these site classes as SI–50, SI–60, and SI–70. The bulk of the commercial forestland on the forest will be in the SI–60 class.

For our analysis we must also distinguish sites by species composition. To do so we consider two extreme cases. One case involves a sale area well stocked with paper birch, which are high-valued, rapid-growing trees of short life. The second case involves a sale area stocked with longer-lived species (ash, beech, maple, and others) that we refer to collectively as northern hardwoods.

Whether a particular site is capable of economic production of timber will depend on the combination of factors identified above. We turn now to the method for evaluating alternative timber regimes, and then to the computation of timber prices and yields and management costs.

The Harvest Management Problem

We begin with a standing stock on an unaccessed area of the forest. The economic harvest scheduling problem, in its simplest form, can be viewed as a variant of the Faustmann harvest-age problem described in chapter 3. The manager must choose the time to enter the existing stand and also the rotation period for the subsequent regeneration harvests. The high initial expenditure on road access may result in the choice of an initial age of harvest that greatly exceeds that desirable for subsequent harvests.

So long as the rate of increase in realizable receipts relative to the asset value of the stocked land exceeds the rate of discount,[10] it will pay to allow the stand to grow. At the time when the annual growth in realizable receipts relative to the asset value declines to the rate of discount, the harvest age has been reached. This is essentially the nature of the analysis that is done below.

The basic elements of our computational exercise can be described in this way. We must find an initial harvest date t and a subsequent rotation period F to maximize net benefits, as follows:

$$\underset{t,F}{\text{Max}} \left[[Y(A+t)-C] + [V(F)/((1+r)^F-1)] \right] \Big/ (1+r)^t$$

where

$$A = \text{existing stand age}$$
$$t = \text{time until harvest of existing stand}$$
$$Y(A+t) = \text{net receipts/acre from harvest of existing stands at age } A+t$$
$$V(F) = \text{net receipts/acre from harvest of regenerated stands at age } F$$
$$C = \text{cost/acre of road construction}$$
$$r = \text{interest rate}$$

The actual computational effort is similar in spirit. However, because of a White Mountain National Forest policy to stagger entry to an area to be clear-cut so as to mitigate the visual impact, only about a quarter of the area will be harvested at one time.[11] This adds a slight complication to the treatment of costs and receipts because we must take note of the staggered harvesting schedule to ensure properly discounted present values. Except where we specifically note otherwise, a maximum of 25 percent of the area to be harvested in any one period and a minimum of ten years between harvest entries are taken to be required.

We calculate too whether intensified management, with precommercial and commercial thinning, is justified. In the actual analysis a richer accounting of costs is also used, including maintenance costs, road opening and closing costs, and the costs of administering harvests and thins.

Timber Prices, Yields, and Management Costs

In this and the following two subsections we present the results of our quantitative analysis, mainly in tabular form. Because the price and yield data are extensive, they are presented in the appendix to the chapter. In the tables, the figures for the existing stands reflect a likely poorer composition of species and lower log quality than could probably be obtained on future managed stands. For regenerated stands, separate tables are given for yields from a managed regime with precommercial and commercial thinnings, and for yields from an unmanaged regime with no thinnings. The economics of managing unaccessed paper birch and northern hardwoods are examined in detail in the sections to follow.

In looking at the tables on prices and yields, it should be noted that both the price and volume of a given stand increase with the age of harvest. Because of growth, the harvestable volume will be greater for later harvests. This growth will be reflected in larger log diameters, which bring more favorable market prices.[12]

The various road costs, which are dependent in part upon the kind of service a road provides, are listed in table 7–7. All road costs are based on an equal mix of class III and IV land. Although presented on a per-acre

Table 7–7.   Road Costs per Acre, White Mountain National Forest
(in dollars)

		Road type	
Activity	Constant service	Intermittent service (winter use)	Intermittent service (all-season use)
Construction	80.06	31.75	63.00
Reopening	0.0	9.19	10.58
Closing	0.0	1.69	1.69
Maintenance (harvest period)	0.475	0.0	0.0
Maintenance (between sales)	0.713	0.713	0.713

Note: These costs are averages based on 1 mile of road per 800 acres accessed on class III and IV lands.

Source: Costs were obtained from the White Mountain National Forest staff.

basis, the costs reflect those typical of a normal sale area. Obviously no reopening and closing costs will be incurred for constant service roads, but for these roads the maintenance cost will be higher than on intermittent service roads. For convenience in our analysis, an average intermittent road, with costs an average of those given in table 7–7, is compared to a constant service road.

Overall road costs will be dependent upon the scheduling of harvests as well as on the type of road. The area to be accessed must be opened at the time of first entry, even if only a quarter of the area will be harvested. Because of this, the total road construction cost is incurred in the year of first entry. All maintenance costs and road opening and closing costs are counted at the time of occurrence and discounted to the present in order to be combined with the construction costs. As an example, with four staggered entries, no thinnings, a 4 percent discount rate, and immediate road construction, the present value of the overall road costs per acre is about $96 for permanent roads, or about $69 for average intermittent roads.

Sales preparation and administration costs are about $139 per acre. For managed stands, there are additional improvement costs. Thinning costs amount to about $60 per acre for precommercial thinnings and about $92 per acre for commercial thinnings. The rather high preparation costs include some expenses incurred to ensure that the harvest is compatible with other multiple uses of the forest. While actual logging and hauling costs do figure into the economics of timber management, they are typically borne by the purchaser and are reflected in the price bid for the timber.

Short-Lived Paper Birch

We now look more specifically at the economics of the unaccessed short-lived species, paper birch. Since we stagger the harvest of a given area over a thirty-year period, the object is to select a level of management intensity, a schedule of the four initial harvest entries, and a subsequent rotation period that will maximize the present value from the land area.

Assume first that we are considering an intermediate productivity site (SI–60) and that the discount rate is 4 percent. Given these conditions, the greatest net present value is obtained with an initial entry at stand age 60, with the three reentries to harvest the initial stock occurring at the minimum required ten-year intervals.[13] Following the initial harvests, subsequent harvest ages, given the discount rate and site index, will be at 70 years. Intensified management with thinnings is not justified. If we suppose the existing stand is aged 40 years, with 4 percent discounting the value of the yield in perpetuity is about $45 per acre if an intermittent road would suffice.

If for some reason a permanent road of high standard were required, the net present value per acre would be reduced to about $34. The selection of a higher-quality road would undoubtedly be motivated by multiple-use objectives such as recreation, and the value of these would have to be considered along with the value of the timber to get the appropriate value to compare with the costs.

Suppose now that all of the conditions remain as stated above, except the discount rate, which is changed to 7 percent. The regeneration harvest period would then be reduced to 60 years and the net present value would be about $11 and $4 per acre, with intermittent and constant service roads respectively.

The assumption made with respect to the present age of the existing stand does not affect the harvest-age solution (unless, of course, the existing stand age exceeds the desirable harvest age). If the age of the existing stand were 60 years and we have assumed it was 40, we would simply have to increase our present values by 4 percent (or 7 percent, as appropriate) for each of the twenty years that we have understated the true age of the existing stand.

The conclusion that regenerated stands would not warrant intensive management needs to be examined. Comparing managed and unmanaged stand yields will show that a significantly higher volume of timber can be obtained from managed stands.[14] Yet in order to achieve higher yields per acre, a number of entries must be made for both precommercial as well as commercial thinnings. These tend to occur at approximately twenty-year intervals, the earliest substantially in advance of final harvest. Moreover, with staggered harvests and periodic thinnings, regular access must be provided at frequent intervals almost indefinitely. Both the recurrence of road costs and early thinning management costs result in substantial capital outlays that are avoided in the case of unmanaged timber stands. It appears that these differences in costs are greater than can be defrayed by the final value of the increment in biological yields induced by intensified management.

In order to put this and related questions into better perspective, we present data in table 7–8 that illustrate how our numerical conclusions will change with changes in assumptions about the discount rate, site productivity index, road standard, and management regime.

At this time we should take up the questions of already roaded sites, low-productivity sites, and single- or staggered-entry harvests. To begin with the first, our analysis has employed data relevant to unroaded areas for which access costs had to be incurred. From our calculations it would appear that for already roaded areas, for which road construction costs of $44 to $80 per acre need not be incurred, there would be some positive return even from low-productivity (SI–50) lands. If these lands were unroaded, it is likely that returns to the low-site productivity class would be too meager to warrant reserving such lands for timber production.

We turn now to the feasibility of staggered harvests for short-lived species. The analysis above was done on the assumption that short-lived species were

Table 7–8. Present Value of Paper Birch per Acre, Initial Harvest in Four Entries

Management regime	SI–50		SI–60		SI–70	
	Roads		Roads		Roads	
	Inter-mittent	Perma-nent	Inter-mittent	Perma-nent	Inter-mittent	Perma-nent
Discount rate = 4 percent						
Unmanaged						
Present value ($)	0	−11	45	34	96	81
Initial entry (years)	60–90	60–90	60–90	60–90	50–80	50–80
Rotation age (years)	80	80	70	70	60	50
Managed						
Present value ($)	−15	−20	34	29	86	80
Initial entry (years)	60–90	60–90	60–90	60–90	50–80	60–90
Rotation age (years)	70	70	70	70	50	50
Discount rate = 7 percent						
Unmanaged						
Present value ($)	−6	−13	11	4	35	23
Initial entry (years)	60–90	60–90	60–90	60–90	50–80	60–90
Rotation age (years)	70	70	60	60	50	50
Managed						
Present value ($)	−8	−15	9	2	31	22
Initial entry (years)	60–90	60–90	60–90	60–90	50–80	60–90
Rotation age (years)	70	70	60	60	50	50

initially harvested over a thirty-year period in the interest of minimizing the visual impact. When the motivation for such scheduling is partly to achieve other multiple-use objectives, the opportunity cost of delaying harvests should not be attributed exclusively to timber. Further, if losses due to natural causes were to be too large among the aging residual stems, a staggered initial harvest schedule might not be feasible for short-lived species. It should also be noted that birch stands may be dispersed within a tract in such a manner that a staggered harvest is not required. Accordingly, we should consider harvesting the area in a single entry.

For a single harvest entry on SI-60 land, the solution calls for an initial harvest at age 80, with regeneration harvests at seventy-year intervals. This gives us a maximum net present value of $71 per acre, assuming intermittent roads and 4 percent discounting. If a higher-standard permanent road were required, the present value would be about $62. Table 7-9 presents data on the present value of paper birch harvested in a single harvest entry; other data in table 7-9 are comparable to those in table 7-8. While our calculations suggest that the savings in costs and the benefit from earlier harvest receipts

Table 7–9. Present Value of Paper Birch per Acre, Initial Harvest in One Entry

	SI–50		SI–60		SI–70	
	Roads		Roads		Roads	
Management regime	Inter-mittent	Perma-nent	Inter-mittent	Perma-nent	Inter-mittent	Perma-nent
Discount rate = 4 percent						
Unmanaged						
Present value ($)	27	18	71	62	138	125
Initial entry (years)	80	80	80	80	70	70
Rotation age (years)	80	80	70	70	60	60
Managed						
Present value ($)	18	11	66	59	135	125
Initial entry (years)	80	80	80	80	70	70
Rotation age (years)	70	70	70	70	50	50
Discount rate = 7 percent						
Unmanaged						
Present value ($)	7	5	25	20	68	55
Initial entry (years)	80	80	70	70	50	60
Rotation age (years)	70	70	60	60	50	50
Managed						
Present value ($)	6	4	23	18	62	53
Initial entry (years)	80	80	70	70	60	60
Rotation age (years)	70	70	60	60	50	50

with single harvest entry would eliminate the negative returns for SI–50 forestland, the yield per acre would in any event not be particularly large. It is not clear, moreover, whether under current regulations adopted pursuant to the National Forest Management Act of 1976 the clear-cutting of large areas of land would be permitted.

Long-Lived Northern Hardwoods

The distinguishing feature of long-lived northern hardwoods is that they do not perish if not harvested in a relatively short period of time after maturity. The relatively long period of time in which harvest decisions may be reached provides greater management latitude and some interesting results.

A mixed hardwood stand might have an economic rotation age of 70 or 80, depending on the site index, discount rate, and other related factors in a typical road-accessed timber management environment. It would not be un-common, however, for a stand that would be ready for harvesting under a roaded condition to be too "young" to harvest economically if access has not yet been provided. The market value of the timber in such a stand might be insufficient to cover the cost of road access in addition to harvest and administrative costs. This will not necessarily mean that such timber is uneconomic.

Deferring harvests until the future permits the volume and value of the inventory to increase with time. It is not unusual, therefore, that a stand of hardwoods which would not be profitable to harvest earlier in time would eventually become profitable. In table 7–10 the present values per acre, the optimal ages of entry for harvesting a previously unaccessed stand (in four staggered entries), and the optimal rotation age for the accessed stand are given for various site productivities, discount rates, road standards, and management intensities. Observe that only for the highest site index with a low rate of discount do we find other than negligible net values.

Again the staggered harvests are predicated on the assumption that aesthetic and recreational considerations would require a succession of small cuts so as not to leave a large clear-cut area exposed to view. The question then arises of the additional cost that may be incurred to mitigate damage to

Table 7–10. Present Value of Northern Hardwoods per Acre, Initial Harvest in Four Entries

	SI–50		SI–60		SI–70	
	Roads		Roads		Roads	
Management regime	Inter-mittent	Perma-nent	Inter-mittent	Perma-nent	Inter-mittent	Perma-nent
Discount rate = 4 percent						
Unmanaged						
Present value ($)	3	1	8	6	25	20
Initial entry (years)	120–50	120–50	100–30	100–30	80–10	80–10
Rotation age (years)	100	100	70	70	70	70
Managed						
Present value ($)	2	1	6	5	22	19
Initial entry (years)	120–50	130–60	110–30	110–40	80–10	80–10
Rotation age (years)	80	90	80	90	50	50
Discount rate = 7 percent						
Unmanaged						
Present value ($)	0	0	1	0	4	2
Initial entry (years)	120–50	130–60	100–30	100–30	70–00	80–10
Rotation age (years)	90	90	70	70	60	60
Managed						
Present value ($)	0	0	1	0	4	2
Initial entry (years)	120–50	130–60	100–30	110–40	80–10	80–10
Rotation age (years)	80	80	70	70	50	50

nontimber values by these restrictions on harvest. In table 7–11 we show the results of assuming that the harvest is made as a single clear-cut in lieu of the staggered harvest. The differences in value are not great. They vary from less than a dollar to a maximum of fifteen dollars per acre harvested. And again it is not clear that the regulations adopted pursuant to the National Forest Management Act would permit the single-entry harvest option. The significance of this finding is that even if the regulations prohibited the single entry harvest, the cost of not taking this option would not be great in any event.

7. Conclusions on Timber Management

Having taken a look at the economic dimensions of forest management on the White Mountain National Forest, we can ask what we have learned. We learn that the least economic timber will come from the biologically lowest productive sites of the commercial forestland, other factors being equal, and

Table 7–11. Present Value of Northern Hardwoods per Acre, Initial Harvest in One Entry

Management regime	SI–50 Roads Inter-mittent	SI–50 Roads Perma-nent	SI–60 Roads Inter-mittent	SI–60 Roads Perma-nent	SI–70 Roads Inter-mittent	SI–70 Roads Perma-nent
Discount rate = 4 percent						
Unmanaged						
Present value ($)	5	4	15	11	40	31
Initial entry (years)	120	120	100	100	80	80
Rotation age (years)	100	100	70	70	70	70
Managed						
Present value ($)	4	3	12	9	36	29
Initial entry (years)	120	130	100	100	80	80
Rotation age (years)	90	90	90	90	50	50
Discount rate = 7 percent						
Unmanaged						
Present value ($)	1	0	3	2	13	8
Initial entry (years)	100	120	80	100	70	80
Rotation age (years)	90	90	70	70	60	60
Managed						
Present value ($)	0	0	2	2	11	8
Initial entry (years)	120	120	90	100	70	80
Rotation age (years)	80	80	70	70	50	50

that a timber management program is nearest to being economic on biologically high-productive sites. However, bringing timber under management approaches economic feasibility only where intermittent roads would be adequate to access unmanaged stands. If the staggered entries for harvesting and periodic entries for stand improvement needed under management would require continuous access provided only by a permanent road, the proper comparison is between the unmanaged stand/intermittent road option and the managed stand/permanent road option. A comparison of these options in otherwise identical circumstances reveals a significant difference in net returns in favor of unmanaged stands.

Now, although it is easy enough to demonstrate such a result, a more complete evaluation in multiple-use terms would have to consider potential recreation and other uses in the unroaded areas of the forest that might require permanent all-weather roads. We do not pretend to have such an intimate knowledge of the White Mountain National Forest that we can volunteer a useful judgment on the matter. But in considering any land and resource allocation decision, these factors need to be taken into account. In any event, we have learned that the difference management makes in biophysical yields may have to be qualified so stringently that we cannot be certain that intensive management, as compared with a more custodial level of care, would be generally economic in the presently unroaded areas of the forest.

As we look at individual species, or species groups having common characteristics, we note how much better the short-lived paper birch does economically than do the northern hardwoods. For example, the birch seems to do well enough to be profitable to harvest from both high and intermediate site classes and for either of the two discount rates that the OMB suggests for evaluations. The northern hardwoods show positive net returns at 7 percent for the SI–70 sites, but although there is a positive net return for the intermediate site class when discounted at 4 percent, northern hardwoods cannot pass the test at 7 percent.

For our purposes it has been convenient to have data on designated site classes and corresponding productivity available. It has also been convenient to have data on the various species found on the forest. In the actual forest, however, the site productivity classes are not all sorted out and aggregated in convenient parcels. Nor do we find all of the species occupying mutually exclusive tracts. The White Mountain National Forest is likely to have intermingling long- and short-lived species and site productivity classes. This suggests that although the bulk of the commercial forestland on the forest occupies SI–60 sites, there is such intermingling that some SI–50 and SI–70 sites would be accessed and harvested incidental to entering the SI–60 sites for harvest. Although such a harvest will look less tidy on the ground than it does on paper, our conclusions are not likely to be affected materially.

8. The Relation between Recreation and Timber: Summary and Conclusions

The purpose of this chapter has been to apply a model, developed by research, to a practical forest planning problem—the problem of estimating the value and implicit price that persons recreating on the White Mountain National Forest would pay rather than do without. As a result, we have estimated the average implicit price at about $14 per person per visit or about $9.33 per visitor day, which appears modest enough when compared to the cost of a good lunch or a mediocre haircut.

With an estimated 2.4 million visitor days per year, the gross recreation value for the forest is about $22.6 million. When the estimated costs of about $2.8 million annually are deducted from the estimated benefits, the annual net value is about $19.8 million, or approximately $26 an acre. If capitalized, this amounts to a present value of $650 (at 4 percent; $370 at 7 percent) per acre of land, which dwarfs the value of timber even on the productive sites.

For vehicular-related forest recreation, roads are an essential facility for recreation access, as they are for timber, and to the extent that the road network can be used for both recreation and timber there is an element of common costs. There appear to be possibilities, also, for flexible timber management policies that minimize the impact of timber activities on recreation. There are opportunities in managing the recreational plant that can accommodate a wider scope for timber management. As an example, developed campsites must undergo renovation after twenty-five to thirty years in any event, so that relocation to other equivalent sites seems possible. If relocation of developed campsites is acknowledged as an option, much greater scope in timber management is possible. We do acknowledge that the distinctive features of some sites are the attraction for locating a developed campground, and we would not argue that these sites should be disturbed. But on the White Mountain National Forest many, if not most, of the campgrounds offer primarily a pleasant forest environment for which there will be numerous substitute sites throughout the forest. A policy implementing proper spatial rotation of developed campgrounds to accommodate a proper timber rotation schedule would not appear to be very difficult to achieve.

There are some sites along the Kancamagus highway (Swift River) and its scenic gorge where the natural attraction of distinctive sites is acknowledged. What manipulation of vegetation is undertaken there should probably be motivated by recreational rather than timber objectives. There are doubtless scenic vistas from vantage points in the backcountry that also should be honored in management objectives. Our review of timber management prospects suggests that we should not expect that many, if any, opportunity returns would be forgone in providing for the preservation of scenic vistas.

Much of the forestland that would provide for attractive vistas does not host commercial timber stands, and where it might we find that except for the SI–70 sites the net value of timber is modest, at best.

We have touched on the importance of roads to access recreational sites and also on scenic vistas in the backcountry that might be frequently observed by hikers and backpackers. In order to clear up any misunderstanding, we note that for backpackers and hikers some mode of transportation to trailheads is essential, that for positioning themselves to engage in pedestrian activities roads are essential. But large areas of roadless country are also as important to them, as are developed camping facilities for others. This consideration militates against accessing areas where the timber is marginal and its quality indifferent. For reasons such as these we reiterate our observation that the forest growing on unaccessed low-productivity sites will doubtless have more value if it is kept in place than if managed and harvested uneconomically.

In summary, we comment on the useful role of recreation valuation models in establishing sensible management objectives. The student of this problem may wince at this observation when recalling the tenuous data on which the valuation here was based. We remind ourselves, however, that virtually all of the information needed to greatly improve confidence in recreation valuation model results is either collected on the forests in the National Forest System but is not tabulated, or is requested on various administrative forms but is not preserved for want of motivation, in the apparent belief that no use will be made of it.

Although incentives to compile data may be lacking for these reasons, we undertook this exercise with two objectives in mind. One was to propose an approach or provide a model that would accommodate to an existing data base that could be accessed through the National Forest System. The other objective, also motivated by a perceived need to improve information, was to provide an approach or valuation model that could assimilate better data when and as it became available. With this as a fallback position to fend off the naysayers, we will experience more pleasure than anyone on seeing results from an application of the model that has assimilated all the information that would convert the wincers to believers.

Notes

1. We remind readers that a visitor day represents twelve hours, whether that total is enjoyed by a single recreator or represents a composite of shorter periods enjoyed by more than one recreator.

2. It is to be remembered that jointness in production can be characterized as a negative as well as a positive relation between two outputs.

3. To be noted are the breakthrough papers by A. H. Trice and S. E. Wood, "Measurement of Recreation Benefits," *Land Economics* vol. 34 (August 1958), and M. Clawson, "Methods of Measuring the Demand for and Value of Outdoor Recreation," Reprint no. 10, Resources for the Future (Washington, D.C., 1959), leading to the so-called travel-cost method. The original work by R. K. Davis, "Value of Outdoor Recreation: An Economic Study of the Maine Woods" (Ph.D. dissertation, Harvard University, 1963), was the precursor to the contingent valuation method. See also R. C. Mitchell and R. T. Carson, *Using Surveys to Value Public Goods: The Contingent Valuation Method* (Washington, D.C., Resources for the Future, 1989).

4. For general discussions of the travel-cost method, see V. K. Smith, "The Estimation and Use of Models of the Demand for Outdoor Recreation," in *Assessing Demand for Outdoor Recreation* (Washington, D.C., National Academy of Sciences, 1975), pp. 89–123; J. F. Dwyer, J. R. Kelly, and M. D. Bowes, *Improved Procedures for Valuation of the Contribution of Recreation to National Economic Development*, WRC Research Report no. 128, University of Illinois Water Resources Center (Urbana, Ill., September 1977). For a discussion specific to forestry, see F. R. Johnson, J. V. Krutilla, M. D. Bowes, and E. A. Wilman, "Estimating the Impacts of Forest Management on Recreation Benefits," Discussion Paper D-115, Resources for the Future (Washington, D.C., 1983).

5. USDA Forest Service, Land Management Planning Staff, "ROS User's Guide" (Washington, D.C., n.d.).

6. This is an estimate of total dispersed and developed recreational use, excluding use of winter sport facilities. See USDA Forest Service, *Report of the Forest Service: Fiscal Year 1983* (Washington, D.C., February 1984). The level of use in 1983 was down about 15 percent from the peak years 1980 and 1981.

7. See J. R. Vincent, *White Mountain National Forest Timber and Its Users: Results of a Survey* (Petersham, Mass., Harvard Forest, 1982).

8. Vincent, *White Mountain National Forest.*

9. The characteristics of the forestland and information concerning timber management and harvesting issues are based on correspondence from and discussion with Edgar B. Brannon, Deputy Forest Supervisor, and the timber management staff of the White Mountain National Forest.

10. The discounts used in this study correspond to the rates required by the Office of Management and Budget. Since there are some differences in the practices of the OMB and the federal agencies, analyses are required to employ both 4 percent and 7 percent rates. We will follow this practice.

11. Because of the very high recreation use of the forest, a harvest policy adopted to mitigate aesthetic damage restricts the clear-cut area to about 20 acres, and the entire harvest of larger tracts is spread over 30 to 45 years.

12. Timber prices used here are the same as those used by the White Mountain National Forest in its FORPLAN study, and reflect the difference in prices of timber from managed and unmanaged stands. See the appendix to this chapter.

13. As a short-lived species, paper birch is usually harvested at 50 to 80 years, but some of our optimizations on lower-productivity sites require longer harvest intervals. Although yield tables that we used permitted so long a rotation interval, experienced

foresters harbored some doubt concerning the condition of the stand and hence the prices and volumes used in the analysis. We have continued to use the information while acknowledging that it may be suspect.

14. D. A. Marquis, D. S. Solomon, and J. C. Bjorkbom, *A Silvicultural Guide for Paper Birch in the Northeast,* USDA Forest Service Research Paper NE-130 (Upper Darby, Pa., Northeastern Forest Experiment Station, 1969); W. B. Leak, D. S. Solomon, and S. M. Filip, *A Silvicultural Guide for Northern Hardwoods in the Northeast,* USDA Forest Service Research Paper NE-143 (Broomall, Pa., Northeastern Forest Experiment Station, 1969).

Appendix 7-A

Prices and Per-Acre Timber Yields for Paper Birch and Northern Hardwoods

Paper Birch, Existing Stand Yields and Prices, Unmanaged

Age (years)	SI–50 Prices $/mcf	SI–50 Inventory mcf	SI–60 Prices $/mcf	SI–60 Inventory mcf	SI–70 Prices $/mcf	SI–70 Inventory mcf
30	0.0	0.0	63.0	0.0	63.0	0.890
40	0.0	0.0	121.0	0.940	121.0	1.580
50	63.0	0.970	148.0	1.350	148.0	2.130
60	121.0	1.300	160.0	1.880	169.0	2.610
70	148.0	1.650	200.0	2.290	200.0	2.990
80	148.0	2.060	196.0	2.570	196.0	3.200
90	169.0	2.260	196.0	2.990	196.02	3.200

Paper Birch, Regeneration Stand Yields and Prices, Unmanaged

Age (years)	SI–50 Prices $/mcf	SI–50 Inventory mcf	SI–60 Prices $/mcf	SI–60 Inventory mcf	SI–70 Prices $/mcf	SI–70 Inventory mcf
30	0.0	0.0	0.0	0.0	63.0	1.145
40	0.0	0.0	63.0	1.206	111.0	2.023
50	63.0	1.243	111.0	1.715	165.0	2.727
60	111.0	1.670	165.0	2.407	190.0	3.350
70	165.0	2.116	190.0	2.935	209.0	3.830
80	190.0	2.636	198.0	3.300	209.0	4.103
90	198.0	2.900	209.0	3.830	209.0	4.103

Paper Birch, Regeneration Stand Yields and Prices, Managed

Age (years)	SI–50 Prices $/mcf	SI–50 Inventory mcf	SI–50 Cut mcf	SI–60 Prices $/mcf	SI–60 Inventory mcf	SI–60 Cut mcf	SI–70 Prices $/mcf	SI–70 Inventory mcf	SI–70 Cut mcf
30	0.0	0.0		63.0	1.063		63.0	1.870	
40	63.0	1.268	0.400	63.0	1.750	0.480	123.0	2.723	0.552
50	63.0	1.700		123.0	2.560		198.0	3.434	
60	123.0	2.372	0.561	198.0	3.300	0.675	233.0	3.979	0.775
70	198.0	2.888		233.0	3.830		258.0	4.302	
80	198.0	3.200		258.0	4.064		258.0	4.521	
90	233.0	3.300		258.0	4.064		258.0	4.521	

Northern Hardwoods, Existing Stand Yields and Prices, Unmanaged

Age (years)	SI–50		SI–60		SI–70	
	Prices $/mcf	Inventory mcf	Prices $/mcf	Inventory mcf	Prices $/mcf	Inventory mcf
40	0.0	0.0	0.0	0.0	0.0	0.0
50	63.2	1.235	63.2	1.482	63.2	1.729
60	63.2	1.377	113.0	1.654	113.0	1.929
70	63.2	1.477	138.0	1.774	138.0	2.069
80	63.2	1.570	163.0	1.885	163.0	2.200
90	113.0	1.663	180.0	1.996	180.0	2.329
100	126.0	1.759	213.0	2.111	213.0	2.463
110	138.0	1.847	219.0	2.218	219.0	2.587
120	163.0	1.935	230.0	2.323	230.0	2.710
130	172.0	2.026	230.0	2.432	230.0	2.838
140	180.0	2.318	230.0	2.549	230.0	2.974
150	180.0	2.318	230.0	2.667	230.0	3.119

Northern Hardwoods, Regeneration Stand Yields and Prices, Unmanaged

Age (years)	SI–50		SI–60		SI–70	
	Prices $/mcf	Inventory mcf	Prices $/mcf	Inventory mcf	Prices $/mcf	Inventory mcf
40	0.0	0.0	0.0	0.0	0.0	0.0
50	63.0	1.583	63.0	1.900	63.0	2.217
60	63.0	1.766	63.0	2.120	155.0	2.473
70	63.0	1.894	155.0	2.274	198.0	2.653
80	63.0	2.013	165.0	2.417	220.0	2.820
90	155.0	2.132	198.0	2.559	233.0	2.986
100	198.0	2.255	220.0	2.707	268.0	3.158
110	198.0	2.368	226.0	2.843	268.0	3.317
120	220.0	2.481	233.0	2.978	268.0	3.474
130	226.0	2.597	268.0	3.118	268.0	3.638
140	226.0	2.722	268.0	3.268	268.0	3.813
150	233.0	2.848	268.0	3.419	268.0	3.999

Northern Hardwoods, Regeneration Stand Yields and Prices, Managed

Age (years)	SI–50 Prices $/mcf	Inventory mcf	Cut mcf	SI–60 Prices $/mcf	Inventory mcf	Cut mcf	SI–70 Prices $/mcf	Inventory mcf	Cut mcf
40	0.0	0.0		0.0	0.0		0.0	0.0	
50	63.0	1.848		63.0	2.218		171.0	2.588	
60	63.0	2.008	0.560	63.0	2.411	0.675	201.0	2.813	0.785
70	63.0	2.170		171.0	2.605		239.0	3.039	
80	171.0	2.338	0.660	201.0	2.807	0.780	273.0	3.275	0.920
90	201.0	2.474		239.0	2.970		291.0	3.465	
100	201.0	2.638	0.725	273.0	3.167	0.870	308.0	3.695	1.010
110	239.0	2.811		291.0	3.375		311.0	3.938	
120	239.0	2.976		308.0	3.573		311.0	4.169	
130	273.0	3.091		311.0	3.711		311.0	4.329	
140	273.0	3.091		311.0	3.711		311.0	4.329	
150	291.0	3.091		311.0	3.711		311.0	4.329	

8

The Flip Side of Joint Production

1. Introduction

Jointness in production is not always attended by positive effects on one forest output when another output is expanded. Stated differently, an increase in the level of production of one forest output may lead to an increase in unit costs of producing a joint output—or, in the limit, may preempt the opportunity to produce the second (joint) output. We see this intimated in our examination of various alternatives for the joint production of timber and water in chapter 5. There we found that timber was not economically feasible considered by itself. When timber harvests and stand improvements were conducted in combination with a complementary water augmentation program, the value of the two outputs combined exceeded the cost of the two programs over a reasonable range of road-access costs. This occurred because some of the cost incurred for one (road access) served equally for the other. But pressing the optimization a step further, we learned that under the most favorable economic outcome (in terms of maximizing returns to the land) the watershed would be managed for water alone, since the relative value of water dominated, requiring a cutting cycle too short to produce merchantable timber. By approaching production as an optimization problem, we learned that a program that provided the maximum net benefits eliminated timber in favor of greater water-yield augmentation.

In this chapter we look at another case that more directly addresses this problem.[1] The setting is in the White Cloud Peaks located across the Sawtooth Valley from the spectacular Sawtooth Mountains and Wilderness in Idaho. During the early 1970s, before the decisions had been taken to establish the Sawtooth Range as a wilderness area and the Sawtooth Valley as a national recreation area which included the White Cloud Peaks, there was a need to consider various options for all of these resources. The Sawtooth Range and Valley have since been dedicated, respectively, to highly dispersed recreation use and to higher-density recreation use consistent with the national recreation area (NRA) concept. But the status of the White Cloud Peaks has remained somewhat indeterminate. In the spring of 1984 it was included in the Idaho Wilderness bill proposed by Senator James A. McClure of Idaho. We will review the salient considerations affecting the decision on the White Cloud Peaks.

2. Conflicts in Land Use in the White Clouds Peaks

The White Cloud Peaks, lying about thirty miles north of Sun Valley, Idaho, in the Challis National Forest, were little known nationally before the spring of 1969 (for the location of the forest, see figure 8–1). In that year, having purchased some existing mineral claims (for molybdenum) in the vicinity of Castle Peak and staked out additional claims, the American Smelting and Refining Company (ASARCO) applied to the Forest Service for a permit to build a mining access road across national forest land in the expectation of more intensive activities in the area. Notice of the application attracted immediate and widespread attention among local conservation groups. And, as the road in question was to be located in a then roadless area along Little Boulder Creek to the base of Castle Peak, it soon attracted national attention through articles in such periodicals as *Living Wilderness* and *Smithsonian Magazine*.

Quite aside from the fact that the road would have penetrated a roadless area (adversely affecting one qualifying criterion for addition to the National Wilderness System), the keen interest expressed by conservation groups was related to the characteristics of the area itself. The White Cloud Peaks, anchored at the southern end by Castle Peak, extend northward for about ten miles of rugged peaks and razorback ridges. While the total area regarded as the White Cloud Peaks administrative unit by the Forest Service was put at 157,000 acres, the area actually occupied by the peaks and related cirque basins is scarcely 20,000 acres in size. In this belt there are well over a score of peaks rising to elevations in the range of 10,000 to nearly 12,000 feet. Viewed from the east, the White Cloud Peaks give the impression of a semicircular, self-contained "pocket" mountain range. Individually the peaks are not more majestic than the peaks in the Pioneer Mountains to the southeast, for example, nor are they more numerous. What distinguishes them is their quality of self-containment, clustered as they are in their peculiar arrangement. They appear physically isolated from other peaks and ranges; this isolation, along with their concentration and distinctive configuration,

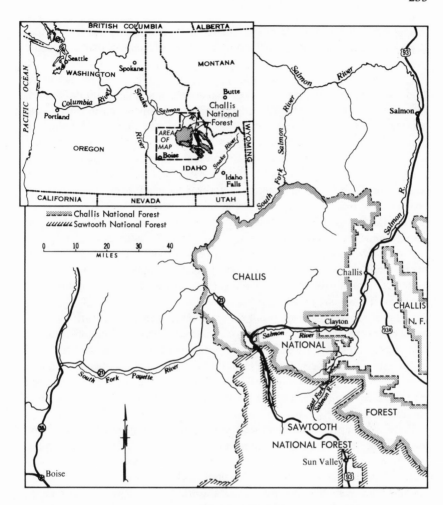

Figure 8-1. Location of Sawtooth and Challis national forests. *Source:* J. V. Krutilla and A. C. Fisher, *The Economics of Natural Environments: Studies in the Valuation of Commodity and Amenity Resources* (Baltimore, The Johns Hopkins University Press for Resources for the Future, 1975).

distinguishes them from the Boulder and Pioneer mountains and collectively lends them a uniqueness which the peaks viewed individually probably would not merit.

The White Cloud Peaks are also unusual in that they and the adjoining lands support a remarkable variety of the larger mammalian wildlife species. Indeed, only within this general region can one find both the mountain goat and the bighorn sheep along with mule deer, elk, antelope, bear, and mountain lion, and numerous smaller mammals and nonmammalian species.[2] Moreover, by reason of the glacial action that occurred in these mountains, the lakes tend to be deeper, with higher energy and oxygen budgets than in the neighboring mountain ranges.[3] Of the 120 lakes in the White Cloud Peaks, about half have sufficient depth to support fisheries. These are principally rainbow, cutthroat, and brook trout. Some of the lakes have had California golden trout and grayling introduced, and the golden trout has flourished when competitive species have not been introduced. As with mammalian wildlife, the White Cloud Peaks can support an unusual variety of fishes, making the area interesting to anglers who wish to fish for new varieties of trout and char.

The geomorphologic features of the area, which provide magnificent scenery; the presence of rare species among the large variety of wildlife; and the opportunity for fishing a wide variety of game fishes make the White Cloud Peaks a prime outdoor recreation area. Although two decades ago the hunting was reported to be light, about a third of the total visitation was motivated by an interest in fishing, and presumably still is. Backpacking, photography, and rock collecting are also recreational activities enjoyed in this area.

There is competition for the resources among the several possible uses, but with one notable exception none are mutually exclusive. Grazing on the Big Boulder Creek Allotment inflicts some cost on recreationists, as the cattle graze in the Frog Lake and Quicksand Meadows areas, which are popular with recreationists, and up along Little Boulder Creek, also popular with fishermen. A range management plan was proposed for the allotment that would, with moderate adjustments at modest expense, eliminate the major sources of conflict. While the plan remained firm as of the spring of 1984, the conflict has not been completely resolved.[4] A somewhat more serious conflict involved the effects of prospective mine-mill operations on utilization of range forage by wildlife, as well as by domestic stock, in the Germania Creek, Wickiup Creek, and Little Boulder Creek drainages. Even so, the better part of all four of the units making up the Big Boulder Creek Allotment, along with the Big Lake Creek Allotment, could remain available for domestic livestock grazing under the proposed plan.

The operation of an open-pit mine and ore beneficiation mill, however, could result in sufficient disturbance to a large area of the existing environment near the heart of the recreationally attractive portion of the White

Clouds to be fatal to the amenities of a wilderness environment. Immense amounts of material would have to be removed and disposed of, tailings from the beneficiation operations would require disposal, and to move so much mass a transport facility would have to be built that would be grossly out of character with the pristine conditions now attracting the recreation activity.

Accordingly, the most intense competition for the resources in the White Clouds involves the incompatibility of mining and a wilderness environment for recreation in an undisturbed natural setting.

3. An Estimate of the Wilderness Recreation Value

Various forms of recreation could be proposed for the White Clouds, but because provision has been made for road-accessible, more intensive recreation in the Sawtooth Valley between the mountain ranges and because of the appearance of the White Clouds in Senator McClure's Idaho Wilderness bill, we analyze the recreation potential in use as a wilderness resource.

There were in the neighborhood of 4,500 visitor days' use in the White Clouds in 1971 when the earlier assessment for wilderness potential was made. This use was not evenly distributed. There was a large concentration at Frog Lake, part of which was accounted for by the reputation of the lake for yielding large fish. But a large part was also accounted for by day-use trail bike traffic permitted by the Forest Service. Trail bike recreation continues to the present, but given the proposed wilderness legislation, it does not seem likely that this will remain a permitted use. In the early 1970s use of the area around Frog Lake already exceeded the capacity of the area to sustain that use without causing unacceptable environmental degradation. A second large concentration occurred at Walker Lake, with much of the remainder distributed among the Little Boulder Chain lakes. Here the usage appeared heavy enough to raise the question of potentially adverse sanitary conditions in the basin.

With the level of use in the White Clouds becoming incompatible with the fragile environment at the time decisions about the area were being made, it was obvious that once the Sawtooth Range and Valley got statutory protection—and thereby national recognition—additional use could be expected and would require attention. The heavy concentration of use at Frog Lake was partly accounted for by the convergence of several trails in the vicinity of the lake, and the existence of a through trail running adjacent to it. Perhaps some extension, relocation, or elimination of segments of the trail system should be considered. Relocation or even elimination of a section of trail (no. 4179) between the lower of the Boulder Chain lakes and its junction with the trail running along Red Ridge would be feasible (see figure 8-2). Frog Lake could then be serviced by a spur trail in a manner similar to the

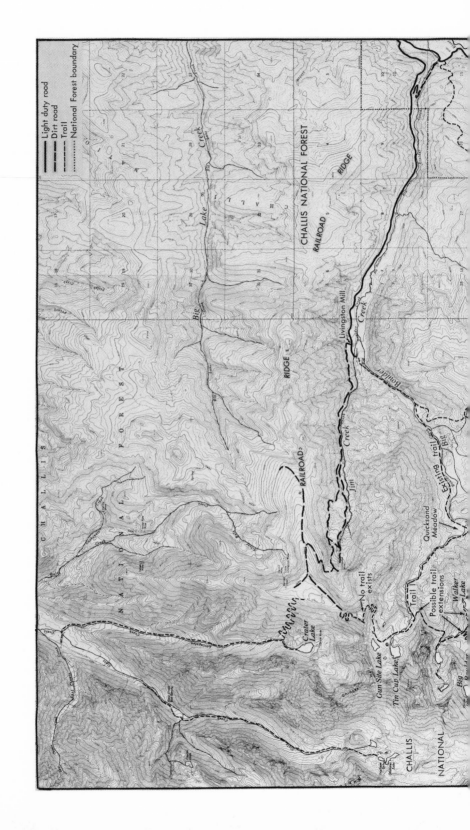

Figure 8-2. Trail system of the White Clouds Peaks. *Source:* Krutilla and Fisher, *Economics of Natural Environments.*

access to Little Redfish Lake.[5] Both because of the concentration of use in a relatively small area and the need to eliminate those uses which most tax the environment, horse traffic should probably not be permitted above Frog Lake and Little Redfish Lake, dispersed corrals should probably be built to accommodate stock, and the upper cirque areas should be reserved for foot traffic alone.

Given the projected use of this area and the potential for unsanitary conditions—or at least unsightly litter associated with defecation—there was discussion in the Forest Service of "wilderness compatible" toilet facilities to deal with the problem. It now appears that the idea of building such sanitary facilities has been rejected in favor of a simpler alternative. Plastic trowels are provided by the national recreation area management to facilitate the burying of excrement and related paraphernalia.[6]

Would the recreation visitation justify expenditures of the sort implied by the above proposals? (See tables 8–1, 8–2.) When this issue came up in the early 1970s a rather complex estimation model, nonetheless good only to a first approximation, was employed. It took into account the estimated increase in recreation capacity, value per visitor day, and its relative increase

Table 8–1. Cost of Trail Extensions for White Clouds Threshold Wilderness Recreation

| | Present value of cost | |
| | Discount rate | |
Trail extension	4 percent	7 percent
Capital outlays		
Trail I $11,000	$ 11,000	$ 11,000
Trail II 12,000	12,000	12,000
Trail III[a] —	—	—
Trail IV[a] —	—	—
Maintenance outlays		
Trail I 5,500	130,242	78,079
Trail II 6,000	142,082	85,170
Trail III 90	2,131	1,278
Trail IV 322	7,625	4,572
USFS 4179 180	4,262	2,555
Present value of total trail costs	$309,342	$194,654

Sources: Letters from Donald Nebecker, November 26, 1971 and January 21, 1972, and discussions on April 14, 1972 with recreation specialists and the coordinator of the USDA Forest Service White Cloud-Boulder-Pioneer Mountains Comprehensive Land Management Planning Study.

[a]Existing trails which would link up with segments of trails I and II to provide loop routes (see figure 8–2).

Table 8–2.   Cost of Sanitary Facilities for White Clouds Threshold Wilderness Recreation

Year	Capital outlays	Present value of cost	
		Discount rate	
		4 percent	7 percent
1	$6,750	$ 6,750	$ 6,750
30	6,750	$ 2,081	887
60	6,750	642	116
90	6,750	198	15
Present value of capital outlays		9,671	7,768
Maintenance outlays, $100/year		2,368	1,419
Present value of total sanitation costs		12,039	9,187
Present value of total recreation costs		$321,381	$203,841

Sources: Information was originally obtained from letters from Donald Nebecker, November 26, 1971 and January 21, 1972, and discussions on April 14, 1972 with recreation specialists and the coordinator of the USDA Forest Service White Cloud-Boulder-Pioneer Mountains Comprehensive Land Management Planning Study.

with time. Since that analysis is published elsewhere,[7] we provide a short-hand sample here to give perspective. Recall that in chapter 7 we estimated a value of $9.33 per visitor day for recreation in the White Mountains. Our estimate for the White Clouds in the early 1970s was a bit more—$10.00, based on estimates of wilderness users in western wilderness areas. The level of use was estimated at between 4,500 and 5,000 visitor days, on the basis of reports by the wilderness patrolman in 1970. Even at that time this level of use, given unchanged capacity, pressed the recreational carrying capacity of the area. Given the facilities and level of use in the early 1970s, and taking the 4,500 visitor days per year estimate at $10.00 per visitor day, for $45,000 per year in perpetuity, we arrive at a present value of $1,125,000 when capitalized at 4 percent, and $642,847 when capitalized at 7 percent.

With an investment in additional trails, and a more uniform distribution of trails throughout the area to reduce the traffic on overused facilities, a trebling of capacity can be estimated as possible without deterioration of the biophysical environment. If capacity is only doubled and there is no increase in the relative value of future recreation, the incremental value of recreation is equivalent to the calculation in the preceding paragraph. In short, the value of the increment in recreation would exceed by a factor of three the costs of facilities required to make it possible.

The purpose of this exercise has been to ascertain whether the level of use and its value merited management inputs even if some capital outlays were involved. But, one needs to ask, is it not relevant also to ascertain whether the present worth of the recreational benefits is at least as great as the opportunity cost, reckoned in terms of the value of molybdenum forgone? That is a good question, but we defer responding to it until we have looked at some of the other uses that we would expect to yield value.

4. The Place of Grazing Stock in the White Clouds

Second to recreation in economic importance in the White Clouds in the early 1970s (as it is today) was the grazing of livestock on portions of the Challis National Forest and on adjacent lands belonging to the Bureau of Land Management (BLM) and the state of Idaho. There were (and are now) two grazing allotments on the national forest either bordering on, or within, the scenic and recreationally attractive area of the White Clouds' eastern slopes. One was made up of three units situated in the Big Lake Creek drainage, and appeared to be free of any appreciable conflicts with recreation activities; the other, the Big Boulder Creek Allotment, was made up of four units (see figure 8–3). The first unit was a combination of Forest Service, BLM, and state of Idaho land located along the bottomlands of the East Fork of the Salmon River. The second unit occupied the Bluett Creek drainage, while the remaining two occupied the major portions of the Little Boulder and Big Boulder creek drainages, respectively. Since only the Big Boulder Creek Allotment overlapped portions of the choice recreation areas, and in turn would be affected by prospective mining operations, we confine our analysis to this allotment.

Years of sheep and cattle grazing, sometimes to excess, were responsible for range conditions which were in places quite poor—conditions which persist today. Some of the most overgrazed range occurred on public land adjacent to private lands along the East Fork of the Salmon River. This area was vital winter range for deer and bighorn sheep.[8] One of the priority objectives of the management plan of the early 1970s was to improve the condition of this critical winter range. This was feasible in part because the Bluett Creek drainage which abuts the land along the East Fork of the Salmon River was getting little use. Because there were no fences, cattle left on this unit tended to stray into the adjacent allotments used by a neighboring permittee. Fencing of these units, along with scheduled rotation and related management practices, promised to improve the winter forage in this area.

Elsewhere range conditions were variable, tending to reflect overgrazing on favored areas and inefficient utilization of the forage on less choice areas. Achievement of a more uniform forage harvest would have required periodic confinement of the grazing stock to restricted areas, thus encouraging uniform foraging within a given fenced unit before rotation to the next unit. It was believed that in the absence of an explicit management program of this

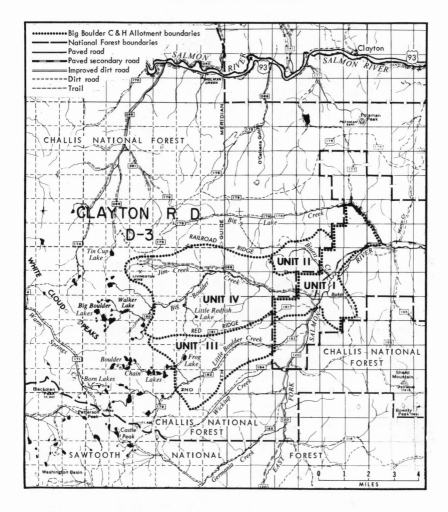

Figure 8-3. Big Boulder C&H Grazing Allotment. *Source:* Krutilla and Fisher, *Economics of Natural Environments.*

sort to upgrade the range conditions, the number of grazing animals would have had to be reduced by approximately 40 percent.[9]

Conflicts between grazing and recreation occurred at Frog Lake, Quicksand Meadows, and Big Meadow in the Little Boulder Creek drainage. Heavy grazing of meadows around the lakes that were used for recreation tended to trample the areas and adversely affect the aesthetic conditions. The conflict between recreation and grazing took on another aspect as well: competition for forage when pack animals were used in recreational travel. If the rest-rotation program for a permittee's livestock was to have the desired effect, recreational livestock could not be allowed to graze on the favored areas of the unit being rested. A coordinated effort, manipulating access to differentiate between recreationists moving on foot without pack animals and those traveling on horseback, would permit restriction of access to allotment units being grazed, with foot traffic directed to resting units. The effect of this would have been to separate horse and foot traffic along trails, a desirable objective from the standpoint of minimizing conflict between recreation participants using different modes of travel.

In summary, the objective of the range management program was the improvement of plant vigor, plant density, and species composition to increase the grazing capacity of the range; melioration of the conflicts between grazing and recreational uses of the White Clouds in areas on the allotment; and provision of forage for wildlife, particularly in the critical winter range along the East Fork of the Salmon. The program would also have had the effect of improving ground cover essential to watershed conservation.

Investments in various improvements would have been necessary to achieve these results. About nine miles of fencing would have been required to obtain the desired control over the movements of livestock. By itself, fencing would have provided the necessary separation of basic units to ensure a feasible rest-rotation program. This was considered to be the first increment of investment. A second level of investment would have involved water development to pick up some additional forage on a section of secondary range in one of the units on the allotment. Also considered was the merit of removing undesirable species of plants as part of the range improvement program. In the early 1970s herbicides were generally used for plant removal, but this would have involved the removal of important cover for wildlife, particularly young antelope and sage grouse. Accordingly, consultation with wildlife biologists was recommended before a decision was made on removing undesirable species. Since that time prescribed burning has come into greater use. Improvement of the winter range for bighorn sheep has been accomplished in this way.

We can assemble the costs of the proposed program by project type—that is, fencing, water development, and spraying—to compare the estimated costs and gains for each undertaking individually. First, however, we should discuss how we reckon the costs. A plan of this sort would be intended to restore the range to an improved condition and then maintain it in such condition in perpetuity. We thus need to cumulate the discounted benefits and costs in perpetuity. Since perpetuity is quite a long time, however, as a practical matter we cumulate the discounted values only while they seem nonnegligible. As a convention, then, we terminate the process when a dollar of future benefit has a discounted present value of one cent.[10] Accordingly, the present-value formula for a program with N projects with economic lives of T_n years can be given as

$$PV = \sum_{n=1}^{N} \sum_{t=1}^{T_n} \left[\frac{B_n(t) - O_n(t)}{(1+i)^t} - C_n \right] / (1+i)^{y_n}$$

where N = number of discrete capital outlays (projects)

T_n = project life in years

y_n = year in which n^{th} project goes into operation

$B_n(t)$ = gross benefit of project n in year t

$O_n(t)$ = operation and maintenance costs of project n in year t

C_n = the investment cost in project n

i = rate of discount

The results of our computations for discounting at 4 and 7 percent are shown in table 8–3.

The conventional query next is, would the improvement in range justify the additional costs? When this question was posed in the early 1970s, the prospects for increasing demand and hence the value of rangeland in the intermountain West were good. Several factors accounted for this. The demand for beef was income-elastic, and this, coupled with a growing population and increasing per-capita income, suggested strongly that the demand for grazing would increase into the indefinite future. Public rangeland, however, is in fixed supply. The private land often associated in use with the public range frequently lies in the bottomlands and is suitable for the production of hay and other higher-valued crops. Accordingly, the opportunity cost of this land for grazing is prohibitive. Because the public range in the intermountain region is suitable largely only for wildlife and livestock grazing, in the 1970s it was deemed likely to remain in grazing but at increasingly higher fees to permittees as demand grew over time.

Since that time, probably in part as a result of increasing awareness of the association of various vascular and cardiac diseases with personal histories of high consumption of beef, the quantity of beef consumed per capita has been declining. This may be reflected in a reduced demand for grazing. The analysis done earlier with the 1970 data postulated a secular rise in the price of an animal unit month (AUM) of forage. In retrospect, the assumptions which served as the basis for those projections have proved to be incorrect. Accordingly, we now recalculate the value of the increased product for each increment of range improvement, taking the 1970 value per AUM ($6.10)

Table 8–3. Estimated Costs of Management Programs for Big Boulder Creek Allotment

	Costs	Present value of costs at 4 percent	Present value of costs at 7 percent
Fencing			
Year			
0	$4,700	$ 4,700	$ 4,700
1	1,175	1,130	1,098
2	7,000	6,472	6,114
30	4,700	1,449	617
31	1,175	348	144
32	7,000	1,995	803
60	4,700	447	81
61	1,175	107	20
62	7,000	615	106
Fence maintenance $100/year		2,368	1,414
Total present value of fencing		$19,631	$15,096
Water development			
Year			
0	$1,900	$ 1,900	$ 1,900
30	1,900	586	250
60	1,900	181	33
Water development maintenance		1,163	707
Total present value of water development		3,830	2,890
Herbicidal spray			
Year			
0	$5,394	$ 5,394	$ 5,394
17	5,394	2,769	1,708
34	5,394	1,422	541
51	5,394	730	171
68	5,394	375	54
Total present value of spraying		10,690	7,868

without assuming an increase in the relative value of forage.[11] The estimated increase in animal unit months and the present value at discount rates of 4 and 7 percent are presented in table 8–4.

Comparing the incremental gain for each measure with the corresponding costs from table 8–3, we see that except for spraying with herbicides with a discount of 7 percent, all of the proposed measures would be justified by the returns. This is especially important in connection with the fencing. Fencing

Table 8–4.   Estimated Forage Increase and Value on Big Boulder Creek Allotment

		Present value of forage	
Improvement	Gains, AUMS	4 percent discount	7 percent discount
Fencing	388.6	$55,228	$33,573
Water development	141.3	20,085	12,210
Spraying	88.5	12,586	7,651
Total	618.4	87,899	53,434

had a double purpose, one of which was to facilitate the routing of stock to prevent stock intrusion into recreationally sensitive areas, the other being to segregate foot from horse traffic. It is seen that these objectives could have been achieved while the facilities were being used for increased forage production. To the extent that the fencing would serve a recreational purpose as well as promote the production of forage, all or part of the cost would be incurred for a common purpose, and represents a case of joint production. In this instance, however, the returns from only one of the joint products would cover total costs. Were this not so, we would not be able to render an economic judgment until after a comparison was made between pooled costs and returns.

5. The Prospect of Mining

The geology of portions of the White Clouds is propitious for mineral deposits. The interfaces between granitic and sedimentary rocks are places where silver, lead, zinc, and molybdenum are frequently found. The east slope areas of the White Cloud Peaks abound in such zones. There are other known deposits of silver, lead, zinc, copper, gold, and antimony within or near the White Cloud Peaks.[12] The old Livingston Mill, in fact, worked an area on the upper Jim Creek drainage, a tributary of Big Boulder Creek, producing lead, zinc, and silver. The East Fork of the Salmon (including the Germania Creek drainage), Boulder Creek (including both streams), and Lake Creek represent the main mineralized districts within the White Clouds. Unfortunately, these are the areas that have the most attractive recreational opportunities as well.

The molybdenum occurrence in these areas is of low grade, and would be open-pit mined. Molybdenum is a comparatively rare mineral, occurring in the earth's crust in a ratio of about 5 parts per million. The concentration in commercial ores is from 1 to 3 parts per thousand, or 2 to 6 pounds of molybdenum per ton of ore. With low-grade deposits, then, one could expect the removal and disposal of a half ton of ore and tailings for every pound of molybdenum concentrate, exclusive of the associated overburden.[13] Mining a large deposit of low-grade ore such as the one found in the vicinity of Castle Peak would have an enormous impact on the landscape. It was suggested that initially the stripped overburden could be used to create a 400-foot-high earth-fill dam in the Wickiup Creek drainage, which was planned by ASARCO as a site for solid waste disposal. The reservoir behind the dam would serve as a settling pond and general tailings disposal area. The sites of the proposed major mining activities are outlined in figure 8–4 against a photographic background of the heads of Little Boulder, Wickiup, and Germania creek drainages.

Until recently there were only two primary producers of molybdenum in the United States: Molybdenum Corporation of America (MOLYCORP) and the Molybdenum Division of American Metals Climax (AMAX). Approximately a quarter of total production, however, is the result of by-product or co-product production through the processing of molybdenum-bearing copper, tungsten, and uranium ores. Kennecott Copper Corporation was the largest producer of by-product molybdenum in the early 1970s, although the Duval Corporation and the Magma Copper Company also were significant by-product producers. The by-product production is significant because it represents a source of supply that meets more than half of the total U.S. industry demand. It thus offers additional sources of supply in what otherwise would be a very noncompetitive market.

The United States has been the principal producer of molybdenum outside the Socialist bloc countries, and has been producing about double its domestic requirements. Being supplied with molybdenum has never been a problem for the United States. On the contrary, in the early 1970s there developed a condition of excess supply. In spite of this the industry undertook to enlarge greatly its productive capacity. The most significant of these developments was the Henderson Mine of Climax Molybdenum, in Colorado; Climax Molybdenum invested a putative quarter of a billion dollars in the mine and the associated production facility. Upon completion, the mine is reported to have been able to double the existing molybdenum production of the United States, which was itself double domestic consumption.[14] In addition, a co-product copper-molybdenum mine having development costs of $165 million was placed into operation near Tucson, Arizona, in the early 1970s by the Duval Sierrita Corporation, creating greater excess capacity.

Perhaps equally important at the time were exploratory and developmental activities in Canada, the second-largest molybdenum producer outside the Socialist bloc. Several large developments were undertaken in British Columbia, which were oriented toward the Japanese market. And Chile and Peru were emerging as sources of competition for U.S. producers in foreign markets.

Given the indications of growing excess capacity in the domestic industry and commitments to further expansion, investment in facilities based on the use of low-grade ore appeared remote. The competition was expected to be so intense, from both by-product and primary producers as well as in foreign markets, that commodity analysts expected pressure to be exerted on prices out to the turn of the century.[15] Looking back some dozen years later, the expectations seem almost prophetic.[16] Demand for molybdenum appears to have peaked in 1973 at about the time that Krutilla and Fisher's original analysis of the economic prospects for mining molybdenum in the White Clouds was undertaken.[17] Demand has been generally below the 1973 level from then up to the present. One reason for the persistence of excess capacity may be the presence of co-product production, and production that occurs incidental to primary outputs that provide by-products. These sources provide from a third to almost 40 percent of molybdenum production. With demand down and by-product production governed largely by primary production of other metals, the state of the industry and market has been quite depressed, and there is a great deal of idle capacity.

Of course, the decision to evaluate the economics of the ore bodies in the White Clouds did not have the benefit of hindsight. Mineral exploration in the Castle Peak area had not been sufficient to determine precisely either the extent of the ore body or the exact molybdenum content of the ore. The ore appears as an outcropping to which only limited access could be gained without removing overburden, but stripping to any depth would result in increased overburden as a function of ore removed. The estimated molybdenum content, based on the sampling that was carried out, ran within the range of 0.17 to 0.20 percent. If this range is typical of the ore body, it suggests a marginal deposit with a high waste-to-metal ratio, which would result in a great deal of landscape disfigurement if development were to take place.

Because of the limited data available for evaluating the worth of the proposed molybdenum mining and milling venture, it was not possible in the early 1970s to develop a precise estimate that could be compared with the values for grazing and recreation that would have been precluded by the development of the mineral resources. In their analysis at that time, Krutilla and Fisher were not persuaded that the prospective mining operation would even qualify as a marginal undertaking without the occurrence of a complex set of unlikely factors, which included unjustified expectation of a long-term trend of rising relative prices for molybdenum. Given the conclusions of the commodity specialists, this appeared a dubious possibility at best until several decades into the future, if then. Because of that assessment it was then,

268

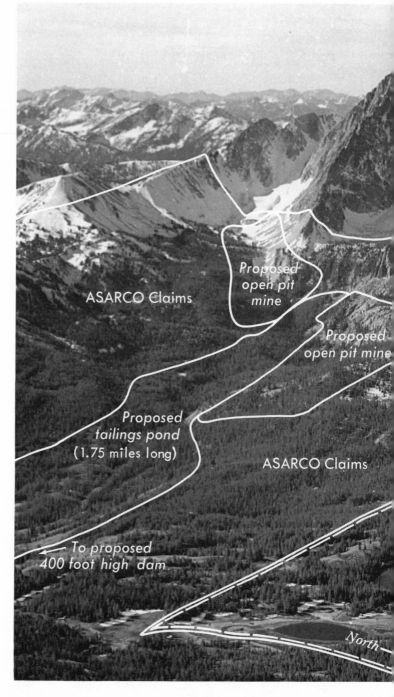

Figure 8-4. Sites of proposed molybdenum mining and milling operations i the White Clouds area. *Source:* Photo courtesy of Ernest E. Day, reprinted wi permission from Krutilla and Fisher, *Economics of Natural Environments.*

and still is for the same reason, difficult to visualize any profitable molybdenum mining being undertaken in the White Clouds for the indefinite future. This follows because if mining activity must be deferred for a lengthy period of time, the present worth of any yields that would not begin until some decades into the future would be small. Moreover, even if the present value of the stream of future yields was confidentially expected to be nonnegligible, it does not follow that *eventual* profitable mining activities should foreclose the *current-* and *intermediate-term* use of the White Clouds as a recreational resource. By saying this we do not intend to suggest that the ultimate use of the area should be expected to be molybdenum mining. Indeed, given the numerous other sources of molybdenum, by-product as well as primary, it is doubtful that the White Cloud Peaks' ore will ever qualify as an economic source for the industry. We intend only to make the point that even were the present value of the future mining operation to exceed the present values of the opportunity returns forgone, the total gain to society would be increased by using the White Clouds as a recreational resource until conditions in the molybdenum industry and the economy are such as to permit the economic mining of the White Clouds ores.

6. Summary and Conclusions

Forestland provides many examples of interrelated biophysical relations, so that the land management activities that will influence the production of one forest product or service will almost invariably alter the output, or cost, of various other forestland resources and the services they yield. In our illustrations of joint outputs we have demonstrated how an activity that may be found to be uneconomic when viewed as an independent management activity might be seen as economically sound when its value is combined with the values of other outputs and compared with costs. We have observed how, on the Black Hills National Forest, pine beetle control activities coincidentally improved recreational deer hunting at or near the site of the clearing activity. Other timber management activities, if carried out so as to jointly improve wildlife habitat, can be expected to do the same in other circumstances. We saw this in the management of subalpine forests for water-yield augmentation. Here we found that there was indeed a joint relation that could be affected by watershed treatment programs. In this case it was possible that a watershed would justify entry for water and timber where it would not for timber alone. But for a program of water augmentation where the relative value of water and timber favors water so heavily, the most economically efficient program would not produce timber with water. The value of water forgone would exceed the value of the gain from timber in this particular instance.

It is possible that in the light of other considerations—for example, aesthetics—the vegetation on watersheds would have to be manipulated to produce both timber and water. However, the prospect for mutual accommodation of mining and wilderness recreation in the White Clouds is out of the question. Here is a case where mine and mill activities affect negatively the opportunity to provide wilderness recreation. Somewhere between carefully regulated mineral exploration and full-scale mining and milling operations, the one use necessarily excludes the other totally—a point that must be realized, because intractable conflicts do arise. In some instances, as in the case of the White Mountain National Forest where most of the recreation is at sites where a pleasant forest environment is provided without any spectacular scenic vistas, the spatial rotation of timber and campsites provides opportunities to manage efficiently both timber and recreation. This is not so where the recreational attraction includes some spectacular scenery to which the mining industry seeks entry, as in the White Cloud Peaks.

Notes

1. This chapter draws heavily on the analysis of the White Cloud Peaks performed a decade or more ago in J. V. Krutilla and A. C. Fisher, *The Economics of Natural Environments: Studies in the Valuation of Commodity and Amenity Resources* (Baltimore, The Johns Hopkins University Press for Resources for the Future, 1975; rev. ed., Resources for the Future, 1985), and portions of the present chapter were previously published in that work.

2. F. Gunnell, "Wildlife Analysis of the Proposed Sawtooth National Recreation and Wilderness Areas," USDA Forest Service, Intermountain Region, 1972.

3. E. Schlatterer, "White-Cloud-Boulder-Pioneer Mountains Comprehensive Land Management Planning Study, Interim Ecological Evaluation," USDA Forest Service, February 1971.

4. Communication to the authors from David Hoefer, assistant superintendent, Sawtooth National Recreation Area, March 1987.

5. For a more complete description of the geomorphology of the White Cloud Peaks and the proposed trail system, see Krutilla and Fisher, *Economics of Natural Environments,* rev. ed., chap. 7, p. 158.

6. Communication to the authors from David Hoefer, March 1987.

7. The interested reader may consult Krutilla and Fisher, *Economics of Natural Environments,* rev. ed., chap. 7, pp. 162–170.

8. D. Pence, "Environmental Analysis Report, Big Boulder Creek C&H Allotment Development Plan," Clayton Ranger District, Challis National Forest, 1971.

9. Pence, "Environmental Analysis Report."

10. For discounting, the original work took a time horizon that would yield for the terminal year a present value of one cent per dollar of future values. See Krutilla and Fisher, *Economics of Natural Environments.*

11. The USDA Economic Research Service has computed the values per AUM for each national forest. For the Challis, an estimate of $11.20 per AUM during 1984 was provided. In the absence of a more detailed analysis, it appears that the increase in the estimate is readily accounted for by the rise in the general price level, rather than by a position change in relative value.

12. U.S. Department of the Interior Task Force, "White Cloud Mountains Idaho Status Report," 1972.

13. E. T. Sheridan, "Molybdenum," in U.S. Bureau of Mines, *Mineral Facts and Problems* (1970).

14. A. Kuklis, "Molybdenum," in U.S. Bureau of Mines, *Minerals Yearbook* vol. 1 (1970) pp. 727–736.

15. Sheridan, "Molybdenum."

16. J. W. Blossom, "Molybdenum," in U.S. Bureau of Mines, *Minerals Yearbook* vol. 1 (1982) pp. 609ff., 618.

17. Krutilla and Fisher, *Economics of Natural Environments*.

9

Funding Nonpriced Resource Services

1. Introduction

In addressing the general problem of managing the public forestlands for multiple uses, we have stressed the issue of *public* forestland because many of the resource services of the forest are public goods or include externalities for which payment, and hence revenues, are not available in exchange for supplying such services. Alternatively, some activities on the land—say, mining—are destructive of certain features of the landscape that would otherwise provide nonmarketable amenities. These amenities would not be given consideration in the private sector calculus. We can understand why private resource owners would not be motivated to provide services the costs of which cannot be recovered. Provision of the full range of resource services of which the public forestlands are capable, therefore, requires some revenue source in addition to user fees (at least for the pure public good, and most of the third-party side effects or externalities). This aspect of public land management involves public budgeting and appropriations processes.

In this chapter we address the budget and the appropriation processes from two points of view. First, we describe the processes as faithfully as we can in a summary treatment—and a summary is all that is possible because of the great complexity of the processes. Second, we review the adequacy of these processes for allocating the budget in a way that is consistent with an efficient allocation of resources in the direct management of forestlands—that is, at the forest level.

2. The Budget and Appropriations

The U.S. Forest Service is comprised of three major resource divisions. In terms of size of budget and personnel, the National Forest System is the largest. But the State and Private Forestry and the Research divisions can be distinguished for reasons other than size—they are not *directly* involved in the management of forestlands. Since our emphasis is on the National Forest System, our attention here will be confined to the budget and appropriations processes as they relate to that system. We will also confine our treatment to the fiscal year 1985, because greater or lesser changes in these processes will have occurred between the time of this writing and the time of publication. While some procedures will doubtless also have changed in that time, it is nonetheless useful to review these processes in their larger dimensions.

The budget process is virtually a continuous activity at one administrative level or another. The instructions for initiating the fiscal 1986 budget, for example, were transmitted to field units from the chief of the Forest Service by a cover letter dated September 1983. This is not to suggest that everyone involved in the process, from developing budget proposals to answering questions of the congressional appropriations committee, is continuously involved in that process. Nonetheless, the extent and volume of work involved in funding the Forest Service is technical, complex, and time-consuming.

The budget process is initiated by the letter of Budget Development Instructions; there are separate instructions for each of the three divisions. In the National Forest System, each forest supervisor is given guidelines for fixing the range of activity levels for which to prepare a budget. At least one of these will reflect an RPA (Resources Planning Act) activity level for the corresponding year.

As there are some 150 national forests, each of which is instructed to prepare a number of different budget proposals for the different levels of activity and outputs postulated, one can appreciate the voluminous data sets that are generated and the computational load that their analysis represents. For this reason a number of sets of codes have been designed to ease computer manipulation. Each forest-level activity is represented by a code. From this set of coded activities a combination can be chosen which will represent the essential inputs for any given forest-level project or program component.

The activities and their corresponding codes are available in the Forest Service's *Management Information Handbook* (MIH), kept at all national forest headquarters. The MIH codes are used to characterize program elements in the preparation of the forest budget proposals. Activities required for recreation management have an A code (for example, A01 represents recreation planning). Activities associated with wilderness, such as inventorying, have a B code designation with a numerical suffix (such as B02); the wildlife and fisheries program has a C designation—and so forth for all of the programs that constitute a multiple-use management regime. Outputs are designated with W codes.

As will be recalled, in earlier chapters we recognize that a given set of activities could, in particular cases, produce more than one kind of output (jointly supplied outputs). We also note that one essential element of multiple-use management is the recognition that increasing or decreasing the level of one output may not leave unaffected the costs of producing some other output. This being the case, it is necessary to capture the essence of these interdependencies. This is attempted by distinguishing among several concepts of costs: *direct costs, resource coordination costs,* and *support costs* (separable and joint).

Direct costs are costs of activities involved directly in the output of a given program component. The cost of timber management activities in preparing a timber sale is an example. A sale or harvest, however, is likely to have impacts beyond the timber resources area. Accordingly, there is need for coordination among resource areas in order to pick up the interdependencies among program components.

Resource coordination costs are the costs incurred outside the program area in question to take into account the interdependencies among resource areas. Among these are the costs incurred frequently to ensure that activities in a given primary component are accomplished in an environmentally acceptable manner (for example, hydrologic and soil-management activities might be required in association with the harvest of timber on fragile sites). While direct costs are expressed in MIH codes, resource coordination costs are expressed in Program Development and Budget (PD&B) codes AZ9 through GZ9 in order to be distinguished from the former and to enable such costs to be traced back to the appropriate resource area.

Support costs are separated into two classes, separable and joint. A *separable support cost* is not unlike a direct cost except that it may be incurred on behalf of a program component that is not immediately producing a forest output.[1] Examples would be the costs of fuelbreaks and fuel treatment to protect a high-valued timber area to be harvested in the future. Another example would be the cost of local or collector road construction activities used to support a timber sale. Part or all of such costs would be attributed to the timber program component. *Joint costs,* on the other hand, would be incurred on behalf of multiple objectives and cannot be attributed to any one objective alone; nor can joint costs be allocated among the several objectives in any nonarbitrary way. In the MIH codes the joint costs are chargeable to special program elements coded from J through T.

Forest-level budget proposals are thus prepared for the use of, and further development by, the service's Regional Office (RO). There they are treated as candidates from which a consolidated set of program alternatives is prepared for the Washington Office (WO). Since there are some 150 national forests in nine separate regions, the levels of detail that field-unit budget proposals provide are consolidated in various ways. It is at the Regional Office that forest-level budgets are converted into Program Development and Budget codes, which tend to aggregate the more detailed MIH entries. For example, the PD&B code AA3 aggregates the MIH codes A05 and A06, collapsing site construction and reconstruction into a single cell. On the output side, the PD&B single cell AW1 represents four MIH cells distinguishing among types of recreation use. Similar aggregation occurs in the various other programs.

At the regional level there is also another technical data management operation. Joint, or common, costs which were identified but not attributed to any particular program component when the field-units budget proposals were being developed are allocated by the RO before the region's alternatives are entered into the region's master data base. (This method purports to employ the "separable cost/remaining benefits" concept first developed for allocating common costs of storage in multiple-purpose water projects.) The RO is required to assign all costs, whether joint or separable, to the "appropriate" program component. The assignment of all costs at this stage is necessary to prepare the data for use by a software package called ADVENT, which is relied upon to analyze and manipulate the data from the field units and convert those data to the form that is used in the Washington-level budget and appropriations processes.

When the budget proposals developed by the field units arrive in Washington, the budgetary data are converted to still other sets of codes. These are the budget line-item codes used by the Washington Office in dealing with the Department of Agriculture, and subsequently with the Office of Management and Budget and the congressional appropriations committees; they are referred to as Appropriation and Function codes. Whereas the field-unit MIH codes are related to activities (kinds of work) that are performed at the field level in undertaking a project or program component, the Appropriation and Function codes differ considerably in concept and orientation, thus making the conversion of budgetary data from an MIH base to the Appropriation and Function base and back again not entirely a crisply satisfactory exercise. Moreover, the MIH codes cover close to 600 land-management activities, whereas there are only about sixty Appropriation codes and fewer than ninety Function codes. Not only is there the problem of translating data between concepts, but there is in addition the problem of preserving identity when aggregation is performed under these circumstances. More to the point, the Appropriation and Function codes will have to be disaggregated into the

large number of program component activities developed along quite different conceptual lines. We will have more to say about this in section 3 of this
chapter, after completing the description of the budget and appropriations
processes.

The budget activities at the Washington Office of the Forest Service consist
of selecting one budget level and getting the budget proposal approved by
the Office of the Assistant Secretary of Agriculture for Natural Resources
and Conservation. The Department of Agriculture, usually the protector of
the incumbent administration's budget philosophy, may have considerable
influence over the proposed Forest Service budget, primarily by emphasizing
certain programs and reducing the size of other programs and thereby affecting the mix of projected field-unit activities and outputs. In due course, the
department's estimate of a budget that is justified by the legislative mandate
of the Forest Service and the policy perspective of the administration is
transmitted to OMB, where, it may be presumed, the estimate receives
searching scrutiny as a matter of principle. It is perhaps worth noting that as
the budget activities migrate upward from the field to Washington, the
relationship of the professionals to the budget process changes somewhat. In
the field there is a much closer and more current association with the actual
management activities on the land, whereas at the Washington level the
professional expertise and the skills and judgment used in analyzing alternative budgets are applied at a macro level more removed from the activities
that take place on, or near, the ground.[2]

After the Forest Service has satisfied the Department of Agriculture and
the OMB staff, what remains of the original submission becomes the President's Budget for the Forest Service. Field-unit personnel levels and resource
management programs must now be reconciled with the President's Budget
preparatory to submission of the President's Budget for the Forest Service to
the Congress.

At this juncture a second formal document concerning the budget for the
fiscal year in question goes from the chief of the Forest Service to the
regional offices; it presents the initial (tentative) allocation of the President's
Budget among the several regions. This allocation is initial because none can
be final pending action by the appropriations committees of the Congress; it
is tentative because there is room for negotiation and some reallocation even
with the initial budget. It is possible, for example, that the initial allocation
by the Washington Office has not adequately covered some region's imperative to pursue a given management activity (such as attending to a fuel
buildup) without interruption. Instances of this sort will lead to some reallocation of the initial budget among regions and forests.

Between the time proposals go up to the region from the forest and the
time the Washington Office sends down initial budget allocations, the original MIH codes have been translated into PD&B aggregate codes, and these

in turn into line items in the congressional budget. It is in the related Appropriation and Function (budget line item) codes (and since 1984 in the MIH codes) that the initial allocations are made to each region. The relationship between the codes and what they represent is presented in table 9-1. The columns in that table represent program components, which reflect the biophysical production relations on the forest; the rows represent various functions and line items—fire control and construction, for example—for which appropriations are made (the matrix is explained below).

The MIH and PD&B codes relate to program components which have a direct relation to the production of forest outputs. The budget line items, expressed in terms of Appropriation and Function codes, do not relate directly to the production relations on forest lands. These latter codes are organized around functions reflecting, on the whole, management specialties. Function code 39, for example, refers to Geological Resource Services, and function code 91 represents Soil and Water Administration.

The OMB and Congress work almost entirely with the budget line items. As a result, the relationship between program components and the other activities that support them is not perceived with great acuity by the OMB and the Congress when the proposed budget is acted on; we treat this issue in section 3 of this chapter. The Forest Service, particularly the Washington Office, must deal with both the line items (in table 9-1, the rows) and the program components (the columns). There are plans to introduce the program component information to Congress and the OMB gradually, perhaps one component at a time, during the annual Forest Service budget presentation.

Unlike table 9-1, the cells that comprise a Forest Service budget matrix are filled with dollar amounts and targets for the interlevel transactions. Thus the various line items not principally associated with a given program component may contain funds for support activities essential to the principal program. (In the cells in table 9-1, P indicates the principal program component to which a budget line item refers, and S the various line items which may support a given primary component. Such support funds, however, do not appear in the principal program component's budget line, or row). From the perspective of Congress, the appropriations intended to fund support activities for a given program component are not visible—that is, these funds do not appear with the primary program component they are intended to support. Consequently, the funding of support activities for a program component is likely to be for either administration or some activity that does not have a current forest-related output or service. We return to this matter after completing our description of the process.

In addition to being informed of the initial allocations in the President's Budget, each Regional Office receives instructions in a document entitled "Fiscal Year Management Attainment Report, Definitions and Targets." This at once establishes forest output targets to be achieved for the level of

Table 9–1. Relationship between Forest Service Functions, Budget Line Items, and Program Components

Organization by function (by skills)	Budget line item (by function)	Land mgmt. planning (Plans)	Fire mgmt. (Prtcn)	Fire mgmt. (Fuels)	Roads (Constr)	Roads (O & m)	Lands (RE mgmt)	Minerals (Opr & mv)	Minerals (Imprv)	Range (Opr & mv)	Range (Imprv)	Recreation (Opr & mv)	Recreation (Imprv)	Timber (Rfrst)	Timber (Tsi)	Timber (Sales)	Soil & water (Opr & mv)	Soil & water (Imprv)	Wildlife & fish (Opr & mv)	Wildlife & fish (Imprv)
Land mgmt. planning	(Multifunction)	P																		
Aviation and fire mgmt.	(Fire $)	S	P	P	S	S		S	S	S	S	S	S	S	S	S	S	S	S	S
Engineering	(Forest, roads, trails $)	S	S	S	P	P	S	S	S	S	S	S	S	S	S	S	S	S	S	S
Lands	(Real estate management)	S	S	S	S	S	P	S	S	S	S	S	S	S	S	S	S	S	S	S
Minerals and geology	(Mineral $)	S	S	S	S	S	S	P	P	S	S	S	S	S	S	S	S	S	S	S
Range management	(Range $)	S	S	S	S	S	S	S	S	P	P	S	S	S	S	S	S	S	S	S
Recreation management	(Recreation $)	S	S	S	S	S	S	S	S	S	S	P	P	S	S	S	S	S	S	S
Timber management	(Timber $)	S	S	S	S	S	S	S	S	S	S	S	S	P	P	P	S	S	S	S
Watershed and air management	(Watershed and soil $)	S	S	S	S	S		S	S	S	S	S	S	S	S	S	P	P	S	S
Wildlife and fisheries	(Wildlife and fish $)	S	S	S	S	S	S	S	S	S	S	S	S	S	S	S	S	S	P	P

Note: **P** = Primary; S = Support (interdisciplinary).

[a] Key to components: Prtcn = protection; O & m = operation and maintenance; Constr = construction; RE mgmt = real estate management; Imprv = improvements; Rfrst = reforestation; Tsi = timber stand improvements.

funding in the President's Budget and explicitly juxtaposes the various codes intended to define the targets in terms of the MIH and PD&B code systems.

Upon receipt of the initial allocations and targets from the Washington Office, each of the regional foresters prepares a document offering "Financial Planning Advice–Tentative" for each of the forest supervisors in his region. Among other things in recent years, this document has implemented a policy of reducing overall personnel levels, on the one hand, and, on the other, increasing the relative and absolute numbers of women and ethnic minority employees on the forests.

At the time the budget proposals are being prepared by the forests, the budget is built up of estimates of funds needed to undertake a set of specific projects. The directions that the forests receive upon allocation of the President's Budget are expressed in terms of output targets, personnel levels (and composition), and funds by budget line item. Recently they have been expressed in MIH codes as well. Changes in the content of budget data returned to the forest may pose difficulties; we discuss these in section 3.

At the Washington level the Forest Service budget, as constrained by the OMB, will go to Congress to be acted upon by the appropriations committees. Interestingly, it is not considered along with other appropriation actions addressing the Department of Agriculture's budgetary requirements. As a result of a historical anomaly, the Forest Service budget is taken up with the Bureau of Land Management and other Interior Department budgets. Throughout the appropriations deliberations the committee members' questions are answered with personal testimony by Forest Service staff and supplementary written replies, as well as with explanatory notes elaborating one or another line item. In due course the committee signs off on the appropriations. In most years, if not all, the congressional power of the purse is exercised either upward or downward on a number of the items on which it does not agree with the executive branch.

When the Congress finishes with the Forest Service appropriations, the Washington Office brings the tentative budget into line with whatever changes are required after the Congress makes its final decisions. The final (adjusted) allocations are sent to the regional offices, which in turn make their allocative choices for the given fiscal year. These choices will determine the work to be done on the national forests. The final Forest Service budgetary data are put through essentially that process described in connection with the President's Budget.

3. Some Critical Observations

Since we have been dealing in this book with nonmarketable and unpriced public forestland outputs, we have naturally looked to the funding process that provides for the supply of nonmarket and unpriced goods and services. The budget and appropriations processes do not necessarily lead to economically efficient results. We have described the mechanics of obtaining budget proposals at the forest level, where program planning and project analyses serve as the basis for budget proposals. But as we have seen, when the budget proposals leave the forest, various adjustments to these proposals are made at higher levels in the hierarchy of budget formulation.

At the regional level, the staff, having information on the various forests in the region, is in a position to work with pooled forest-level data to select the best set of regional program components for any given budget level.[3] Part of the work done on the budget by the Regional Office staff is directed to selecting a strong program. But the regional staff also undertakes an allocation of joint costs and in other ways prepares the regional data to conform to the requirements of the ADVENT software program. ADVENT converts budgetary information coming from field units to a data and code set that is congenial for budget purposes at the Washington level. While the data from the field underlie the data base on which ADVENT operates, and thus the data are essentially preserved, the form in which the data survive is very different. Moreover, when the activities in Washington—whether in the Department of Agriculture, during negotiations with OMB to establish the President's Budget, or in the appropriations committees—alter the budgeted amounts, changes will have been made in both the budgeted *amounts* and the *relationships among them.*

After the budget has been determined by the actual appropriations of the Congress, the appropriations are then distributed to the various units in the National Forest System. Allocations now differ from the data prepared by the field units not only in amount, but also in form. That these changes occur would be a matter of only passing interest if the outcome that is sought would be realized in this way. However, changes in the amounts appropriated from the amounts requested, when accompanied by a change in the form in which the allocations are made, are likely to lead to unintended consequences.

We can illustrate this by a hypothetical example. Assume a region X is planning its program in terms of *program components.* Part of the plan might look like the plan in table 9-2. Since the OMB and the Congress work principally with budget line items, plans are reviewed for the most part on a row-by-row basis. Region X's budget line-item summary is $20 for timber and $20 for fire. The region's program plans, however, call for $23 for timber and $17 for fire, because $3 of the budget line-item funds for fire are programmed in the timber program to coordinate fire management considerations with the management and harvesting of timber.

Let us assume now that congressional appropriations for the Forest Service provide for full timber line-item requests but reduce the fire line item by 25

Table 9–2. Program Planning, Region *X*

	Program components		Total
	Timber	Fire	Total
Congressional budget line			
Timber line item	$20	$ 0	$20
Fire line item	3	17	20
Total	$23	$17	$40[a]

[a]Line item and program components.

percent. Because of the role of fire management in the timber management program, the cut in fire management appropriations may inadvertently affect the timber program as well. For example, if after receiving the Forest Service appropriations the Washington Office were to treat the fire protection funds as fixed rigidly to the fire protection component (see option 1, table 9–3), fire coordination funds would be excluded from the timber budget and such exclusion would adversely affect the achievement of program output targets. On the other hand, a judgment might be made in favor of retaining the fire coordination activities in the timber program. In that event the allocation would look like option 2 in the fire line item in table 9–3. But in this case there would be a reduction of 29 percent in the fire budget, a cut perhaps greater than Congress intended. Another option ostensibly within the spirit of congressional intent would be to reduce the amount originally proposed for fire in each program component by 25 percent, as in option 3. Again this would result in an allocation to the timber program that would be less than originally proposed and perhaps less than Congress intended.

Table 9–3. Allocation for Region *X*

	Program components		Total
	Timber	Fire	Total
Congressional budget line			
Timber line item	$20.00	$ 0.00	$20.00
Fire line item			
Option 1	0.00	15.00	15.00
Option 2	3.00	12.00	15.00
Option 3	2.25	12.75	15.00
Totals			
1	$20.00	$15.00	$35.00
2	23.00	12.00	35.00
3	22.25	12.75	35.00

These illustrations involve only two programs and two budget line items. They are oversimplifications because in reality there are some ten principal program components and roughly an equal number of principal budget line items, each with numerous subdivisions. Tracking the effects of adjustments made in such an interdependent system of programs is a monumental task, and it is an open question how well it can be done. The Forest Service's ADVENT software package was developed to aid that task, but a great deal of supplementary knowledge and human judgment is required to carry out the work.

This is a problem of which the Forest Service has taken note. A task force has been formed to undertake a coding burden study. Preliminary indications are that among other things the study team will recommend a drastic reduction in the number of codes and a sharper definition of the reduced number. But as of the moment there is no indication that the task force will address the problems that arise because forest-level budget proposals are developed in one system of coding and the budget is funded on a conceptually different basis.

If budget reform were to be carried out to the extent that the *system of codes* as well as individual codes had to be justified, one can imagine the direction that could be taken. It is apparent that the MIH coding system addresses the activities that relate to actual forest management. It perhaps ought to be suggested, however, that the MIH codes currently are more loosely defined than they ought to be, and that by use of more discriminating definitions the level of detail, and hence the number of MIH codes, would likely increase. Under the circumstances it is fair to ask if this would not in fact be desirable. The activity codes relate directly to the work that is to be performed on program components at the basic production units (forests) in the system. Proposed projects, in their early stages, have to be cast in terms of their costs and the value of their outputs in order to attain candidate status for the budget proposal. In a real sense they reflect the biophysical production relations on the forest that must be related to the demand—largely local—for most of the outputs of the forest.

From the standpoint of production, or economic efficiency, there does not appear to be an equivalent rationale for the budget line-item format or codes. As the traditional overseer of the funding of government activities, however, Congress may consider them to be time-tested, practical means by which to exercise funding control over agencies of the executive branch. From this point of view, whatever changes are made in the President's Budget might be regarded simply as additional constraints on the agency, which must be accommodated in the field. It is possible that these changes would permit a set of program components that would still maximize the net value of forest outputs, subject to such constraints. However, if there are joint outputs or other interdependencies involving two or more program components, a directive in the appropriations bill to meet specific outputs targets with personnel and funding ceilings (the latter stipulated by budget line item) may simply not be possible. This follows because there is no way of knowing by examining the Appropriation and Function codes whether the assumptions underlying the production targets and the personnel and funding levels are consistent with the production relations on the forests.

One can question whether the consequences of using the budget line-item format and its associated Appropriation and Function codes are compatible with the directives in the Renewable Resources Planning (RPA) and National Forest Management (NFMA) acts. It should be recalled that under NFMA the forests undertake a planning exercise which ultimately results in an approved five-year forest plan for each forest. Budgeting on the forest level, it is to be assumed, reflects the activities and projects implied by the approved plan. If the result of the budget and appropriations processes is to change not only the amounts requested for management but also the relations among program components, these processes unwittingly introduce a distorting factor in forest management that is difficult to reconcile with the planning and management acts.[4]

If in congressional deliberations over agency appropriations the budget line item has become such an ingrained method of working that consideration of an alternative budget form or structure is out of the question, it seems that the function of translating program field codes into their Washington counterparts and back to field codes should be transferred to the regional offices. The Regional Office is the highest administrative level in the Forest Service that works with and retains the forest-level data. If such a transfer were to be made, when the funding by budget line item is made available at the regional level the conversion of the Appropriation and Function codes into MIH codes could be done with the basic forest-level data on hand. These data, because they are cast in terms of program alternatives, are likely to be consistent with the basic production relations on the land. This might be one way of reducing the undesirable consequences of working with budget line-item codes.

4. A Digression on Capital and Cost Allocation

As provisions of the National Forest Management Act have been implemented, more exacting analytic and accounting requirements have come in their wake. During the appropriations hearings for fiscal year 1985, considerable pressure was applied to the Forest Service by the chairman of the Subcommittee on Interior and Related Agencies of the House Appropriations Committee to develop an accounting system that would permit closer accounting of expenditures and receipts for its timber management program. Until that time the Forest Service reasoned that the activities engaged in on behalf of timber also provided benefits for other uses.[5] Moreover, some activities that the Forest Service refrained from undertaking in order to accommodate its environmental and amenity objectives increased the cost per unit of timber volume harvested. This cost cannot properly be compared with harvesting costs on private industrial forests.[6]

This is admittedly a very complicated problem. There are two important issues that must be confronted. First, the budget requirements and appropriations for an agency engaging in the production of goods and services are different from the budget requirements of those government agencies that perform largely transfer or regulatory functions (most agencies fall into this category). The budgets for most other agencies mainly involve personnel services and transfer payments. And while budgets and appropriations looking not much beyond the current year may be adequate for the more typical government agency, such short-term practices are ill-suited for land and water resource management agencies if some sort of economic accounting is expected from the agency. An explicit capital budget and corresponding accounting practices are needed for good management and effective oversight. We will justify these statements after introducing the second issue.

The hallmark of natural resources utilization is the high degree of interdependency that characterizes their production. One aspect of such interdependency is the condition that when one product or output is being produced by the forest it almost invariably affects the cost, quality, or character of another product or service. Thus managing one output—say, timber—will affect other resource commodities or services whether positively or negatively, or even preemptively in the extreme case. Accordingly, when a given bundle of inputs will affect more than one output or resource service, we have a case of joint production. From an accounting standpoint, joint production is not a problem when the objective is simply to evaluate the worth of a given set of inputs—that is, whether the gains from their employment exceed their costs. It does become a problem if there is some reason other than economic evaluation for attempting to determine what the cost of each output of a jointly produced combination is when considered individually. We will take up the question of joint cost allocation after reviewing the need for recognition of capital outlays.

Capital Outlays

At present the Forest Service does not have a capital budget. As far as accounting is concerned, the service's capital expenditures are treated as though they were expenditures on current account. There are good reasons for separating expenditures by current and capital outlays. Perhaps the best is that separation would permit a more rational accounting to the Congress and the public. Unless an accounting system is implemented that would treat capital and current expenditures differently, serious distortion will occur in the results when imputing annual costs to a given project or program. For example, the investment in a road which will yield services over a substantial period of time should be depreciated over its useful life rather than be charged as an operating expense during the current year. Similarly, the cost of a plantation, the yield from which will not occur for many years, should be carried in a capital account.

But the issue of treating capital differently from operations and maintenance expenditures cuts deeper than the sorting out of expenditures after the event to determine relevant costs. Capital is sufficiently different from current expenditures that it also ought to be treated differently in the budgeting and appropriations processes. Just as investments, by definition, are those outlays that are not recoverable in a single year, it may also happen that a capital outlay may have to extend over more than one year if a given project takes more than a year to complete. Completion of the undertaking would thus require the Congress to make a commitment to provide appropriations extending beyond the year in which the project is initiated. Moreover, if the project should be a plantation that would be under management, the total stream of episodic timber management expenditures—that is, costs of thinnings, fuels management, and other activities until harvest—are properly chargeable to capital. If future expenditures are tied to present investment decisions, the Congress must treat investment expenditures knowledgeably and recognize that there are strong time-period interdependencies that must be acknowledged if the output targets on which Congress has signed off as part of the RPA process are to be realized.

Joint Cost Allocation

The mere fact that a given bundle of inputs used in timber management has produced two or more outputs is no reason to feel that the total cost of the bundle must be parceled out to each output to determine if the forest practice in question is economic. The problem involves two issues. First is the issue of whether or not the separable cost of each output is covered by its value. Second, it must be determined whether overall costs are covered.

To determine the separable cost of any output, one needs to know the aggregate cost of the total package producible by the bundle of inputs. Then one individual output is dropped from the package of uses, and the corresponding reduction in the aggregate cost is calculated. The amount of the reduction is the separable cost of that element of the package of outputs. These separable costs can then be used to test whether the values of the individual elements in the package of outputs are at least equal to their separable costs. If any are not, the overall program can be improved by dropping these outputs from the package. Then the sum of separable costs of the surviving outputs is subtracted from the aggregate cost of this set of outputs to get their joint cost. Finally, to ensure that that cost is covered, the net value of the surviving outputs is compared with the joint cost. To establish that the use of the bundle of resources is economically justified, the total net value of the surviving outputs must at least equal the joint cost. Note that the economic evaluation can be achieved without at any time allocating the joint cost.

This general approach has traditionally been followed in financial reporting in the water resources area. For such financial reporting one additional step is taken which does indeed end in allocating joint costs. Since some of the outputs in the water resources field are priced and expected to be self-supporting, it is important to learn how much of the joint cost should be allocated to the paying partner in order to determine what revenue targets the self-liquidating use must achieve to meet its legislated obligation.

Until recently there has been no similar requirement for the Forest Service. During the 1985 appropriations hearings, however, Congressman Sidney Yates, chairman of the Subcommittee on Interior and Related Agencies of the House Appropriations Committee, directed the Forest Service to develop a system of accounts that would permit the Congress to understand how much it costs the Forest Service to produce timber in any given case after netting out the costs associated with the production of other multiple-use benefits. On the surface this looks like the same question as that for which the water resource agencies had to find an answer, and for which the separable cost/remaining benefits method was found acceptable by the Congress. There is a hitch, however, when attempts are made to apply the method to multiple-use forestry. This has to do with the added factor of growth over time in forests, and its effect on site-related benefits and costs.

When a multiple-purpose project is put in place in a river development scheme, it will remain in place unchanged unless some modification is introduced by conscious design.[7] This is not the case with forest management projects where the outputs are a function of vegetation manipulation. Let us say, for example, that a clearing is opened by the harvest of a site and that it is to serve also a wildlife management objective—that is, it is to provide forage. Regeneration may begin almost immediately, or only after some time, depending on local conditions. After some time the site, by virtue of the growth of its standing stock to a higher age class, must be replaced by another site just harvested in order to continue meeting forage needs. Indeed, multiple-use forest management requires, among other things, diversity in age classes.[8] And a given site will tend to fill different roles at different times, depending on the age of the stand that it supports and the mix of age classes on surrounding sites. Aside from the value of the timber that is sold from a site, the value of the site also depends on the value of the forage at one stage, of the cover at another, and of other benefits over an appropriately long period of time. While the vegetation at a given site changes over time, under multiple-use management the vegetation at other sites is manipulated as well to provide complementary services for timber, wildlife, and other multiple-use benefits. In short, there are interdependencies among sites as well as between time periods at a given site.

The two types of interdependencies combine to pose problems in trying to apply the separable cost/remaining benefits method to a given project, such as a timber harvest or any single activity that takes place at one site at one moment in time. The solution lies in finding the present value of the between-harvest services that the site will provide over its entire life. However, since the services the site renders are dependent also on the flow of complementary services from interacting sites, the yield from the site in question cannot be separated from the influence of the interacting sites. In evaluations in the water resources field, where there are interdependent sites (large storage at an upstream site and head at downstream sites), the output from the different sites is combined and interdependence is thus taken into account. Something similar must be done in the forest management case. Some unit for accounting should be defined that would encompass all of the interacting sites, and minimize the spillovers between accounting units. A third- or fourth-order drainage might be needed to contain the interactions. What would be accounted for then would not be a single activity at a given moment in time, but rather the discounted value of the time streams of several forest outputs generated on the management or accounting unit. The plan for the entire unit could then be compared with alternate plans to identify the one having the largest present net worth.

In suggesting how to deal with the hitch that the dynamics of forest growth introduces, we do not intend to imply that implementation of the proposed accounting unit would not be an extraordinarily complex—perhaps nearly infeasible—task. In principle, the current FORPLAN model is attempting to apply such a dynamic interactive accounting model. But as we shall see in chapter 10, application of the model in the forest planning environment, while promising as an initial effort, has so far fallen considerably short of serving as an adequate economic accounting system for the multiple-use management of public forestland.

5. Summary and Conclusions

In this chapter we have addressed the issues associated with the funding of nonpriced resource services. Some of these are public goods which cannot be provided through normal market processes. Others may be technically suited to being supplied by the market, but are not handled as market goods for reasons of tradition, law, or administrative policy. We have also briefly reviewed the public funding of the National Forest System in order to better understand the processes by which those funds are provided, and to address the implications of these processes for economically sound resource management decisions. It is clear that the processes are very complex, if not unnecessarily cumbersome.

Currently a Forest Service task force is addressing this problem, and preliminary indications are that a substantial reduction in the number of MIH codes is in progress, which is expected to aid in simplifying budget preparation. There is no reason to anticipate that the task force will terminate its activities with only that recommendation, but we cannot anticipate what additional recommendations may be made. Even so, it seems that budget preparation problems arise not so much from the number of MIH codes (hence, number of management activities that can be defined), but from the differences between the set of pragmatic codes used for program proposal evaluation and the set of budget line items that has emerged historically. The budget line items may be suitable for some agencies of the federal government; they appear to be misapplied in the case of a public agency that has been directed to undertake programs requiring capital expenditures and related economic operations.

That the appropriations committees of the Congress recognize that the operations of the National Forest System are somehow different from those of most government agencies is reflected in its current interest in having the Forest Service develop an accounting system that would indeed result in the different treatment of expenditures on capital and expenditures on current account. Distinguishing between those expenditures that result in an investment that will yield productive services over an extended useful life and those that are properly treated as current expenditures would be a significant step in providing relevant information with which to review and evaluate various Forest Service programs.

But even with an accounting system that distinguishes between capital and current expenditures there will be the issue of how to handle common costs. One can argue that for estimating the economic benefits and costs of projects that make up a program component, the evaluation should be made on some spatially integrated accounting unit in which the site interactions could be reasonably contained. We have seen that to perform a proper economic evaluation common costs need not be allocated among the several outputs. In the next chapter we undertake a prototype evaluation exercise with a capital accounting model that responds to the accounting problem.

We would like to conclude that the matter of evaluation would be properly disposed of if only the theoretically acceptable evaluation procedure were used. Unfortunately, we do not have good estimates for some of the hard-to-measure benefits, such as the economic benefit of arresting soil erosion. That being the case, some of the nonpriced resource services provided for under multiple-use legislation are entered as constraints on an objective function that would otherwise maximize the benefit from just the priced commodities and resource services. It is to be hoped that in time, with perseverance and ingenuity, methods will be discovered for estimating the value of such hard-to-measure benefits. There are challenging problems here for both earth and life scientists as well as economists.

Notes

1. In conventional accounting, where a distinction is made between current and capital expenditures, these costs would be carried in a capital account.

2. While the above characterization is substantially correct, it should be said that the Forest Service has a policy of rotating personnel between the field and the Washington Office. We do not suggest that personnel engaged primarily in budgeting in the Washington Office have not had extensive field experience.

3. While it is true that the Regional Office is in a position to select the better projects, it does not follow that the budget items it selects are necessarily the most economic. The provision in the National Forest Management Act requiring public participation in the forest's planning process permits an influence which, under the circumstances, could hardly be described as economic. In what is essentially a political theater, expressions of preferences for resource uses are not subject to income constraints as they are in market transactions. See J. A. Haigh and J. V. Krutilla, "Clarifying Policy Directives: The Case of National Forest Management," *Policy Analysis* vol. 6, no. 4 (Autumn 1980) pp. 409–439.

4. In actual practice the requirement to meet production targets as evolved out of the appropriations process leads to an emphasis on physical production goals rather than economically efficient land management. Evidence of this is cited in the appendix to the letter of the director of the General Accounting Office to Chairman Sidney Yates of the House Subcommittee on Interior and Related Agencies, dated April 4, 1986: "Forest Service field personnel advised the subcommittee staff that management decisions are based primarily on attaining the yearly timber volume goals and that their success as managers is based on meeting these targets, not on any type of assessment involving a cost-benefit comparison" (appendix I, p. 7). A Forest Service task force report, "Analysis of Costs and Revenues in the Timber Program of Four National Forests" (n.d.), echoes this observation, saying that employee incentives tend to lead away from, rather than toward, economically efficient management behavior. Finally, at field-level interforest communications, on an electronic bulletin board designated as the "rumors conference," we find on December 17, 1985, message 566 addressed to operations research analysts (ORAs) from Reuben Weisz, R3 LMP (Region 3, Land Management Planning), referring to the " . . . lack of correlation between the budget allocation and the budget proposals." In the message there is also reference to the problem that exists "between budget allocation and the implementation schedule for the Forest Plan."

5. It bears noting that there must be a demand for these joint outputs. Increased browse for winter range has value but is of dubious value on the summer range. To

take another example, if there is adequate or excess capacity for dispersed recreation accessible from present roads, additional road access would not necessarily add anything to net benefit.

6. See R. E. Benson and M. J. Niccolucci, "Cost of Managing Nontimber Resources when Harvesting Timber in the Northern Rockies," Research paper INT-351, USDA Forest Service, Intermountain Research Station (September 1985).

7. Perhaps a qualification should be noted. It is true that a given hydroelectric project will play a different role in a power system when the hydroelectric potential and development can supply all of the power for the system than when it operates as a complement in an essentially thermal power system. Accordingly, if the demand over time grows to exceed the hydroelectric potential, in response to that change there will be changes in the operating policy and likely engineering modifications to the structures. While in developing countries there may still be hydroelectric potential to be developed which could meet the host region's power by itself, this is a situation of the past in the United States. In any event, such river projects do not entail the kind of problem that is presented in the forest by biological growth of the stock and its relation to habitat, forest aesthetics, and other amenity considerations.

8. S. G. Boyce, "Management of Eastern Hardwood Forests for Multiple Benefits (Dynast-MB)," Research paper SE-168, USDA Forest Service (July 1977).

10

Below-Cost Timber Sales and Forest Planning

1. Introduction

In the late 1970s Tom Barlow and his associates at the Natural Resources Defense Council (NRDC) drew attention to the practice of selling timber from the national forests at a price below the costs of production.[1] The NRDC argued that, among other things, the below-cost sales depressed the prospects for profitable forest ownership and timber management among private forestland owners. These claims regarding the prevalence of below-cost sales were later reinforced by studies of the timber sales program from the General Accounting Office (GAO) and the Congressional Research Service (CRS).[2] Recent Forest Service annual reports also confirm that for many national forests—the general exceptions being the highly productive forests of the Pacific Coast and the Southeast—the value of timber products sold each year is often less than the cost of timber production in the same year.[3]

Although the NRDC pressed its concern about below-cost sales continuously,[4] it was not until the spring of 1984, during hearings of the House Appropriations Committee's Subcommittee on Interior and Related Agencies, that Congress took effective note of the issue.[5] At hearings of the subcommittee, Chairman Sidney Yates noted that the Forest Service was projecting large increases in the timber harvests for all national forests; he requested that the service abstain from increasing harvests on those forests which showed sales below cost for three of the past five years. While Congressman Yates subsequently placed this injunction in abeyance, the issue has remained alive because of a provision in the appropriations bill of fiscal year 1985 requiring the Forest Service to adopt an accounting system that would, in spite of the multiple-use objective of many land management practices, permit Congress to learn the costs of producing timber alone and would exclude the costs of the other jointly provided forest outputs. The request by Congressman Yates seems to have been motivated by a desire to gain some perspective on the economic rationality of the Forest Service timber sale program so as to address the question of appropriate levels of timber funding. If so, framing the below-cost sales issue in terms of the need for a financial accounting system was probably an unfortunate event.

As might have been predicted, the request for an accounting system that could provide a useful basis for evaluating the economics of timber sales has led to a great deal of pointless effort. Among other things, the request has embroiled the Forest Service and the General Accounting Office in a costly and perhaps hopeless search for an appropriate way of tracking and apportioning the costs of capital expenditures, shared facilities, and those management activities which jointly affect the production of timber and other multiple-use services. The debate over allocation procedures has inevitably been highly politicized.

The initial NRDC below-cost sale study was based on a cash-flow accounting approach, which compared timber revenues to timber management expenditures over each year of a five-year period. Much of the ensuing debate, including that resulting from congressional concern, has focused on the extent to which particular current expenditures (such as those for roads and general administration) should be considered components of the current costs of timber production. Environmental and conservation groups are no doubt hoping that on the basis of existing cash-flow analyses, timber harvesting can be restricted, and that restriction may result in the protection of much of the unroaded land on the national forests. The forest products industry is pressing to have many of the budgeted timber costs allocated to other multiple-use services.[6]

It should be recognized that any single-product, cash-flow accounting scheme can give only limited information as to the economics of a timber sale program. Current timber management decisions cannot be divorced from the broader program of multiple-use and forest asset management. While cash-flow accounting does strongly suggest that there is an ongoing problem with uneconomic timber sales, the fact remains that the economics of the timber program cannot be fully evaluated without determining the effect of timber management activity on the flow of value from nearby or related forest sites, or without considering the effect of current timber management actions on the future flow of net value from all the multiple-use services of the forest.

A product-by-product, cash-flow accounting scheme has two inherent limitations within the multiple-use forestry setting. There is first the difficulty of reflecting the dynamic nature of forest management while looking only at current receipts and expenditures. The effects of current management activities are not fully, if at all, reflected in current outputs and revenues. The effects of timber management activity, which may be either positive or negative, can continue for many years, or indeed may not even be realized until long after the activity has been implemented. Second, there is the difficulty of assigning costs among jointly provided products. Although many activities are taken to be primarily for timber management purposes, even these may jointly influence the provision of other services. Only a fraction of the total production costs—the separable costs—can ever be unambiguously allocated to individual outputs. The separable cost of timber is the savings in cost that would result from eliminating the current timber harvest while maintaining the planned production of other services. The assignment of separable costs to outputs does not generally result in a full allocation of total costs. The problem of allocating that remaining component of cost, referred to as the common cost, has long been known to have no single conceptually correct solution.[7]

It is significant, however, that there is no need to allocate costs in undertaking an economic evaluation of production alternatives. Indeed, the use of allocated costs proves to be a most unreliable means of determining whether any one output is economic and appropriately included in production, or uneconomic and better excluded from production. Rather, benefit-cost analysis provides the appropriate means for selecting a production plan. Otto Eckstein notes in his discussion of the allocation of joint costs:

It is important to stress that cost allocation and benefit-cost analysis have very little in common. Allocation is an essential part of the financial analysis of projects, while benefit-cost analysis is the main component of the economic analysis. Since, by their very nature, cost allocations must be arbitrary, the introduction of their results into the benefit-cost analysis only serves to obscure the essentials of the economics of a project.[8]

Quite simply, a timber sale is economically justified if the expected net present value of the overall multiple-use benefits from the national forest are higher with the planned sale than they would be for any production plan without the sale. While it may seem appealing to resolve the complexity of multi-output production by allocating costs, acting as if there were a set of independent single-output production processes, the attempt to use these accounting costs to refine plans can result in gross misallocation of production effort, and hence excessive social costs.

The Forest Service, in addressing the below-cost sales issue, has suggested that timber programs are best analyzed within the full framework of the forestland management plans, rather than at the level of the individual timber sale. The linear programs used for developing land management plans provide a mechanism for evaluating alternative timber programs within the context of the wider forest, reflecting the full impact of timber sales on the current and future flow of multiple-use benefits and costs. In this, the Forest

Service is backed by a Society of American Foresters' task force report on below-cost sales. The task force concludes that separate analysis of a single output violates the logic of the multiple-use planning process.[9] The Forest Service has on occasion gone further, arguing, in effect, that any timber sales required to meet the selected forest plan are automatically justified, whether or not individual sales are economic, as long as the planning process has followed legal guidelines.[10] Since the output targets of forest management plans are not selected on economic criteria alone, some losses on timber sales might well be anticipated and consistent with the plan. Not surprisingly, this position has not entirely satisfied Congressman Yates, chairman of the appropriations subcommittee.

We also think that the linear programs used for the land management planning process should, if well-constructed, provide an appropriate framework for an evaluation of the economics of timber sale programs. In this chapter we provide some discussion of the linear programs used for multiple-use planning of the national forests (FORPLAN),[11] and we determine the extent to which forest planning models can be used to address economic and policy analysis questions of the kind raised by below-cost sales. In particular, we assess the possibility of computing the separable costs of individual forest outputs.

In section 2 of this chapter we explore some of the complexities in the evaluation of timber sales that may not be apparent from casual observation. The below-cost sale problem is formulated in an economic multiple-use framework. An essential element here will be the development of a capital accounting procedure for the definition of costs that allows us to reflect both the immediate expenses attributable to timber management and the opportunity costs associated with the impact of a timber program on the flow of future service from the forest. This definition of full current economic costs provides us with a measure which can be appropriately compared to the immediate revenues from timber sales. In section 3 procedures for the measurement of cost are developed and the details of computing separable cost using the FORPLAN model are provided. FORPLAN models, which can be variously structured in practice, should, if adequate for planning purposes, also provide a basis for economic benefit-cost analysis of forest management choices.

In section 4 we look at two applications of our cost accounting procedures. Using two FORPLAN models—those of the Shoshone and the Mt. Baker–Snoqualmie national forests—we demonstrate the practicality of our procedures for the computation of separable costs, and provide illustrations of apparent below-cost sales which prove to be economically justifiable because of their beneficial effects over the longer term. It should be emphasized that we do not assess the actual forest plans, which were not complete at the time of our analysis. In section 5 we comment on the usefulness of the current planning effort as a basis for the analysis of timber sales, recommending changes that would give the effort greater value as a source of information on the economics of multiple-use management practices.

It would be natural to ask why, since we had access to the FORPLAN planning models, we did not simply solve for the "best" overall production plan and compare this to the plan selected by the forests. The primary reason is that we wanted to address the below-cost sales issue in much the same manner as it had been addressed in previous studies and in a form that would reasonably address the accounting concerns of the congressional appropriations subcommittees. Our analysis is best viewed as a natural extension of existing work on below-cost timber sales, such as the early NRDC study. Forest production plans are taken as a given; we consider whether the revenues from planned timber sales cover the costs associated with the harvests. We have simply moved from the financial accounting of costs to a more clearly defined and useful economic cost-accounting framework. This approach does result in a somewhat limited analysis of the economics of forest plans, as will be made clear.

2. Below-Cost Sales in a Multiple-Use Framework

It is worth noting at the outset that there are circumstances under which below-cost sales of timber, even sales that return less than the immediate administration and preparation costs, might be justified on economic grounds. To justify such sales, the harvest must generate benefits beyond the immediate revenues from the sale. These benefits may result from improvement in the flow of other multiple-use services or from increases in the value of the area for future timber harvests. Since the Forest Service operates under a multiple-use mandate, it is authorized to expend funds to manipulate the forest or grasslands vegetation for objectives that provide nonreimbursable resource services.

By the same token, the acceptance of some below-cost sales as part of a general land-management program does not mean that every below-cost sale is economic. When a below-cost sale results in the degradation of other multiple-use services of the land area and cannot be shown to provide significantly enhanced net revenues from future harvests, the sale cannot be justified on grounds of simple economic efficiency. No attempt to partially allocate the costs of such a sale to other services should be allowed to disguise the fact that this would be an uneconomic harvest. Even some profitable timber sales may be uneconomic when viewed from the broad multiple-use perspective. Accordingly, whether circumstances do exist to justify a particular sale is an empirical question.

While it is correct to say that one needs to look at the empirical evidence, in actuality the sheer mass of analytic work that would be occasioned by performing such an exercise for every timber sale would be overwhelming. The Forest Service has never felt it necessary to justify each timber sale individually, and would no doubt find it most difficult and costly to gather the information required to do so. Some sense of the magnitude of the task can be gained by noting that in fiscal year 1985 there were 366,874 Forest Service timber sales. While many of these were small sales, there were almost 1,200 large sales of more than 2 million board-feet each. Apart from the sheer number of sales involved, proving that a particular sale is economic is difficult because of limited knowledge of the effects of timber harvesting on the flow of multiple-use services. The effects of a timber sale are not confined to the harvested stand, but will depend upon the location of the sale in relation to the configuration of facilities, harvests, and residual stands on a wider management area. The appropriate unit of analysis is certainly not the sale area alone, but rather the spatially interdependent set of sites.

Then there are also the dynamics of forest growth to be considered. The flow of services from a sale area will change over time as biological growth occurs and as other sites in the management unit are harvested. This introduces a temporal interdependence in that the value of a current sale is not independent of its impact on the future flow of outputs and services from the sale area and surrounding sites. Each sale must be viewed as one component of a total management program over time, and it must be determined, by comparing the present values of overall benefits and costs, whether an alternative management program which excludes the timber sale would provide a greater net economic return. No approach to the below-cost sales issue which focuses only on the cash flow, comparing current expenditures to current receipts, can ever adequately address the economics of timber sales. To the extent that there are net beneficial impacts from a timber sale, they will often occur over many years following the current harvest.

Below-cost timber sales are often associated with areas into which new roads must be built. The more accessible and more productive timber lands will often have been accessed earlier, and the more difficult terrain and poorer lands will have been left unroaded. When the poorer lands are scheduled for harvest, both the cost of access and the lower quality of the timber stock combine to prevent the receipts from timber sales from covering the cost of harvesting.

The building of a new road into a previously unroaded area will change the character of the area in many ways. As one effect, the area will become ineligible for wilderness designation. The loss of a potential wilderness area represents a stream of future value that is forgone, an opportunity cost of the harvest. In contrast, other forms of recreation will not generally be adversely affected by new roads into the forest. A timber-access road would in all probability open access to opportunities for dispersed as well as developed recreation. A new access road may provide recreational and hunting benefits. However, such recreational benefits may be significant only if existing opportunities are not already available in excess of demand.

Below-cost sales are not confined to previously unaccessed areas, nor are the potential benefits of a timber program associated only with improved road access. Improvements in multiple-use benefits may be a direct effect of man-made openings in the forest. It has been suggested that under some conditions carefully prescribed timber harvests can lead to economically significant increases in the quantity of water available for streamflow (see chapter 5). Well-designed timber sales may result in improved habitat for game species and so increase the value of the forest to hunters.[12] In the Black Hills, White Mountain, and Pisgah national forests, as in other national forests, manipulation of roadside vegetation has been used to open up scenic vistas which otherwise would be screened from view. Such changes in multiple-use benefits resulting from timber harvests would be maintained for some years after the date of the sale, although they would vary over time.

The current harvest may enhance not just the flow of multiple-use amenity services, but also the net flow of revenues from future timber harvests. As Schuster and Jones have illustrated, the high initial cost of roads to open up new lands to harvesting might result in below-cost sales in the early years that would be offset by less costly and more profitable sales from later entries in the same area.[13] In another argument, the Forest Service has often said that harvests from stands which are currently diseased or otherwise in poor condition might be justified by future revenues from subsequent well-managed regenerated stands that could be introduced onto the land. Again, the benefits of the current harvest are reflected in the improved future profitability of the forest in the production of timber. It should also be noted that in order to avoid the risk of even larger future losses, such as those resulting from the spread of wildfire or insect damage, one might occasionally find it economic to harvest from poorly stocked lands that show no potential for positive current or future net revenues.

It is worth pointing out that, because of the effects of discounting, it is unlikely that the net revenues from future harvests of a regenerated stand would themselves justify much of a loss on the current harvest.[14] But that is not the end of the story: the Forest Service operates under an even-flow harvesting constraint, meaning that the current harvest level is constrained to be no higher than the level which can be maintained in perpetuity. The harvesting and subsequent management of poor timber lands can, by raising the potential future harvest levels, allow an immediate increase in harvests from productive lands across the forest. Such an "allowable-cut" effect will be most apparent on forests having significant existing stocks of older-growth timber. The present value of the net revenues that result from the immediate increase in the allowable cut may offset the time-discounted cost of later below-cost sales.

Two points related to the allowable-cut effect must be made. First, a sale-by-sale analysis cannot reflect the effect of forestwide constraints, such as the even-flow constraint. Second, the even-flow constraint itself merits critical attention. Note that the same benefits that would be gained by increasing current harvests could be achieved by simply relaxing the even-flow constraint. The even-flow constraint creates incentives that encourage both the harvesting of poorly stocked unroaded lands and a general intensification of timber management, because these actions result in greater allowable current harvests from the abundant stocks of mature timber typically found on some national forests. It is indeed ironic that the even-flow constraint has in the past been generally supported by those conservation groups that are alarmed over below-cost sales.

The primary economic justification for a below-cost sale must be that there are net beneficial impacts from the timber harvest and the related roading activity. Other arguments have been presented to explain or justify widespread below-cost sales of timber from the national forests. According to one justification, there are social benefits to timber sales beyond the benefits measured in the economic efficiency calculus.[15] Maintaining the stability of communities and industries which are dependent upon the supply of national forest timber seems to be long-established Forest Service policy. To support community stability, the Forest Service has been willing to subsidize local industry by selling timber at prices below the level needed to justify the harvest. This is essentially a political justification for below-cost sales.[16]

Still another explanation for below-cost sales has been that timber-cost data are in many cases misleading, overstating the true costs of Forest Service timber sales.[17] It is argued that these costs, if properly adjusted, might show that below-cost sales are really not widespread. This explanation—something of a denial of the below-cost problem—has resulted in, or at least added to, an unfortunate focus on Forest Service cost-accounting procedures. It is our view that direct tinkering with the day-to-day accounting procedures is both an unproductive and mischievous approach to the below-cost sales issue. The existing accounting procedures were not meant to be, and probably could not be, designed to provide the basis for economic judgments on the mix of multiple-use outputs.

It does seem to be true that some of the money charged to the timber budget goes to cover administrative and planning needs, many of which would continue in the absence of an active timber program.[18] The planning and the design of Forest Service timber sales make them costly by the standards of the private forest industry, but some of this apparent inefficiency results from an attempt to guarantee the protection of multiple-use services. Further, it seems likely that some of the expenses charged to timber are for activities designed to improve other multiple-use services as much as to provide timber. The relative ease of funding the timber program, in contrast to the funding of other multiple-use services, has created incentives to cover many costs through the timber budget.

It proves to be difficult to sort out from budget figures alone an amount that can be unambiguously attributable to timber management. One reason for the difficulty is the jointness in forest production activities. An appropriate allocation of costs under conditions of jointness requires technically demanding procedures that could not be used for the day-to-day accounting of expenditures.

Perhaps a more significant problem is that the Forest Service accounting approach is a yearly cash-flow system, comparing current expenditures to current receipts. The trouble lies in the fact that current revenues result primarily from inherited stocks and past actions, while current costs result in future benefits. The economic analysis of a timber sale requires us to consider not just the immediate expenditures and receipts, but also the change in the whole stream of future multiple-use benefits and costs that results from the sale. The focus on the budgeting and accounting framework, with its single-year look at benefits and costs, has so narrowed the analysis of the below-cost sales issue as to turn attention away from the many potentially good economic justifications for below-cost sales. The attempt to allocate current expenditures alone cannot help the Forest Service justify those legitimately economic sales for which high current expenses are offset by later returns. Similarly, in those instances when otherwise profitable timber sales have a negative effect on the future flows of recreational and amenity services, a focus on the current cash flow would miss the very real opportunity costs of these sales.

Up to this point in this chapter we have not distinguished between capital and current expenses. Some expenditures, such as those for roads and facilities, are easily recognized as capital expenses, yielding their services over more than one year's duration. It should be understood that most other expenses of forest management have a similar capital component, and that any activity leading to alteration of the forest vegetation has long-lived effects, either improving or degrading the future flow of value from the forest lands. If we wish to determine a meaningful measure of economic cost against which to compare the current receipts from timber harvests, the treatment of costs in a capital accounting framework is called for.

3. The Costs of Forest Outputs: A Capital Accounting Approach

We consider now a procedure for computing the full current costs of forest outputs that is based on capital accounting principles. The method of evaluating cost to be described will yield an economically relevant measure of the current harvest costs. Those sales having high current costs which are offset by later returns will be clearly shown to be justified because of the resulting improved asset value of the forest. On the other hand, sales which result in a deterioration of the future multiple-use services of the forest will be, appropriately, charged for the economic opportunity costs of those forgone multiple-use values.

In the capital accounting framework, the true cost of production in the current period includes both the actual expenditures and a measure of the impact of current actions on the asset value of the forest resources. The asset value of the forest corresponds to the present value of the flow of future benefits and costs from all the multiple-use services of the forest. The procedure developed here provides a relatively simple means for determining this full economic cost of current management. Our focus is on the appropriate computation of separable costs of timber production within the capital accounting framework. Our measure of the separable cost of a timber sale is the increase in full economic costs (including opportunity costs of future values forgone) that results from the inclusion of timber output in a current management plan.

Separable Costs

Separable costs are the extra costs that are required by the inclusion of an output in a planned production mix. The separable costs of an output are determined by computing the cost savings that result from excluding that one output while providing the same level of all other outputs, and providing them by least-cost means. It should be noted that the least-cost production of the other outputs is consistent with there being some incidental yield of that output for which the separable cost is computed. Indeed, in the forestry case the reduction of an output level to zero is often not a meaningful or economically sensible possibility. In the discussion here, we will use the phrase "dropping an output" to indicate that production of the output is being reduced to the level (perhaps zero) that incidentally results from the least-cost production of the remaining outputs and services.

An example should illustrate the potential importance of the separable cost idea to forest management. Suppose a timber sale will cost $250 per acre and that the sale will result in a significant increase in the water available for streamflow. Suppose also that there is no means to provide this increase in water flow except through timber harvesting—that is, the least-cost means of providing the improved streamflow would require this timber sale. Despite the expenses of the timber sale, then, the separable cost of the timber would be zero. Even with a very low revenue from sale of the timber, the harvest would be justified if the combined benefits from timber and water exceeded the sale costs. This is very much the case presented in our study of the management of timber for increased water yields in subalpine forests in Colorado, in chapter 5.

Direct versus Separable Costs.   As discussed earlier, separable costs must be distinguished from the direct costs of producing an output (see chapter 9). To be avoided is the intuitive assignment of costs to outputs simply because the activity seems most obviously associated with one output. Currently, the Forest Service budget figures on which the below-cost sales argument relies are based on the service's *Management Information Handbook* codes. The MIH codes, as we have seen, provide the basis for budgeting and accounting, categorize all activities, and assign costs according to the primary intent of the activity. However, the costs most directly associated with timber sales do not necessarily correspond to the separable cost of timber.

Consider a timber sale that provides better opportunities for hunting. For such a sale, the separable cost of timber will be less than the direct costs. Suppose the sale costs were $500 per acre, and that these costs would be the same even were the sale planned without regard to hunting benefits. The direct costs of timber could then no doubt be considered to be this $500, a minimum expenditure necessary for a timber harvest. Now suppose that the cheapest means of providing the same improved hunting benefit from this land area would cost $150 per acre. The separable cost of the timber sale is then only $350 per acre, $150 less than the $500 direct costs. Despite the fact that all of the sale activity is directly required for the timber harvesting, the separable cost of timber is less than the direct costs, because the harvest provides complementary hunting benefits for which the forest would otherwise have had to incur costs.

Why Consider Separable Costs?   Some historical perspective is gained by referring to usage in the multiple-purpose water resources field, in which hydropower has a role similar to timber in forest management. That is, revenues are collected for both timber and power, while the remaining services of both systems are largely nonpriced. Accordingly, we suggest considering for the moment the conventional, officially sanctioned, method of cost allocation—the separable cost/remaining benefits method used in such circumstances for governmental cost accounting and congressional oversight.[19] Under this method separable costs are the basis for cost allocation. Fairness seems to require that the cost allocated to a product at least equal those extra costs that are required to include the product in the selected output mix. This principle of fairness is a basic element in most cost-allocation methods. Other allocation methods differ primarily in their schemes for assigning those common costs which remain unallocated after the assignment of separable costs to each output. The assignment of common costs is to a large degree arbitrary, although bounded.[20]

Separable costs also provide the basis for a simple, although limited, test of economic efficiency in the design of projects. The value of each product and service should be tested against its separable cost. If the benefits from including an output in the production plans are not greater than the resulting increment in cost (that is, the separable cost), the current production plans are not economically efficient. A greater return could be achieved by dropping the output from the production plan.[21]

Alternative Incremental Cost Tests. It should be realized that other than for its relevance in cost allocation procedures, the separable cost test would be just one of many possible comparisons of incremental benefits to incremental costs. Other comparisons, some more easily applied, do allow for a similar focus on economic rationality in the provision of a single product. For harvests to be economic, an intended production plan which includes timber harvesting must provide greater net benefits than any alternative plan with reduced timber harvests. The comparison to one such alternative plan— that which excludes current timber harvests but maintains the currently proposed production levels for the nontimber outputs—provides the basis for the separable cost test.

Another, somewhat more rigorous, test of the economics of planned timber harvesting could be based on a comparison of the net benefits of the proposed production plan to the net benefits of an alternative plan which provides the greatest net present value from nontimber outputs alone. Let us call this alternative the amenity-emphasis plan. If the amenity-emphasis plan provides greater net benefits than the currently proposed plan, the proposed plan is not efficient. The proposed timber harvests could then be said to be uneconomic; that is, economic benefits would increase if we dropped the timber harvest and instead emphasized the production of nontimber outputs.

This incremental benefit-cost test can be stated in a manner analogous to the separable cost test, putting the focus on the decision to harvest. Planned timber harvests are not economic unless the benefits from the timber harvest exceed both the forgone current amenity benefits and the increased total costs (current expenses and forgone future values) of selecting the current proposed plan rather than choosing an amenity-emphasis alternative. The combination of increased costs and forgone amenity benefits might be referred to as the separable opportunity costs of the timber harvest.

Similar incremental benefit-cost tests of the desirability of partial reductions in the overall timber program might be even more useful, since it is unlikely that complete cutbacks in timber harvesting would either be desirable or seriously considered.

Some Computational Concerns. The alternative incremental benefit-cost tests, such as those just described, often prove to be easier to implement than separable cost tests. The computation of separable costs is not always straightforward, in that it will doubtless involve more economic and engineering analysis of management projects than is customary. In particular, to determine the separable cost of each output, projects must be designed in a way that best provides all other uses except the one in question. Such projects may not otherwise be of inherent interest and might not be easily developed. Other incremental benefit-cost tests can be easier to implement because the alternatives against which a planned project is compared are less rigidly defined.

One practical difficulty in computing separable costs is that some outputs are so closely linked that one output cannot be adjusted without altering another output. No meaningful accounting of the separable costs for a single output in such a pair is possible. More precisely, if one attempted to find the separable cost of the single output it would be found to be, trivially, zero. Of real concern is that in some forest plans a similar strong link between outputs may be artificially introduced. For example, increases in roaded recreation may be treated as resulting from timber harvests alone, and no other means for providing the recreation output may be considered. When treated in this manner, the costs of the two outputs cannot be separated.

Forest planning models based on FORPLAN linear programs do allow for choosing among many alternative mixes of output and services. Despite FORPLAN's flaws, nothing generally available on the national forests offers any greater hope than FORPLAN for the evaluation of the economic efficiency of timber harvest programs. With this background, let us now turn to the definition and computation of costs in the capital accounting framework.

Defining Costs in a Capital Accounting Framework

Many forest management actions, such as road construction, will have an unavoidable impact on the future flow of outputs and costs. Roads require large expenditures at the time of construction which may not seem justified by the immediate benefits. However, the construction of a road, if it is built for a justified and continuing purpose, can increase the capital value of the forest. A road may improve the net present value of the future flow of services from the forest by reducing future harvest costs and improving access for recreational use. In making a fair comparison of the current cost of a road against its current benefits, it is appropriate to subtract from the current construction cost an adjustment for the value of those future services from the road that will continue after the end of the current planning period. In effect, this spreads the construction cost of the road over its useful life.[22]

It is not so widely recognized that most other management activities on the forest, especially the activities associated with timber harvesting, have a similar capital cost component. The harvesting of timber reduces the amount of timber that remains standing and—leaving multiple-use considerations aside—will generally reduce the asset value of the forest. Simply put, stocked lands are worth more than unstocked lands. The stocked land provides us with the opportunity for harvesting in the near future, the harvested land does not. It seems likely that timber harvesting would also reduce the future flow of amenity values, further decreasing the asset value of the forest. Put another way, in addition to the immediate costs of a timber sale there are opportunity costs to timber harvests, the future values which are forgone.

There are, as we have noted earlier, exceptions to the rule that harvesting will decrease the asset value of a forest. A well-designed timber sale might result in increased future flows of certain multiple-use services of the forest, sufficient to offset the loss in value associated with the reduced level of timber stock. The improved access that results from the construction of roads for the timber harvest might provide for an increase in future net benefits from both timber and multiple-use services. A harvest of diseased trees or harvests from poorly stocked lands might allow the introduction of vigorous timber stands and result in an increase in the asset value of the forest. Whatever the case, it will never be appropriate to make a harvesting decision simply on the basis of the current year's receipts and expenditures. To do so would be to ignore the impact of the current harvest on the resources of the forest and on the future yield of multiple-use benefits from the forest.

Accordingly, an adjustment must be made in the measure of costs if we are to look at a meaningful comparison of benefits and costs in a single budget period. The full measure of the economic cost of current-period management activity must include both the immediate expenditures and the resulting change in the asset value of the forest over the budget period. A few essential definitions should now be considered and recalled.

Asset Value and Asset Depreciation.   The asset value of the forest or forest area is the maximum attainable net present value from the flow of outputs and services of the land. The depreciation in the asset value of the forest over a budget period is the initial asset value of the forest minus its subsequent asset value at the end of the period. If the timber harvest improves the potential flow of value from the forest, depreciation will be negative—that is, forest value appreciates.

Full Current Costs. The cost of managing the forest asset over a single planning period includes the current operating expenditures plus the depreciation in the asset value of the forest and its facilities. The reader should be careful to avoid confusion between this full economic measure of current cost, always referred to as the full cost or the economic cost, and the actual current expenditure from which it is to be distinguished.

Separable Cost. The separable cost of an output is the overall cost savings that would result from dropping the output in the current period while providing, at least cost, the planned current-period production for all other outputs. Given our definition of full current costs, the separable cost of a product is the increase in current expenditures, plus the increased depreciation in the forest asset value, that results from including the product in the current management plan.

We can get some sense of what a capital accounting view of costs means by looking at an example. Say $520 per acre is spent harvesting timber in the current period, and suppose the harvest results in improved future hunting opportunities, thus partially offsetting the value of the reduced timber stocks, so that the end-of-year asset value of the area is less than the initial asset value by $20. The economic cost of the first-period harvest is then $540. This is not the separable cost of timber, but rather it is the full current cost of providing the supply of timber and hunting in this period.

To find the separable cost chargeable to timber, it is necessary to determine the least-cost means of providing the hunting opportunities alone. Say this can be found to require current expenditures of $200, but would also result in a $50 improvement in the asset value of the area, for a full current economic cost of $150. The separable cost of the first-period timber sale is then just $390, found by subtracting $150 from $540.

Our method of determining separable cost focuses on the immediate harvest plans. The separable cost of timber, as computed by our method, can be compared to the revenues from the current timber harvest—or, more accurately, to the increment in the current harvest revenue over the harvest revenue that might result incidentally from the provision of the other multiple-use services. Two important points should be understood. First, our method is not equivalent to finding the cost savings that would result from dropping timber harvesting for all times. Such an approach would be unfairly restrictive, and could lead us to eliminate timber management from a management unit for all time when perhaps the harvest should simply be postponed. Second, our approach is fully consistent with the Faustmann-type harvest-timing solutions described in chapter 4 and embedded in economic harvest-scheduling linear programs (such as FORPLAN when it is used to maximize the present value of services from a forest). That is, the revenue from the harvest of any single-aged stand will not cover separable costs until the date of economically optimal harvesting.

Computing Costs and Separable Costs

Generally, one of the difficulties in applying a capital accounting procedure is that there will not be an accurate means for determining the current value of the assets. The FORPLAN linear programming model can be used for scheduling forest treatments so as to maximize the present value of the aggregate of multiple-use services. In the following pages we describe our procedure for using FORPLAN to compute separable costs. But this requires some initial development of notation.

Represent the maximum present net value of the forest at the beginning of the budget period—the initial asset value—by the term V_0. This is the typical output of a FORPLAN run which uses maximum present net value as the objective function, employing only those constraints required by law or necessary for accurate representation of production and demand.

The current expenditures and value of the forest at the end of the current period will depend upon the specific management actions taken during the period. The current expenditures are represented by the term $C_1(p)$, where p is the mix of outputs provided in the first period. The term $V_1(p)$ represents the asset value of the forest at the end of the first period, given that outputs p are produced in the first period.

A FORPLAN model is typically structured by decades rather than years, and treats most expenditures as if they were incurred at the midpoint of the decade. As a result, separable costs can only be computed for the production over a whole decade; the model does not allow us to compute the separable costs of production for a single year. The long time interval also raises new questions as to the appropriate discounting in presenting the costs. We will follow a convention of computing the present value of costs as of the beginning of the current period. With the annual discount rate used in the FORPLAN model represented by r, $TC(p)$ the full current costs of management for any mix of current outputs p are computed as

$$TC(p) = C_1(p)/(1+r)^5 + [V_0 - V_1(p)/(1+r)^{10}]$$

The current expenditure C_1 is discounted back five years, reflecting the FORPLAN assumption that the costs are incurred at the midpoint of the decade. The second term on the right reflects the gross depreciation in the asset value over the planning period. In this term, the asset value $V_1(p)$ is discounted back the ten years from the end of the decade.

There are other reasonable ways in which we could define the measure of capital cost. Perhaps most logical would be to measure costs in comparison to a no-management alternative. This would require replacing V_0 in our cost definition with $V_1(0)/(1+r)^{10}$, where $V_1(0)$ represents the end-of-period asset value of the forest when there are no management activities undertaken in the first period. That approach would focus on the potentially allocable costs, ignoring fixed costs. Alternatively, we might replace V_0 with $V_0/(1+r)^{10}$, focusing strictly on the change in the forest relative to initial condition. In comparison, our definition may be viewed as further including in the costs an expected return on the initial asset value of the forest. The choice among these measures of full cost proves to be irrelevant to the computation of separable or incremental costs.

The inclusion of the asset value $V_1(p)$ in our definition of full current cost is significant. It means that we consider the forest to have the option of selecting economically optimal management in subsequent periods. In particular, we do not restrict the forest to its stated plans for future management. This approach maintains our focus, appropriately, on the forest's first-period production plans.

We use the term "proposed management plan" to refer to the mix of outputs the forest intends to provide in the first period. To find the separable cost of providing an individual output in this proposed plan, we must develop new management programs in which the current production of the one output is dropped, while all other outputs are provided at least cost and at the same level as under the proposed plan. Programs in which one output is dropped are referred to here as the "nontimber" plan, the "nonrecreation" plan, and so forth; there is one plan for each output for which a separable cost is to be evaluated.

As an example, consider a forest that can provide three outputs: timber harvest, general recreation, and wildlife-based recreation. Suppose the proposed management plan for the forest calls for the provision of amount T of timber, amount R of general recreation, and amount W of wildlife recreation in the first period. The computation of the separable cost of the first-period timber harvest can be described as follows.

First, find the full least cost of providing the proposed levels of outputs T, R, and W in the first period, by evaluating

$$TC(T,R,W) = V_0 - \operatorname*{Max}_{t,\rho,w} \{-C_1(t,\rho,w)/(1+r)^5 + V_1(t,\rho,w)/(1+r)^{10}\}$$

$$\text{subject to: } t \geq T;\ \rho \geq R;\ \text{and } w \geq W \qquad (10\text{--}1)$$

$$= V_0 + C_1(T,R,W)/(1+r)^5 - V_1(T,R,W)/(1+r)^{10}$$

The calculation of the full cost of a production plan is easily accomplished using FORPLAN; the procedure is described in greater detail below. The calculation requires a two-step process, as represented by equation (10–1). First, the initial asset value V_0 is determined in one FORPLAN run, then the maximization that makes up the second part of equation (10–1) is performed as a separate FORPLAN run. The term V_1 in equation (10-1) represents a discounted sum of all future net benefits, and this value is determined optimally within the FORPLAN run.

To get the separable cost of the first-period timber harvest, it is necessary to find the full least cost of a nontimber plan. The nontimber program calls for dropping the first-period timber harvest while producing amount R of recreation and W of wildlife in the first period (with perhaps some incidental timber production). The cost can be represented most conveniently as

$$TC(R,W) = V_0 - \underset{\rho,w}{\text{Max}} \ \{-C_1(t,\rho,w)/(1+r)^5 + V_1(t,\rho,w)/(1+r)^{10}\}$$

$$\text{subject to: } \rho \geq R; \text{ and } w \geq W \qquad (10\text{-}2)$$

$$= V_0 + C_1(R,W)/(1+r)^5 - V_1(R,W)/(1+r)^{10}$$

The separable cost of the current timber harvest is now found as the cost savings from following the nontimber program rather than the proposed first-period program. That is,

$$SC(T) = TC(T,R,W) - TC(R,W)$$
$$= [C_1(T,R,W)-C_1(R,W)]/(1+r)^5 \qquad (10\text{-}3)$$
$$+ [V_1(R,W)-V_1(T,R,W)]/(1+r)^{10}$$

The separable cost of the current harvest is the increase in current expenditure C_1 plus the net reduction in the forest asset value V_1 that results from including the timber harvest in the current plan. It is apparent from equation (10-3) that separable cost tends to be reduced when the inclusion of the timber output in the current period plan results in an improvement in the future flow of value from the forest. When timber harvesting improves the flow of net benefits above the level that would result without a harvest, the term $V_1(R,W)-V_1(T,R,W)$ is negative and separable costs are correspondingly lower than the increase in immediate expenditures required by the harvest.

The separable cost test calls for the comparison of the benefits from the net increment in timber harvest to the separable costs. A planned harvest is uneconomic if these incremental benefits from the planned timber sales are less than the separable costs; that is, the planned timber program is not economic if

$$B(T)-B(T') \leq [C_1(T,R,W)-C_1(R,W)]/(1+r)^5 \qquad (10\text{-}4)$$
$$+ [V_1(R,W)-V_1(T,R,W)]/(1+r)^{10}$$

where $B(T)$ represents the benefits from the planned timber sales and $B(T')$ represents the benefits from any incidental production of timber under the alternative nontimber program.

Similar incremental benefit-costs can be derived to evaluate partial reductions in the timber harvest or for comparing the planned program to alternatives with both reduced timber and differences in the current provision of recreation and wildlife services. In general, a timber program is uneconomic if we can find any alternative program with reduced first-period timber harvest such that

$$B(T)-B(T) \leq [C_1(T,R,W)-C_1(T',R',W')]/(1+r)^5$$
$$+ [A_1(R',W')-A_1(R,W)] \qquad (10\text{-}5)$$
$$+ [V_1(T',R',W')-V_1(T,R,W)]/(1+r)^{10}$$

where T', R', and W' represent the first-period production of timber, general recreation, and wildlife services under the alternative production program, and the function $A_1(r,w)$ represents first-period benefits from recreation r and wildlife services w.

It should be instructive for the reader to apply the harvesting rule of equation (10-4) to the Faustmann harvest-timing problem for a single-aged timber stand. The application of rule (10-4) will correctly indicate that a harvest before the optimal Faustmann age is uneconomic, while a harvest at the Faustmann age will just cover separable costs. One might also compare equation (10-4) to the condition for dynamic optimality in harvesting renewable resources.[23] That very similar condition calls for harvesting to a level at which price equals the current marginal operating cost plus the marginal user cost of depleting resource stocks.

What exactly is required of FORPLAN runs in order to compute the separable cost of timber? First, V_0 the initial asset value of the forest can be found with a FORPLAN run that maximizes the present net value from all the goods and services of the forest.

It is presumed that the forest has already proposed a management plan, which of course may not be the same program as that which would maximize present net value. A second run is used in calculating the full least cost of providing the first-period outputs planned under the proposed management alternative. This proposed management run requires the maximization of present net value in a FORPLAN run that (a) puts no price on the outputs in the first period (we are not interested here in the first-period benefits, only in costs); (b) constrains the level of all outputs in the first period to at least the proposed level; and (c) prices all outputs in the normal manner from the second period on, with only the essential constraints in these later periods (and so accounts for V_1 the present value of the forest outputs from the second period on, an essential component of the capital costs). The proposed management run is equivalent to solving a present-value maximization problem like that represented in equation (10-1). The full cost of the proposed program is found as V_0 minus the objective function value determined in solving this proposed FORPLAN run.

As noted, in order to compute the separable cost of timber we must determine the full least cost of providing the first-period production levels of all other outputs, excluding timber. The nontimber run requires a maximization of present net value, with (a) no prices on any first-period outputs; (b) constraints on the first-period levels of all nontimber outputs, requiring them to at least equal the previously proposed level (with no constraint on the first-period timber, the harvest will drop to zero unless some harvest is necessary in the provision of other services); and (c) prices on all outputs from the second period on. This nontimber run is equivalent to solving the present-value maximization represented in equation (10-2). The full current cost of the nontimber program can be found as V_0 minus the objective function value from the nontimber FORPLAN run.

With the proposed management run and the nontimber run just described, the separable cost of timber can be computed. This is most easily done by taking the objective function value of the nontimber run and subtracting the objective function value from the proposed management run. To find the separable cost of other outputs, it is necessary to make FORPLAN runs similar to the nontimber run, one run for each output, dropping the emphasis on the single output of concern while maintaining the constraint that all other outputs meet the previously proposed level of supply in the first period.

It is our conclusion that only a capital accounting framework can provide the kind of information the Congress and the interested public requires in order to correctly evaluate the below-cost sales issue. Having come to this conclusion, we have sought to test whether the models available for national forest management are adequate to the task. The Shoshone and the Mt. Baker–Snoqualmie national forest FORPLAN models allowed convenient development and testing of our method for determining separable costs. We turn to that application now.

4. Applying FORPLAN: The Shoshone and Mt. Baker–Snoqualmie National Forests

In this section we provide a brief overview of the FORPLAN linear programming models as used for planning on the Shoshone National Forest and the

Mt. Baker–Snoqualmie National Forest. An appendix to this chapter provides more detailed descriptions of these planning models. It should be emphasized again that we do not evaluate the actual production plans for these forests; final versions of the forest plans were not completed at the time this chapter was written. Nor do we investigate the quality of the economic and physical data that make up the planning model. Our study is methodological, an investigation of the suitability of FORPLAN for examining policy analysis issues such as have been raised by the below-cost sales debate.

These two forests have used Version II of FORPLAN for their land management planning.[24] As we have seen, FORPLAN offers a variety of options for modeling the forest management choices. Early applications focused on the choice of treatments for homogeneous timber analysis areas, without regard to the spatial location of a particular timber stand. A homogeneous timber analysis area would be made up of all land in the forest that was approximately uniform with respect to current timber stock and in potential for the growth and harvest of timber. The land in such an analysis area would often be spread across the forest in noncontiguous stands. Those multiple-use outputs which depended upon location-specific features and on the interrelation of vegetation conditions across lands in different analysis areas could not be accurately modeled. Indeed, the early applications of FORPLAN were not altogether adequate for modeling the timber output.

The FORPLAN Version II model, as it is used on the Shoshone and Mt. Baker–Snoqualmie national forests, has been improved by the introduction of a second, overlapping categorization of the forestland. In addition to the timber analysis areas, the Shoshone and the Mt. Baker–Snoqualmie planning models are structured around geographical subdivisions of the forests—the allocation zones. The division of a forest into allocation zones can be based on a variety of factors: the natural features of the land, the road networks and resulting patterns of land use, and administrative boundaries. Allocation zones are for the most part areas of between 5,000 to 50,000 acres. Within each allocation zone there may be portions of several timber analysis areas.

The zone allocation choice and the harvest scheduling decision for timber analysis areas make up two fairly distinct components of a Version II linear programming model. For each of the geographical allocation zones, there is a choice among several general management emphases. For each timber analysis area, the model may choose among many prescriptions for scheduling the timing of timber harvests and treatments. Tying the two components of the model together will be a variety of constraints. Through these constraints the timber program is forced to be compatible with the overall multiple-use emphasis proposed for each zone. The choice of a general management emphasis for a geographical zone results in constraints on the rate of harvesting and on the location of timber management activities within the zone. The model also allows for further constraints to ensure compatibility in the assignment of general management emphases to adjacent, or otherwise linked, geographical allocation zones.

An advantage of this structure is that many of the multiple-use outputs can be directly associated with the choice of an aggregate emphasis for the geographical allocation zone. Typically, much of the recreation, the wildlife, the livestock grazing, and the water-flow and water-quality effects will be modeled as outputs of the overall zone management emphasis. Certain costs, particularly those associated with recreation and grazing management, general administration, and the provision or maintenance of structural facilities (including the basic road network), will also be modeled within the zone management component of the linear program.

The various timber outputs and related costs are determined by the choice of harvest prescriptions for the timber analysis areas within each zone. Associated with each harvest prescription will be a time schedule of the associated harvest volumes and timber management costs. Some road costs will be tied to the timber harvest and will be reflected in the timber prescription. In addition, the direct effect of the timber harvest on other multiple-use outputs and services will be identified here. Each alternative timber management prescription may be associated with incremental improvement or degradation to the base level of those outputs (recreation, wildlife, livestock grazing, and water) which have resulted from the choice of a management emphasis for the allocation zones.

The basic structure of the Shoshone and the Mt. Baker–Snoqualmie models seems to provide an acceptable representation of multiple-use forest production. The geographical zone structure has also proved to be of significant convenience in our analysis. We were able to partition the forestwide linear program and work with a model for a small area of the forest, thus greatly reducing the potential cost of generating linear programming solutions. The forest areas that we selected for analysis were large enough to allow the full multiple-use effects of a timber sale to be captured, and so to qualify as appropriate accounting units.

The Shoshone National Forest

The Shoshone National Forest is in the northern Rocky Mountains in Wyoming just to the east of Yellowstone National Park. The Shoshone is not a major timber-producing forest. About 25 percent of the forest is above the timberline, and 20 percent of the forest is classified as rockland. Those lands with timber cover are often either unproductive, difficult to access, or poorly stocked. Historically, the Shoshone has had many below-cost sales. Low sale prices have often been a result of the large quantity of insect-damaged timber to be sold. On the other hand, wildlife is an important resource, abundant in population and diversity.

The Dunoir/Doby Cliff Allocation Zone. The Dunoir/Doby Cliff planning unit of the Shoshone forest is a diverse area of about 50,000 acres containing some of the Washakie Wilderness and having a potential for high levels of forage and timber production. The zone has important wildlife habitat, including crucial preferred winter range for moose and bighorn sheep. There are elk calving areas, and a major elk migration route passes through the zone. In this area, well-designed timber sales are expected to improve both the wildlife services and the roaded recreation opportunities, mainly at the expense of primitive recreation. Any increases in livestock grazing would directly compete with these potential wildlife services. The Dunoir/Doby Cliff unit is more productive and heavily timbered than much of the Shoshone forest. Given the prices and costs used in the planning process, it appears that timber management in this zone will return a positive present value. Even so, it is likely that certain timber sales will be below cost, with negative cash flows in the first few decades.

The Shoshone Linear Programming Model. The FORPLAN model for the Dunoir/Doby Cliff allocation zone considers the choice among eleven aggregate emphasis prescriptions, which differ mainly with respect to the nature of the recreation management they prescribe, and especially the extent to which they allocate land to wilderness. The prescriptions also offer a range of choices from emphasis on domestic livestock grazing to emphasis on the provision of wildlife habitat. The timber scheduling component of the model offers (for the predominant single-aged stands) a choice of no timber management, shelterwood harvesting, small-area clearcuts (5 to 10 acres), or standard clearcuts. There is also considerable choice as to the intensity of a thinning program and the age at which a regeneration harvest occurs.

The outputs included in the planning model are timber, with separate values by species for sawtimber, roundwood, and firewood; livestock grazing, with separate values for cattle AUMs and sheep AUMs; recreation, with separate values for pristine wilderness, primitive, semi-primitive nonmotorized, semi-primitive motorized, roaded natural, and developed recreation visitor days; fish and wildlife services, with values for consumptive and nonconsumptive user days; and water, with a positive value for increases in water flow. Both deterioration in visual quality and increased sedimentation are considered to result from timber harvesting. No explicit monetary values are included in the model for these last two "outputs." The discount rate used in this FORPLAN model, and in all forest planning models, is 4 percent.

We have found it convenient to compute the separable costs for three major groupings of outputs: timber and water, recreation (excluding fishing and hunting), and fish and wildlife-related services. Since no livestock grazing is provided in our solution, it was not necessary to consider the separable cost of that output. The combined separable cost for timber and water-flow increments was computed, since timber harvests provide the only means for increasing water flow above the base level. It was impossible to consider altering the level of water flow while maintaining a constant level of timber harvest. Accordingly, when we refer below to the separable cost of timber, it should be understood that the timber/water combination is meant. Although outputs were grouped for convenience in the presentation of separable costs, all the separate pricing and production details for each of the individual outputs were maintained in the model.

With little modification, we have used the model developed by the Shoshone planning staff for the Dunoir/Doby Cliff allocation zone. However, the Shoshone FORPLAN model for the forest as a whole includes a variety of constraints that do not seem applicable at the level of the single management unit. For example, the imposition of an even-flow constraint on timber harvests from such a small area does not seem appropriate. The forestwide model also includes constraints which reflect the limited demand for certain outputs; in particular, a forestwide limit is set on the number of recreation visitor days of capacity that should be valued. Although it seems likely that recreation capacity might exceed the demand for recreational use of the Dunoir/Doby Cliff unit, no separate estimate for demand by analysis units was available. Therefore, all recreational capacity produced by this unit has been valued, with no demand constraints applied.

It should also be noted that in our analysis we have chosen to use timber prices that were at 75 percent of the level used in the Shoshone plan. This adjustment has been made to provide more striking illustrations of below-cost sales. Even with this lower timber price, most stands could provide a positive net return to timber management, although often with periods of negative cash flow. In any case, the prices used in the actual Shoshone planning effort seem to be rather optimistically determined.

The Shoshone FORPLAN Runs and Results. The computation of the separable costs for the first-period supply of timber, recreation, and wildlife services required that five FORPLAN runs be made. First, we required a management plan to evaluate. For purposes of illustration, we have evaluated the separable costs of the outputs that would be produced in the first period under the solution to a run (PNV) which maximized present net value of outputs from the Dunoir/Doby Cliff analysis zone. The objective function value for the PNV run also gave us our estimate of V_0, the initial asset value of the Dunoir/Doby Cliff allocation zone. It should be stressed that forests have typically not chosen present net value maximizing solutions as their proposed management plan, and that output levels proposed under any other plan would be rather less likely to cover their separable costs.

Our choice of the PNV run as the "proposed" plan guaranteed that all outputs do cover their separable costs; if they did not, present net value could be improved by dropping some output—which would contradict the fact that present net value had been maximized. Beginning with the knowledge that all outputs do cover their separable costs is convenient because it allows us to clearly illustrate some of the faults inherent in those analyses of the below-cost sale issue that focus on current cash flow alone. It should be noted that by our choice of the PNV solution as the basis for analysis, it follows that any timber harvests which appear in the solution are economic. That is, eliminating the harvest would result in a lower present value of net benefits from the management area. Nevertheless, it will be seen that not all of our chosen timber harvests result in immediate revenues sufficient to cover the costs of current harvest-related activity.

Once the PNV management program was chosen for analysis, four further FORPLAN runs were made. The run PLAN1 was used to determine the least economic cost of providing the first-period outputs that were proposed under the PNV solution. This is the "proposed" management run described in the preceding section and corresponding to the maximization problem represented in equation (10–1). In this FORPLAN run, we maximized present net value, with first-period levels of all outputs constrained to be at the planned levels, with first-period outputs unpriced, and with all outputs in the later periods priced at their normal levels and unconstrained as to the level of production.

The method for generating the other FORPLAN runs has also been described in the preceding section. The nontimber (TIM) run was used to determine the least economic cost of producing all current outputs except timber. Similarly, the nonrecreation (REC) run and the nonwildlife run (WLF) were used to determine the least economic cost of providing all first-period outputs except recreation and except wildlife, respectively. Various other runs were made for special purposes, as we shall describe.

The results of our FORPLAN runs are given in table 10–1. The first row in the table gives the objective function value for the five linear programming runs (PNV, PLAN1, TIM, REC, and WLF). The PNV objective function value is not directly comparable to the other values, since the others do not include the value of first-period outputs. The second row gives the present value of the expenditures planned for the first decade under these various programming solutions. The third row presents the total first-period costs of each program. This is a full measure of the economic costs of management, reflecting both the current expenditures and the changing asset value of the forest. The separable costs for the three outputs timber, recreation, and

Table 10–1. Shoshone Results: Objective Function, Costs, and Separable Costs (values in thousands of dollars)

	FORPLAN runs				
	PNV	PLAN1	TIM	REC	WLF
Objective function	15,405	9,080	9,235	9,080	9,080
Current expense	—	1,575	1,490	1,575	1,585
Full cost	—	6,325	6,170	6,325	6,325
Separable cost	—	—	155	0	0

wildlife services are given in the fourth row, under the headings TIM, REC, and WLF. The separable cost of timber production in the first decade has been found by subtracting the objective function value for the PLAN1 (proposed) run from the objective function value for the TIM (nontimber) run. The other separable costs have been similarly calculated.

The total cost of the planned (PLAN1) first-period production, including the capital costs, was found to be $6,325,000. This amount is considerably greater than the actual first-period expenditures of $1,575,000, primarily because it includes a $5 million opportunity cost for holding the forest asset. This amount is the present value of the interest income (at a 4 percent discount rate) that might be expected to be earned over the decade on the initial asset value of the forest unit. Apart from that fixed cost, the capital cost also reflects some improvement in the condition of the forest resulting from the planned management actions. Separable costs—the costs that can be unambiguously attributed to each output—are seen to be relatively low, with little of the total cost allocated to the outputs.[25] In particular, no separable costs are associated with the inclusion of either recreation or wildlife services in the overall program.[26]

Separable Costs of Timber. We can conclude that dropping the current timber production would save only $155,000 in total cost. Of this amount, $85,000 can be thought of as representing an allocation of the current expenses. The remainder, $70,000, represents the increased depreciation in the forest's end-of-period asset value that results from including timber harvesting in the planned first-period program, an allocation of the capital costs. The separable cost amounts to $41 per thousand cubic feet ($41/mcf) of planned timber production, which can be compared to the average price of $156/mcf that the model indicates would be received for all wood products. The present value of the benefits from planned first-period timber sales was $590,000. However, the nontimber run itself resulted in a substantial incidental timber harvest in the first period, with a present value of $383,000 from this harvest. That timber harvest proved to be essential for maintaining the planned supply of wildlife and roaded recreation. Inclusion of the timber emphasis in the planned program results, therefore, in a net increase in current benefits of just $207,000, an amount to be compared to the separable costs of $155,000. Thus by the separable cost test it does not pay to drop the timber program. This is not surprising, since our management plan had been selected to maximize net present value, and therefore all planned harvests were guaranteed to be economic.

Separable Costs versus Direct Costs. The separable costs of timber are seen to be fairly small. Such low separable costs can be expected when products are complementary, as many are on this forest. For comparison, the present value of the direct budget expenses associated with timber activity in the first period was $466,000. This direct-cost figure includes costs of timber monitoring, reforestation, site preparation, timber stand improvement, sale preparation, harvest administration, timber product administration, and fuels management, as well as any other expenses (some increases in road costs, project planning expenses, and fixed costs) that the linear programming model indicates will result from increased timber harvests.

The Separable Opportunity Costs of Timber. The separable cost test is just one of many possible incremental-benefit cost tests, as we have mentioned earlier. A test that better evaluates the economics of the current proposed timber program is a comparison of the present value of net benefits from the proposed management program to the present value of the alternative program that best emphasizes nontimber outputs in the first period. A FORPLAN run called AMENITY was made to generate this first-period amenity-emphasis solution.[27] Our proposed plan (PNV) was found to provide $140,000 more in net present value than could be attained from the amenity-emphasis program in the first period. Put another way (see equation [10–5]), a decision to maintain the planned first-period timber production rather than specialize in amenity services results in extra first-period timber receipts of $590,000, which more than offset the $450,000 cost. This cost includes increased current expenses, forgone increments to the current amenity services, and forgone future flows of net value.

Below-Cost Sales. Not all the timber harvests on this forest return immediate benefits sufficient to cover the immediate expense. All sales are of course economic, because our choice of a management program maximizes the present value of net benefits from the multiple uses. There seem to be three reasons why we see below-cost sales. The primary reason is a requirement that harvest entries into an area be staggered over time. The expenses of the first entry are offset by the revenues from later harvests in the already accessed area. The second reason is that there are declining timber stands. Some stands on this forest are of sufficient age and in such poor condition that if they are not harvested soon, they will not be available to harvest later. It proves better to get what timber there is now and promptly regenerate the stand rather than to incur much the same expense later while getting nothing and delaying the regeneration of the stand into productive timber. The multiple-use benefits of the timber harvest—roaded recreation and improved wildlife habitat—are the third reason for accepting below-cost sales.

Some 1,000 acres of a 6,719-acre lodgepole pine stand are scheduled in our PNV solution to be harvested in the first decade. This is stand AKW in the Shoshone model, 150-year-old lodgepoles on a low-productivity site. The harvest costs and other timber-related expenses in the first decade would be $69,000, the timber benefits would be $36,000. Partially offsetting this "loss" would be first-period multiple-use benefits resulting from harvesting, giving total first-period benefits of $53,000. Further losses of a similar magnitude would be incurred in the second decade. However, the considerable returns from later harvest entries would more than make up for these early losses. Our method of determining the separable cost of a timber sale in a capital accounting framework, reflecting the future benefits that result from current action, shows that this first-decade sale would indeed be economically justified.

The Mt. Baker–Snoqualmie National Forest

The Mt. Baker and Snoqualmie national forests, combined by the Forest Service for planning purposes, run south from the Canadian border for some 190 miles along the western slopes of the Cascade Mountains of Washington state. This forest is an important source of timber supply, with significant stocks of large old-growth conifers. Most of the lands below the timberline are very productive. The forest also offers considerable recreational opportunities, especially to residents of the Seattle area, providing dispersed recreation, developed campsites, and facilities for winter sports. The potential importance of the forest as a recreational area is somewhat diminished by the great number of attractive alternatives available in this region of the country. Both the North Cascades and Mount Rainier national parks adjoin the forest, and the Olympic National Park is similarly close to Seattle. In addition, designated wilderness areas within the Mt. Baker–Snoqualmie National Forest tend to attract dispersed recreational use, in competition with general multiple-use areas of the forest.

The Nooksack River Area.   For the location of our second test of FORPLAN for capital cost accounting, we selected an area at the northern tip of the Mt. Baker–Snoqualmie forest, in the region of the Nooksack River. The land unit is made up of six adjacent allocation zones from the Mt. Baker–Snoqualmie FORPLAN model.[28] With a total area of 83,000 acres, the unit is bounded by Canada to the north and the Mt. Baker Wilderness to the east and south. None of this wilderness is within our study area. North Cascades National Park is just to the east. The land is crossed by the Nooksack River, and state highway 154 runs beside the river. Several campgrounds are provided along the road corridor. Lakes, picnic areas, and winter recreation facilities can be found toward the eastern end of the highway. Maintained gravel roads run to the north of the Nooksack River along Canyon Creek and south along Glacier Creek.

About 45,000 acres of our study area are considered to be available for timber management. Almost half of this is unroaded, although very little of the land is more than a mile from an existing road. While the unit is generally productive for timber, steep slopes and difficult access could make harvesting expensive in some areas. As a result, not all lands within our study area return a positive net value to timber management.

The Mt. Baker–Snoqualmie Planning Model. The forest planning staff has developed five alternative aggregate emphases for each of the allocation zones in our study area. These choices range from a nonmarket service emphasis to an intensive timber management regime. The choice of an aggregate emphasis results in constraints on the location and timing of timber harvests within a zone. Recreation outputs and the cost of recreation management are completely determined by the choice of an aggregate emphasis for zones, with no incremental effect from the timber harvest. Much of the cost associated with road construction and maintenance is also determined by the selection of an aggregate emphasis.

In practice, the Mt. Baker–Snoqualmie staff has run its FORPLAN model with preselected management emphases for each zone. Several forestwide alternatives are then generated by altering these chosen patterns of land management in different FORPLAN runs. The linear programming model has been used only to schedule the timber management activity within the bounds set by the overall land management emphasis. This approach offers considerable savings in the cost of solving FORPLAN models, as compared to the cost of the option of allowing the linear program to select among zone management emphases. The approach also has the great advantage of confining attention to reasonable and feasible programs for the overall management of the forest. A disadvantage is that it narrows the range of management options considered in the planning exercise.

For our purposes, it seemed best to modify the Mt. Baker–Snoqualmie FORPLAN model so that the zone emphasis was chosen within the linear program. Without this modification, the limited flexibility in the means of providing different potential output mixes would have made it difficult to compute separable costs. One risk in allowing the linear program to select the optimal management emphasis for each zone is that the choices might be incompatible with the treatment selected for an adjacent zone. We were prepared to constrain the allocation choices across adjacent forest zones in order to ensure consistency in the chosen land use. However, with our FORPLAN solutions, such constraints have not proved necessary.

Separable costs have been computed for three output groups. All timber products were grouped together, while the several recreation outputs were formed into two separate groups. One recreation group combined all recreation activity on lands classified as unroaded. Our second group combined all recreation activity on roaded lands. As with the Shoshone study the considerable underlying detail, within the FORPLAN model, on the production of individual timber and recreation outputs was maintained. The grouping of outputs was done for convenience in the presentation of separable costs.

Demand limits, constraining the amount of each recreation output to be valued within the model, were determined for our unit from the forestwide constraint. The constraint levels were based on the proportion of each recreational land type within our zone as compared to the forest as a whole. There was no demand constraint limiting the amount of timber which could be sold from our unit. As in the actual Mt. Baker–Snoqualmie model, timber prices were assumed to be increasing over time at a rate of 10 percent per decade.

In our runs, we have included a harvest-flow constraint that allows at most a 25 percent fluctuation in the harvest levels between successive decades. The inclusion of the flow constraint distinguishes the Mt. Baker–Snoqualmie study from the Shoshone example. With its productive timber land, Mt. Baker–Snoqualmie is not a forest one would expect to be associated with below-cost sales. However, the harvest-flow constraint, coupled with the presence of large stocks of mature timber, does result in below-cost sales in later decades of the planned harvest program. These later harvests from otherwise uneconomic land allow a higher current and sustained harvest level. It is because of the importance of this effect that we include a flow constraint in our Mt. Baker–Snoqualmie study. Except as described, our model is the same as that used by the Mt. Baker–Snoqualmie planning staff.

The Mt. Baker–Snoqualmie FORPLAN Runs and Results. As in the Shoshone example, separable costs have been evaluated for the first-period output levels that would be produced under the present net value maximizing solution for the study area. By definition, all timber sales that occur in the solution to the present net value maximizing (PNV) run are economically justified, given our acceptance of the flow constraints we have imposed on this harvest solution.

Four other FORPLAN runs were made. First, we needed a proposed management run (PLAN1) to determine the least cost of providing the first-period outputs from the PNV solution. Then, in order to compute the separable costs of these outputs, three other runs were made. The nontimber

(TIM) run was used to determine the least cost of providing all first-period outputs except timber. The first nonrecreation run (REC1) determined the least cost of providing the planned levels of all first-period outputs except recreation from roaded lands; the second nonrecreation run (REC2) was used to find the least cost of providing the planned first-period production of all outputs except recreation from unroaded lands. The PNV run provided our estimate of V_0, the asset value of the forest unit.

The results of the Mt. Baker–Snoqualmie FORPLAN runs can be found in table 10–2. The first row gives the objective function value for the solutions of each of the five runs. The PNV objective function value corresponds to V_0 the initial asset value of this land area. The second row gives the present value of actual expenditures made in the first decade under the various alternative FORPLAN runs. The third row presents a measure of the full economic cost of the first-period programs, including both current expenditures and the depreciation in the asset value of the forestland over the period.

The total economic cost of the planned program is $109,046,000, higher than the actual first-period expenditures. The extra expense includes $48,762,000 of opportunity cost (forgone interest income) on holding the forest asset, plus $8,268,000 in additional depreciation resulting from first-period management activity. The separable cost of timber can be found by subtracting the objective function value for the PLAN1 program from the objective function value for the TIM run. This is equivalent to subtracting the total cost of the nontimber program (TIM) from the total cost of the proposed plan (PLAN1).

Separable Costs of Timber. The separable cost of the first-period planned timber harvests is $76,830,000. This cost includes a $50,504,000 allocation of current expenditures (most of the current expense) as well as $26,326,000 of capital costs chargeable to timber. The capital costs primarily reflect a reduction in the volume of standing timber stock rather than a loss of other future multiple-use values. The timber program results in a tremendous increment to the current management costs. It also requires us to forgo the

Table 10–2. Mt. Baker–Snoqualmie Results: Objective Function, Costs, and Separable Costs
(values in thousands of dollars)

	FORPLAN runs				
	PNV	PLAN1	TIM	REC1	REC2
Objective function	150,299	41,253	118,083	41,572	41,401
Current expense	—	52,016	1,512	51,465	51,985
Full cost	—	109,046	32,216	108,727	108,898
Separable cost	—	—	76,830	319	148

large increase in the asset value of the forest that would result from the growth of unharvested timber stock.

The separable cost can be expressed as $1,070 per thousand cubic feet (mcf) of merchantable timber, which can be compared to an average received price of $1,570/mcf. In contrast to the Shoshone example, no significant incidental production of timber occurs under the Mt. Baker–Snoqualmie nontimber program TIM. The planned timber program results in an increase in present net benefits, and the timber program as a whole passes the separable cost test, as must be expected with our choice of an overall program PNV that was designed to provide the maximum present net value.

Below-Cost Sales. Were it not for the flow constraint in our model which limits fluctuations in timber-harvest levels across successive decades, much of the timber on our study area would be scheduled for harvest in the first decade. The timber is for the most part of substantial size and value, financially very desirable to harvest. The effect of the harvest-flow constraint is a smoothing of the harvest supply over time, which must result largely through the delayed harvesting of mature timber. However, simply forcing the delay of harvests from economically mature timber stands would greatly reduce the potential present net value attainable from the forest. To the extent possible, the linear programming model searches for cheaper ways to smooth the flow of harvest—ways that would allow the forest to maintain a high current harvest from the valuable mature stands.

The scheduling of future harvests on poor timber lands provides one way of maintaining both a smooth harvest flow and a fairly high allowable harvest from existing mature stands. The future flow of timber harvests from poor lands could, through the harvest-flow constraint, allow an immediate increase in the allowable current harvest. The future economic losses that would result from treating poor timber lands might be more than offset by the immediate increase in receipts from the higher current harvest. The tradeoff of future costs for current revenues is desirable because future costs are discounted in the overall present value calculation. That is, future below-cost sales may be used to justify high current levels of planned harvests. That is what we find in our Mt. Baker–Snoqualmie harvest scheduling solution.

In the Mt. Baker–Snoqualmie PNV solution, some very poor timberlands are scheduled for harvest from the fourth decade on. The most striking example is a timber analysis area of which roughly half is scheduled for harvest in the fifth decade. This 3,000-acre area (analysis area AA061) has stands of mixed age and species and is inaccessible for harvest except by expensive aerial logging systems. The proposed harvest in our solution for this analysis area—14,832 thousand cubic feet—makes up two-thirds of the scheduled harvest for the fifth decade. That harvest would result in a net loss of $3.4 million, as evaluated at the time of sale. Discounted to the present, the timber activity on this analysis area would result in a net loss of $589,000. This loss in present value is justified within the model because the harvest has the effect, through the harvest-flow constraint, of increasing the allowable cut from better stands in the early decades. In a FORPLAN run with no harvest-flow constraint, none of this analysis area would be scheduled for harvest.

It is questionable whether the forest would actually want to harvest such poor land when the time comes to do so. Certainly the large losses could not at that time be economically justified unless there were strong equity arguments to be made for sustaining the harvest level. The implicit claim that current high harvest levels could be sustained is probably not true. The current widespread occurrence of below-cost sales on the less productive forests of the Rocky Mountains may portend the situation that may arise later in the Northwest. On the poorer forests of the Rocky Mountains, past harvesting of the more accessible timber was sustained at high levels in anticipation of future harvests from less productive or less accessible stands. Now those forests are faced with the actual harvesting of the poorer stands.

5. The Use of FORPLAN in Program Evaluation

In practice, multiple-use planning at the level of the national forest has come to mean FORPLAN. As the basis for forest plans, FORPLAN models should logically provide grounds for justifying the planned harvest program. Indeed, there is no generally available alternative.

Limitations of FORPLAN

The inability of the Forest Service to use its forest plans to directly justify timber sales suggests that the planning process is not yet everything it ought to be. If a plan has led to the choice of a below-cost sale, it should be a simple matter to note this and to compare such a management action to the best alternative plan having no below-cost sale. An economic accounting to justify a below-cost sale requires only that the chosen action be shown to lead to a greater present value of net benefits than do the available alternatives.

The fact is that the forest planning models are not very well structured to permit economic evaluation of production decisions. The models are not designed to realistically reflect localized projects, such as a timber sale. And as now structured, the models have very limited flexibility. They do not have sufficient detail to allow more than a sketchy consideration of incremental changes in the scale and scope of the production program.

The Shoshone and Mt. Baker–Snoqualmie models are among the better FORPLAN efforts. They stand out because of their zone structure and the obvious attention to multiple-use values. The many other models, which are structured solely around timber analysis areas having no spatial definition, would appear to be of little practical use. Yet despite the quality of the Shoshone and the Mt. Baker–Snoqualmie models, it has become obvious that they are not perfectly suited to the type of incremental analysis that is needed to evaluate forest plans.

While all FORPLAN models allow considerable choice as to the level and timing of timber harvests, nothing even comparable to this flexibility is considered for the other multiple-use outputs. The Shoshone model, for example, does not provide means of providing roaded recreation, water-flow increments, or wildlife opportunities other than through timber harvesting. This inflexibility accounts for our probable underestimation of the separable cost of timber for the Shoshone. In the Mt. Baker–Snoqualmie model, few multiple-use services are considered to be directly affected by the level of the timber harvest; the model thus allows independent adjustments for each output, but perhaps at the cost of some loss in potential model accuracy.

A second problem comes from the lack of flexibility over time. The models are not structured to allow adjustments in land-use emphasis over time—a useful response to changing demands. In the Shoshone model, the choice of a zone allocation prescription results in a once-and-for-all determination of the land-use emphasis. We found that our attempt to determine separable costs for the first period alone was restricted by this inflexibility. It was not possible to choose a multiple-use emphasis in the first period and then shift toward a timber emphasis in later years. The result was that much the same management plan was proposed irrespective of our first-period emphases, and the separable cost of timber was again probably understated. The Mt. Baker–Snoqualmie model is somewhat less restrictive. It allows for a one-decade delay in the introduction of a management emphasis, with the zone held to its current management emphasis until implementation occurs.

A third problem is that some of the multiple-use services of the forest are either not represented or not valued in the FORPLAN model. This is quite understandable in many cases, but must be considered in evaluating the model solutions. The Shoshone model, for example, does not attempt to value the visual quality of the forest or the sedimentation that results from timber harvests, and the reduction in the quantity or quality of these services was not reflected in our computed asset value of the forest. As one result, the true cost of a timber harvest that results in deterioration of visual quality was understated.

There are other reasons why the structure of these FORPLAN models may lead to erroneous values for the separable cost of timber. The treatment of road costs in the Shoshone model provides an illustration. In the Shoshone model much of the road cost ($566,000) was associated with the aggregate emphasis prescription and thus was effectively independent of the chosen timber harvest level. While the allocation of road costs between the zone and timber scheduling components of the model might have been accurate at anticipated harvest solution levels, it probably is not accurate once we make the model choose a harvest level much lower than that anticipated. No doubt much of the zonal road expense could be avoided if it there really were no harvest activity. It has to be expected that a FORPLAN model will be designed to be most accurate in the range of the expected and familiar solutions.

Improving FORPLAN

We should acknowledge that criticizing the use of FORPLAN as we have above is a little unfair, since the planning effort has accomplished a great deal under difficult circumstances and the process is quite obviously improving over time. And after all, for adequate planning not every contingency can be or must be modeled. Limited data and the great expense of constructing and running FORPLAN models have most restricted the quality of the planning effort. Nevertheless, some practical improvements might be suggested.

The greatest deficiency of the FORPLAN models, as used in practice, is the lack of specific detail on current projects and timber sales. Often as much effort goes into modeling and solving for management decisions 150 years into the future as into the current choices. The models should be restructured to encourage a more detailed evaluation of current options. At the same time, the detail in which future harvests and treatments are planned should be greatly reduced. In terms of model size and solution costs, these two adjustments could offset each other, and would result in more useful models.

These changes in the FORPLAN models should be fairly easy to implement. We can view the modified linear program as having two components. The first is the zone allocation choice for the current decade. The model should offer choices among many different programs, including variations on timber sales, to be considered for implementation in the first decade. One output of each such prescription should be a classification of the condition of the zone resources at the end of the decade. All lands ending the first decade in the same condition class would then be aggregated. The second component of the model considers land use from the second decade on. For each of the condition classes a few broad prescriptions could be considered. In this component there should be limited specific detail in the planning choices and highly aggregated land classifications. Because of the effects of discounting, the inevitable inaccuracy in the later planning choices should have little impact on current choices. The second component of the model would serve mainly to ensure that the future sustained yield of the forest is taken into consideration when selecting the current program.

6. Summary and Conclusions

This chapter has been written largely in response to demands for the development of cost allocation procedures to be used as a basis for judgments on the Forest Service timber sale program. It seemed that the procedures most likely to be used would take into account only the current cash flow from each sale, neglecting the impact of timber harvests on the multiple-use services and on the future flow of value from the wider forest area. But no approach to the below-cost sales issue which focuses only on current expenditures and receipts from the sale can adequately address the economics of the timber program. In order to compare current benefits to a measure of current costs, it is necessary to develop a capital accounting view of costs. This is the approach we have taken, adding to current expenditure a capital cost, an amount which summarizes the impact of current actions on future flows of value from the forest as a whole. Such an approach provides the basis for the computation of economically relevant separable costs and the meaningful comparison of these incremental costs to the current benefits from individual outputs.

Our essential conclusions may be summarized as follows. First, an analysis of the timber sale area alone does not account for the spatial interdependence of forestlands. The timber sale itself is the wrong unit for analysis. One has to investigate a larger area in order to reflect all the interactions among the multiple-use outputs. Second, an analysis based on a single year's cash flow will fail to account for the temporal interdependence in forest management. A one-year cash-flow approach is irrelevant to an economic evaluation of a timber sale and the associated road construction, which may have long-lived effects on the forest and on the flow of resource services from the forest. Third, the desire to gain some perspective on the economic rationality of Forest Service timber sales should be the essential element in the concern with below-cost sales. In our view, this perspective will not be best gained by searching for cost allocation methods or by modifying Forest Service accounting procedures. A timber sale is economically justified if the expected net present value of benefits from the services of the forest would be higher with the planned sale than they would be from a production plan which excluded or reduced the harvest. A below-cost sale is not necessarily uneconomic. Whether a particular sale is uneconomic is an empirical question. Each sale should be viewed as a component of the total management program, to be evaluated on the basis of the present value of overall benefits and costs. However, it should be clear that a below-cost sale, with negative current cash flow, can never be justified on grounds of economic efficiency unless it results in an improvement in the flow of nontimber multiple-use values or in the net flow of value from future timber harvests.

Notes

1. T. J. Barlow, G. E. Helfand, T. W. Orr, and T. B. Stoel, *Giving Away the National Forests: An Analysis of U.S. Forest Service Timber Sales Below Cost* (Washington, D.C., Natural Resource Defense Council, June 1980).

2. U.S. Comptroller General, "Report to the Congress: Congress Needs Better Information on Forest Service's Below-Cost Timber Sales," GAO/RCED-84-96 (Washington, D.C., U.S. General Accounting Office, June 1984); R. E. Wolf, "State-by-State Estimates of Situations Where Timber Will Be Sold by the Forest Service at a Loss or a Profit: A Report to the Subcommittee on Interior Appropriations" (Washington, D.C., Congressional Research Service, Library of Congress, 1984).

3. See, for example, USDA Forest Service, *Report of the Forest Service: Fiscal Year 1985* (Washington, D.C., February 1986).

4. G. E. Helfand, F. K. Benfield, J. R. Ward, and A. E. Kinsinger, "Reform of Uneconomic Federal Timber Sale Procedures Is Badly Needed and Long Overdue," testimony of the Natural Resources Defense Council before the Subcommittee on Public Lands, U.S. House of Representatives (Washington, D.C., Natural Resources Defense Council, June 1985).

5. Committee staff returned a study under a covering memorandum, "Memorandum for the Chairman, Re: Timber Sales Program of the U.S. Forest Service," February 27, 1984.

6. M. Rasmussen, "Below-Cost Timber Sales," *American Forests* vol. 91, no. 1 (January 1985).

7. O. Eckstein, *Water Resources Development* (Cambridge, Mass., Harvard University Press, 1958); see also H. P. Young, N. Okada, and T. Hashimoto, "Cost Allocation in Water Resources Development," *Water Resources Research* vol. 18, no. 3 (June 1982) for a description of various cost allocation procedures.

8. Eckstein, *Water Resources Development,* p. 262.

9. Society of American Foresters, Below-Cost Timber Sales Task Force, "Fiscal and Social Responsibility in National Forest Management" (Bethesda, Md., 1986).

10. J. H. Beuter, "Federal Timber Sales" (Washington, D.C., Congressional Research Service, Library of Congress, February 1985). This useful description of federal timber sale practices provides several quotes from Forest Service testimony to the Appropriations Subcommittee regarding forest planning and below-cost sales.

11. K. N. Johnson, "FORPLAN Version 1: An Overview" (Washington, D.C., USDA Forest Service, Land Management Planning Systems, February 1986); K. N. Johnson, T. W. Stuart, and S. A. Crim, "FORPLAN Version 2: An Overview" (Washington, D.C., USDA Forest Service, Land Management Planning Systems, August 1986).

12. S. G. Boyce, "Management of Eastern Hardwood Forests for Multiple Benefits (DYNAST-MB)," Research paper SE-168 (Asheville, N.C., USDA Forest Service, Southeast Forest Experiment Station, 1977); J. W. Thomas, *Wildlife Habitats in Managed Forests: The Blue Mountains of Oregon and Washington*, Agricultural Handbook no. 533 (Washington, D.C., USDA Forest Service, 1979); E. A. Wilman, "Valuation of Public Forest and Rangeland Resources," Discussion paper D-109, Resources for the Future (Washington, D.C., August 1984).

13. E. G. Schuster and J. G. Jones, "Below-Cost Timber Sales: Analysis of a Forest Policy Issue," General Technical Report INT-183 (Ogden, Utah, USDA Forest Service, 1985).

14. For similar comments, see U.S. Comptroller General, "Report to the Congress," pp. iii–iv and p. 21.

15. USDA Forest Service, *Report of the Forest Service: Fiscal Year 1983* (Washington, D.C., February 1984).

16. We are not inclined to address this issue; but see J. A. Haigh and J. V. Krutilla, "Clarifying Policy Directives: The Case of National Forest Management," *Policy Analysis* vol. 6, no. 4 (Autumn 1980).

17. R. E. Benson and M. J. Niccolucci, "Cost of Managing Nontimber Resources When Harvesting Timber in the Northern Rockies," Research paper INT-351 (USDA Forest Service, Intermountain Research Station, September 1985).

18. A Forest Service report, USDA Forest Service, "Analysis of Costs and Revenues in the Timber Program of Four National Forests" (Washington, D.C., USDA Forest Service, Land Management Planning Systems, n.d.), discusses some implications of cost accounting practices.

19. Inter-Agency Committee on Water Resources, "Proposed Practices for Economic Analysis of River Basin Projects" (Washington, D.C., Government Printing Office, 1958); Eckstein, *Water Resources Development.*

20. Young, Okada, and Hashimoto, "Cost Allocation in Water Resources Development."

21. However, note the possibility that it might be even better to maintain the timber program with cheaper means of harvesting and less attention to amenities.

22. This is not a blanket endorsement of depreciation accounting for roads. Many forest roads do not improve the asset value of the forest. Some roads are obliterated after a timber sale, some cause severe environmental damage, and others will provide access only to future below-cost sales.

23. See F. M. Peterson and A. C. Fisher, "The Exploitation of Extractive Resources: A Survey," *Economic Journal* vol. 87 (December 1977).

24. Johnson, Stuart, and Crim, "FORPLAN Version 2."

25. For a variety of reasons, discussed in section 5, we suspect that the separable costs of timber are underestimated. In a paper by M. D. Bowes, J. V. Krutilla, and T. B. Stockton, "The Economics of Below-Cost Timber Sales" (Washington, D.C., Resources for the Future, July 1986) a possible upper bound on the true separable cost is determined to be $480,000.

26. The separable cost of the combined recreation and wildlife services was found to be only $7,000.

27. The setup for the AMENITY run was identical to that for the PNV run, except that no value was placed on timber outputs in the first decade. The present value of the amenity-emphasis solution was found by adding the value of any incidental first-period harvests to the actual objective function value.

28. Our six zones are ZO3, ZO4, ZO5, ZO6, ZO8, and ZO9 from the Mt. Baker-Snoqualmie FORPLAN model.

Appendix 10–A

The Structure of FORPLAN Linear Programs

In this appendix we describe the essential elements of the Shoshone and the Mt. Baker–Snoqualmie forest planning models, but do not pretend to have precisely or fully captured the structure of each model. In table 10–A1 a stylized example of a FORPLAN model is provided. Opportunities for the allocation of land and the scheduling of harvests over a two-decade planning horizon are represented.

This stylized "forest" consists of two geographical allocation zones identified as ZO1 and ZO2. There are also two timber analysis areas, AA1 and AA2. Zone ZO1 is an area of 6,000 acres, of which 1,500 acres are in the low-productivity analysis area AA1 and 4,500 acres are in AA2, the moderate-productivity timber class. Zone ZO2 has 4,000 acres, with 1,000 acres in timber analysis area AA1 and 3,000 acres in analysis area AA2.

Allocation Zones. In table 10–A1, look first at the block of columns titled Allocation Zones. For each allocation zone we have modeled two alternative management prescriptions or aggregate emphases. For zone ZO1 there can be a choice of either aggregate emphases AE1, with a high amenity emphasis, or AE2, with a high timber alternative. For zone ZO2 we can choose between the aggregate emphases AE1, an amenity emphasis, or AE2, a somewhat restricted timber emphasis alternative.

Timber Analysis Areas. Now look at the second block of columns in table 10–A1. For each timber analysis area there are five alternative timber management prescriptions or harvest schedules modeled. These are: MN, a minimum level or no-harvest alternative; LO_1, a low-intensity management alternative with the harvest in the first decade; LO_2, a low-intensity timber management alternative with a harvest in the second decade; MD_1, moderately intensive timber management with a harvest in the first decade; and MD_2, moderately intensive timber management with the harvest in the second decade.

For convenience, only two time periods have been considered in the example. More typically, the planning choices and the resulting outputs and costs would be modeled over fifteen decades. The size of the timber scheduling component of the linear program increases greatly in such an expanded time-horizon model. In particular, there would be many possible choices for the timing of thinnings and harvests from the existing timber stands, and each choice could be combined with various options for the management and harvest of regenerated timber on the same land area.

Outputs and Cost Columns. The third block of columns, Outputs/Costs, identifies the cumulated outputs and costs that result from the choice of management prescriptions. We include here the timber harvests in decade 1 (H_1) and decade 2 (H_2),

Table 10-A1. Example of a FORPLAN Linear Program for Multiple-Use Forest Planning

Column groups: *Allocation zones* — ZO1, ZO2 (Aggr. emph.[a]); *Timber analysis areas* — AA1, AA2 (Harvest schedule); *Outputs/costs* — H1, H2, R1, R2, C1, C2.

	ZO1 AE1	ZO1 AE2	ZO2 AE1	ZO2 AE2	AA1 MN	AA1 LO1	AA1 MD1	AA1 LO2	AA1 MD2	AA2 MN	AA2 LO1	AA2 MD1	AA2 LO2	AA2 MD2	H1	H2	R1	R2	C1	C2	Right-hand side
Zone acreage	1	1																			= 1
Zone acreage			1	1																	= 1
Analysis area acreage					1	1	1	1	1												= 2,500
Analysis area acreage										1	1	1	1	1							= 7,500
Access limits (first decade)	−.075	−.125	−.075	−.125		1	1														≤ 0
Access limits (first decade)	−.225	−.375	−.225	−.375							1	1									≤ 0
Emphasis restrictions			−.25		1																≥ 0
Emphasis restrictions			−.75							1											≥ 0
Emphasis restrictions				−.125	1					1											≥ 0
Emphasis restrictions											1										≥ 0
Emphasis restrictions													1								≥ 0
Timber output						90	110	100	120						−1						= 0
Timber output											95	115	110	130		−1					= 0
Recreation output	.167	.100	.150	.900		.01	−.04	.01	−.04		.01	−.02	.01	−.04			−1				= 0
Recreation output	.183	.108	.157	.905		.02	−.02				.02	−.02						−1			= 0
Cost	.500	.917	.625	1.125		2	3	2	3		2	3	2	3					−1		= 0
Cost	.533	.833	.675	1.000		.2	.4				.2	.4								−1	= 0
Objective function	A11	A12	A21	A22	B11	B12	B13	B14	B15	B21	B22	B23	B24	B25							= PNV

Note: See text of appendix 10–A1 for explanation of figures.

[a] Aggr. emph. = aggregate emphasis.

recreation output in each decade (R_1 and R_2), and the overall management costs in each decade (C_1 and C_2). In actual FORPLAN models there could be many more outputs and there might be a detailed tracking of costs for specific management activities. An output or cost column need not be generated unless there is a specific constraint associated with the output or cost. The columns here are to highlight the multiple-use nature of the production activities.

Zone Acreage Constraints. The first block of rows in table 10–A1 represents constraints indicating that all acreage within each allocation zone must be assigned to a management emphasis. For example, the land in zone ZO1 could be assigned to either aggregate emphasis AE1 or to emphasis AE2. The emphasis variables are measured here in terms of the fraction of the zone assigned to the emphasis. A value of zero would indicate that none of the zone was allocated to that emphasis, while a value of one would indicate that the zone is fully allocated to the emphasis. As the model is specified, it is possible that a zone might be fractionally allocated among more than one emphasis. Such a solution would violate the logic of the model. Fortunately, this type of solution does not often result. If an area were to be assigned to more than one emphasis in a solution, the model would usually be tinkered with to ensure the selection of a single emphasis in each zone.

Analysis Area Acreage Constraints. These constraints indicate that the acreage in each timber analysis area must be fully assigned to the various timber management prescriptions (MN, LO_1, MD_1, LO_2, and MD_2). It is quite likely that a timber analysis area will be assigned to more than one timber prescription. This occurs because of the direct restrictions on harvesting imposed by the aggregate emphases chosen for each management zone. More generally it will occur because of the desirability of spreading harvest levels evenly over time.

Access Limit Constraints. The access constraints give an upper bound on the acreage of timber that can be harvested in the first decade from each timber analysis area. Such constraints may be used to represent the effect of gradual construction of a road network or may reflect environmental concerns over the rate of harvesting. In national forest FORPLAN models accessibility constraints are often included for the first four decades.

In table 10–A1, the first access constraint indicates the maximum acreage in analysis area AA1, across all zones, that is available for harvesting in the first decade (in other words, that is to be allocated to harvest schedules LO_1 or MD_1). The second access constraint indicates the maximum acreage in analysis area AA2 that may be harvested in the first decade. In each case, no more may be harvested than is made accessible by the chosen zone management emphases. In our example, emphasis AE1 makes 30 percent of a zone available for harvesting in the first decade; it is assumed that each analysis area is equally accessible. Emphasis AE2 makes 50 percent of the land in a zone available in the first decade. Since, in our example, one-fourth of each zone is in analysis area AA1 and three-fourths are in AA2, each acre assigned to emphasis AE1 makes .075 acres of AA1 and .225 acres of AA2 accessible for first-period harvest. Similarly, each acre assigned to emphasis AE2 makes .125 acres of AA1 and .375 acres of AA2 available for early harvesting.

Emphasis Restrictions. Emphasis constraints allow for restrictions on the choice of harvest emphasis to ensure that timber treatments within a zone are consistent with the chosen zone management emphasis. In table 10-A1, the choice of zone aggregate emphasis AE1, the amenity emphasis, on zone ZO1 can be seen to restrict timber management to low-intensity prescriptions. In particular, all 1,500 acres of timber analysis area AA1 present in this zone (.25 of the total zone area) must be allocated to minimum-level, no-harvest (MN) management. The 4,500 acres of timber analysis area AA2 which are in this zone (.75 of the zone area) can be allocated only to minimum-level (MN) or low-intensity harvest management (LO_1 or LO_2). For zone ZO1 the choice of aggregate emphasis AE2, the timber emphasis, results in no constraints on timber harvesting practices.

Outputs and Costs Rows. Each prescription generates a flow of outputs and costs over time. For the zones, the choice of an aggregate emphasis prescription will result in recreation visits and some expenditures in each of the two decades. For example, the choice of the amenity emphasis AE1 for zone ZO1 would give us 1,000 recreation visitor days (.167 RVD per acre) in the first decade and 1,100 visitor days (.183 RVD per acre) in the second decade. This amenity emphasis would also require expenditures of $3,000 ($.50 per acre) in the first decade and $3,200 ($.53 per acre) in the second decade. The choice of the timber emphasis AE2 for this same zone would give us less recreation, but because of the greater roading would require higher expenditures.

The choice of a harvest scheduling prescription for the timber analysis areas results in a flow of timber harvests as well as incremental changes in the recreation outputs and the expenditures in each decade. For analysis area AA2, the choice of low-intensity timber management with a harvest in the first decade (LO_1) would result in a per-acre harvest of 9,500 cubic feet in the first decade, per-acre increments in recreation of .01 and .02 visitor days respectively over the two decades, and an increase in costs of $2 per acre in the first decade and $.20 per acre in the following decade. Thus if all 4,500 acres of timber analysis area AA2 in zone ZO1 were allocated to harvest prescription LO_1, first-period recreation from the zone would be increased by 45 visitor days above the level resulting from the zone aggregate emphasis prescription, while first-period costs in the zone would be increased by $9,000.

As with the output and cost columns, these rows need not be explicitly generated in the linear program unless there are to be constraints associated with the output or cost.

The Objective Function. The final row of our linear programming example in table 10–A1 gives the objective function. Our objective function calls for the maximization of PNV, the present net value of the flow of services from the chosen land allocations. Each prescription, whether for a zone or an analysis area, has a number entered in the objective function row. This number gives the discounted value of the time stream of outputs and costs that results from implementation of the prescription.

To illustrate, let us represent timber harvest prices by P_h and recreation values by P_r, and let δ represent the discount factor applied to values in the second decade. Our objective function value A_{11}, associated with management emphasis AE1 on zone 1, is then equal to P_r $(1,000+1,100\delta) - 3,000 - 3,200\delta$. Similarly, our objective function value B_{13} associated with the choice of harvest schedule MD_1 for timber analysis area AA1 is equal to $100P_h + P_r(.01+.02\delta) - 2 - .2\delta$.

Most forests put a value on all those outputs for which one can reasonably be determined. For outputs for which no value is available, the forest may choose to set constraints to ensure a certain level of production. Step demand functions, with lower output prices at higher output levels, can be accommodated in FORPLAN models. The step demand function option is most often used in situations where the capacity to provide an output exceeds any likely level of use. Any output provided in excess of the maximum demand level would be valued at zero.

11

Conclusions: An Interpretive Synthesis

1. Introduction

Two matters, one empirical and one conceptual, are of critical importance in understanding the economics of multiple-use management of public forestlands. The first concerns one widely acknowledged and generally well-known fact: the extent to which recreation resources have emerged as the dominant resources on the public forestlands during the second half of the twentieth century.[1] Except on the Pacific slope forests and the southern forests, the value of national forest recreation resource services substantially exceeds the value of national forest timber resources. For the National Forest System overall, the value of annual recreational services is roughly double the value of the annual timber sales (see chapters 2 and 7). Curiously, the implications of this development have at times been overlooked in analytic policy studies. Failure to recognize this development often leads to flawed policy recommendations by those whose analyses implicitly assume that systemwide national forest policy prescriptions can be made largely on the basis of national forest timber data.[2]

The second matter, of equal importance, is to understand the analytic implications of multiple-use forestland management for policy studies. The analytic approach to multiple-use management requires a much different mind-set from that needed for the approach often taken when public forestland policy is addressed. The reason for this may lie in the history and development of forest economics.

The economic principles of forest management were first illustrated by Martin Faustmann by means of a simplified capital theoretic formulation. For this purpose a single-stand model, adequate where timber is the sole use of the land, is sufficient to indicate the economic principles for determining the commercial maturity of a timber stand (see chapter 4). Indeed, in the case of a single-use forest plantation, the classic Faustmann formulation will suffice, and the solution, in principle, is not difficult. A difficulty arises when the congenial single-stand model, its use fostered by almost a century and a half of familiarity in the discourse of forest economics, becomes part of a default mind-set and is unwittingly used to derive management principles—or worse, to treat national forest policy issues—as though a model for the management of a single-use forest would suffice. While resource analysts generally perceive that public forestlands are intended to provide a mix of forest outputs, some credentialed policy analyses of forestland management issues have not been entirely liberated from the world of Martin Faustmann.[3]

To appreciate the operational significance of multiple use, one needs to go beyond the management of a set of single and independent stands to accommodate in part the biological diversity on which multiple forest outputs depend. This is by now second nature to forest ecologists, wildlife biologists, and especially practicing land managers on the national forests themselves. Consequently, if we are to provide a conceptual framework that is adequate to deal with the economics of multiple-use management, or with policy regarding multiple-use management issues, we must introduce explicitly into our analyses an economic model that is capable of handling multiple interdependent stands (or, more properly, sites) (see chapter 4). For adequate analysis, it is important to recognize the significance of the linkage or interactions among stands in providing the biophysical environment in which other outputs, in addition to timber, can be produced. From chapter 5, recall the interdependence between the patch-cut sites and the surrounding sites that supported residual stands to promote increased water for streamflow. We see other, perhaps not so precisely specified, relationships between forage and cover in wildlife habitat prescriptions.[4] Of course, the relationship between harvests and habitat can also be negative, as when private, cost-effective logging results in the silting of spawning beds of trout and salmon. The relationship between forest recreation and timber harvests is quite complex, depending in part on the recreation activities in question, yet it cannot be ignored in the light of the economic importance of forest recreation. Failure to recognize these site interdependencies has led some to advance policy prescriptions that could eliminate much of the multiple-use enjoyment of national public forestlands.

As an example, it has been advocated that the public timber lands be separated from the recreation lands, divesting the former to the private sector and retaining the latter in public ownership. The rationale for this prescription is that by "concentrating on what each type of land does best, the total output value of recreation and commodity production would be increased."[5] Such a prescription would require that the forests of the Pacific slope and the

South concentrate on timber production under private ownership and management, and that the national forests of the Southwest, the central Rockies, the Ozarks, and the Appalachians concentrate on the production of forest recreation. Implied is that some participants in Pacific slope forest recreation would have to be accommodated by regions other than their own, and that if this were either not feasible or more costly, there would be welfare losses in the region for those whose weekend recreation opportunities had been pre-empted. The prescription also suggests that timber would not be harvested in Ozark and Appalachian forests even though both have areas where logging and recreation can be accommodated.

Overlooked in the policy study in which this prescription appeared is that forest recreation in the Pacific slope region depends in part on the rather specialized forest environment there—an environment as good for spawning salmon and steelhead as it is for growing timber. Further, in the California region of the National Forest System (region 5), which contains a significant share of the Pacific slope forests, the value of the annual forest recreational services exceeds by a comfortable margin the annual value of timber. It is at least partly because of the high value of these prime recreational resources on the national forests that the annual value of recreation services provided on all of the national forests combined greatly exceeds the aggregate value of the annual timber harvests. When it comes, then, to classifying what is timberland and what is forest recreation land, a taxonomic approach will not find the mutual exclusivity implied by the policy prescription.

Timber and forest recreation lands below timberline are usually coextensive. Stream sides and riparian zones hold both attractive recreational sites, quality sites for growing trees, and important habitats for fish and wildlife. Where the object is only to enjoy a pleasant forest environment, the careful location and relocation of campsites can accommodate both recreation and timber management objectives in the same area of the forest, if not in close proximity (see chapter 7). Moreover, in some cases the pursuit of multiple-use production takes advantage of the interaction among sites when the objectives are pursued in a properly coordinated fashion. In the light of these interactions, a policy prescription to convert national forests into single-purpose units is akin to arguing for the separation of orchards and bees so that each might concentrate on what it does best.

One of the interesting practical advantages of the multiple-interdependent-stand approach is the opportunity provided to avoid the pitfalls of an unwitting independent-stand, single-product mind-set when addressing public land management policies. In one important current example, the single-stand formulation leads inevitably to a conclusion that the harvest of any stand is uneconomic under conditions where the proceeds from the sale fall short of covering the costs of preharvest management, sales preparation, and harvest administration. But from the perspective of a multiple-stand, multiple-use model, it is clear that such a sale is not *necessarily* uneconomic (see chapters 4, 5, and 10). That is to say, when more than one stand or site is considered, it is possible to appreciate that even recently harvested stands may provide enhancements in the value of resource services from the complementary or adjacent sites. Recall again the relation between the patch cuts and the residual stands in a watershed water augmentation program (chapter 5), or the relation between forage and cover for the improvement of wildlife habitat (chapter 6). The importance of such site interdependencies in providing amenity and aesthetic services is well attested to by the powerful opposition to large clearcuts, extensive roading, and the harvesting of timber by those who look to the public forestlands for outdoor recreation. Because of the important resource interactions among sites, the change in value of the between-harvest resource services from the full area affected by a harvest must be reckoned, as well as the value of the timber harvest from the sale area alone. Accordingly, we cannot assert that the value of the total yield of resource services of a managed site does not exceed the management costs even if the value of the timber harvest—an integral part of site management— alone does not. It must also be acknowledged, however, that not every below-cost sale is *necessarily economic* simply because it serves a multiple-use objective; nor will *all* conclusions using a single-stand model necessarily be incorrect. But the issue is open to resolution by a proper empirical analysis (see chapter 10).

Again, with a single-stand model or with any model that does not take into account the interdependence among stands, managing for old-growth stands will almost never be justified. Unless harvesting is extremely costly, the economic rotation age will be reached long before a stand reaches old-growth status. If an old-growth stand is inherited from the past, it will be scheduled for harvest as quickly as is convenient. We have seen this occur on private timberlands in the Douglas fir region as well as on private industrial forests throughout the country. Now it is true that we would not expect management for old growth, even in an interdependent-stand forest model, if we were addressing the problem of economic timber management *starting from bare ground*. The carrying cost of waiting for the old-growth forest to emerge makes such management uneconomic under normal circumstances, whether a single-stand or multiple, interdependent-stand model is used. However, because of the value of old-growth groves in a multiple-objective manage-ment regime, under which these stands would potentially augment the flow of resource services from a much wider forest area, it is possible to conceive of conditions where retention of inherited stocks of old growth would be economically justified, although perhaps not on the most productive timber sites.

While we can conceive of situations in a multiple-stand, multiple-use management regime whereby inherited old-growth groves would be retained under efficient management, there remains the legitimate question of how much old growth should be retained. At best we would have to know a great deal more than we currently do to fix an economic value on increments or decrements in stocks of old growth. Given the irreversibility attending the elimination of an old-growth grove and an irreplaceable ecosystem, however, one would in principle require a premium above an uncomplicated benefit-cost comparison to warrant the destruction of an old-growth grove.[6] We do not see much prospect of acquiring such information from either natural or social scientists before the issue is resolved politically. And perhaps under the circumstances the political arena is where the issue ought to be resolved.

2. Concise Review of Differences in Model Results

Before proceeding further we should summarize some of the differences we may expect between single-use timber management and multiple-use management of interdependent stands. First, it is important to recognize that multiple stands, as well as single stands, can be managed for a single use (timber, for example).[7] However, economic multiple-use management cannot be effectively achieved in the absence of consideration of multiple stands and the appropriate distribution of age classes and other attributes of a biologically diversified forest. The findings summarized below, produced from a dynamic optimization multiple-stand model, contrast the results from management for timber alone with results from the same model when applied to management for multiple objectives. In both cases the same set of sites and the identical initial economic conditions (price, discount factor, and others) are used.[8] While the results are specific to particular applications of the model, they are suggestive of a potential richness in multiple-use management that has not been made apparent in previous analytic treatments of the economics of forest management.

In applying the model, management for timber alone was found to be uneconomic at low timber prices. More precisely, under single-use timber management, at sufficiently low timber prices, we found that all initially stocked stands would be promptly harvested and then abandoned, with no artificial regeneration and with no further timber management. Under multiple-use management there was a greater likelihood of regeneration and continuing harvest. The amenity value of standing timber and the value from stand age diversity encouraged the continuing management of the timber resource. Even at very low timber prices, some level of timber harvest was desirable, as harvesting represented an effective means of manipulating vegetation for improved wildlife habitat. In general, on public forestlands, below-cost timber sales and timber management were often found to be justified economically on the basis of the long-term improvement in multiple-use value that might result from harvesting.

At high timber prices we found, as might be expected, that the full multiple-use solution corresponded quite closely to the timber management solution. However, even with high timber values we often found that some subtle and interesting effects on the harvest schedule resulted from our consideration of the full mix of multiple-use values. If we began with a forest area of fairly uniform stand age, we found that it could prove advantageous, particularly in the initial decades, to delay some harvesting or to delay the regeneration of a stand in order to introduce a more diverse age distribution. These means of introducing diversity in the age mix of stands were found to impose few financial costs, and sometimes improved the flow of amenity values significantly.

In general, the effect that consideration of multiple-use values will have upon the scheduling of timber harvests cannot easily be described. Analytic work based on the study of single stands has led many to conclude that multiple-use values will be best met by the implementation of longer rotation periods than would typically be desirable under management for timber alone. This is said to be true at least when older stands are to be preferred for their amenity value. In contrast, in our multiple-interdependent-stand example the benefits of maintaining some balance in the mix of age classes were seen to make the optimal harvest timing dependent upon the current conditions (that is, the specific age classes in the set of stands) of the forest area. Harvest schedules, especially during the initial few decades of a multiple-use management regime, could be very complex. Younger stands might be indicated for harvest cuts, while older stands were left unharvested. Also, a higher relative recreational value did not necessarily lead to a longer rotation, even when older stands were generally preferred for their amenity values. Instead, we might see an increasing amount of the area set aside as protected old growth, while a shorter timber rotation cycle was instituted on the remaining sites in the management unit. We found such solutions even when the management area was perfectly homogeneous. Such harvest solutions could not be found if the stands were treated independently.

We found too that the benefits from diversity of standing stock might lead to the allocation of areas of the forest in a manner that looked like the specialization of function by land area. This seemed to be the case, for example, when we chose to indefinitely preserve some stable stands of old growth while harvesting adjacent stands on short timber rotations. However, it is important to note that such specialization resulted within the context of the multiple-use model and does not correspond to the solution that might result if land were initially allocated to primary functions. It seems likely that specialization of land use, such as we found, is often apt to result in more effective production of such services as wildlife and increased water flow than would be possible from uniform management of a land area.

An interesting related result of the application of our model was the potential advantage from the specialization of use over time, rather than location. We found that multiple-use values were sometimes best provided by periodic high levels of timber harvesting, with intermediate periods of little or no harvesting. Such a policy of uneven harvest flow could be indicated for areas where high levels of harvesting promote wildlife values and low levels have little benefit. They might also be indicated for areas where amenity values are greatly degraded by even the lowest levels of harvesting. In such areas we might choose to have periods of heavy harvesting, followed some years later with periods of high amenity values. All of the complex harvest solutions described above contrast strongly with the simplistic harvest schedules that result from single-stand or independent-stand models.

The differences we have seen in timber management under the economically efficient multiple-use management of interdependent sites and under the single-use timber solution can be expected to be found where the various multiple-use values depend on the diversity of the forest. Where sufficiently high values are attached to the nontimber services, the improvement of the age mix or other attributes of the forest may dominate completely the timing of the harvest (see chapter 5). The complexity of the harvest solution results from the nonlinearity of the nontimber benefit function across the individual sites. Such nonlinearity seems generally to be the case.

To encapsulate, the general multiple-use harvesting policy is seen to be complex; no simple rule of thumb is likely to describe it. We can expect that sometimes younger stands should be harvested while older ones are left uncut. We can expect that rarely will there be a uniformly optimal harvest rotation age, especially over the early years of management. We can expect that some sites may be managed as set-asides for special purposes—stable old-growth groves in one case, permanent clearings in another (see the discussion of water augmentation in chapter 5). We might sometimes expect a policy of specializing over time to be optimal, with forestland producing high timber harvests in some periods and high recreational benefits in others. We find that even-flow policies and uniformity of management standards are not inherently desirable goals; coupled with allowable-cut harvest incentives, they can be detrimental to sound management. The optimal harvest age is not likely to be at the age of maximum sustained yield. Indeed, long rotations, far from being a desirable compromise in multiple-use management, may simply provide both uneconomic timber and a poor balance of age classes for nontimber uses. These conclusions flow from a more realistic perception of the forest condition for multiple-use management than is possible from the perspective of a single-stand model. Perhaps partly as a result of that more realistic perspective, the harvesting decision can be quite sensitive to factors about which we have little empirical knowledge.

3. Further Observations on Realistic Multiple-Use Management

The linkage between sites and time periods makes it impossible to obtain a discerning view of the economics of multiple-use forest management when the analysis is encumbered by a single-stand economic framework. Nor is it meaningful to render an economic accounting on the basis of the financial results of a single sale. For competent economic analysis and policy prescription, it is essential to address a realistically scoped management or accounting unit that spatially contains the major site interactions in a designated collection of sites for capital accounting purposes. We have suggested that for conducting an analysis of a management plan an appropriate unit might be a drainage—perhaps a fourth-order drainage—as was done in chapter 10. Although such an analysis does not provide a system of accounting in a financial sense, it does provide a mechanism for economic accounting—that is, it is designed to address the external effects of harvests (whether positive or negative) and the spatial and temporal interdependencies through changes in forest asset values. Such a design is important for the budgeting and appropriations processes, and is beyond the capability of a commercial or financial accounting system. Moreover, this approach builds on the positive accomplishments of Forest Service personnel who have worked on the first-round forest plans under the National Forest Management Act of 1976. Through the use of FORPLAN, all current and capital expenditures are converted to a capital stock value which can then be tracked through time to yield information on the economic performance of meaningful forest management tracts, in lieu of specific timber sales.

The proper analysis of the below-cost sales issue is of prime importance to the economic management of public forestlands because it lies at the heart of the concept of using tracts of forestlands for more than one purpose. Moreover, with the explosion of demand for outdoor recreation in the postwar years and the growing reliance of the construction industry on softwood sawtimber from the national forests of the Pacific slope, the need to accommodate as many uses as are consistent with efficient management is an imperative in public lands management.

Having argued for an analytic model similar to that being used in the current forest planning exercise directed by NFMA, we hasten to add that great improvements in the data and the model's application will be required in the future to provide the flexibility needed for the results to be fully acceptable as a form of accounting to Congress and the public. Two types of deficiency must be addressed: (1) the quality and the extent of the data that represent the nonmarketed or nonpriced resource services, and (2) the way in which the FORPLAN model is structured and applied.

A preliminary point should be made about the factual information used by forest planners. It is generally not the estimate of total nonpriced resource services that is of relevance to most forest planning situations (albeit such information is useful for obtaining perspective on the relative value of different forest resources); rather, what is normally to be sought is the change in the value of an output, or combination of outputs, in response to incremental changes in an input bundle or management activity (see chapters 5 and 6). Generally this is a harder set of data to develop; partly because this is so, substantial ignorance exists concerning precise biophysical responses to management activities, under specific conditions, for several multiple-use outputs.

In some cases, as in the relationship between habitat and wildlife, the response to management of vegetation is just now becoming sufficiently well understood to serve in the development of quantitative response models for the economic analysis we have in mind.[9] Improvement of our knowledge of biophysical responses to incremental changes in management inputs should be a high priority if we expect to get a proper capital accounting of forestland management.[10] When the immediate resource response is not known, the relevant information, which is the economic value of the resource response, can sometimes be inferred directly from the behavior of the consumers of the resource services (see chapter 6). We might be fortunate in being able to evaluate the increment in management activity in the absence of knowledge of the direct biophysical response, but this will not always be possible, particularly if the product or service is consumed off site. For example, the determination of benefits from increased water flow resulting from management of vegetation would prove impossible without knowledge of the hydrology of the watershed. Without such technical knowledge, we could not determine the increment in water flow to be valued.

Another area of deficient information has to do with the economic valuation of the biophysical product of forestland management. Getting a monetary estimate of the incremental benefits from incremental changes in management inputs depends in great measure on being able to make realistic estimates of the resulting biophysical change in the forest condition. Realistic estimates now seem in prospect in the areas of silvicultural treatment, forest hydrology, and wildlife biology. It is also possible that research can broaden and deepen the current incipient effort to model demand behavior in response to management activities of specific kinds, as in the recreational hunting case treated in chapter 6. But this has been a little-researched area, and knowledge of incremental recreational response to increments of management activity falls far short of providing the quantitative information one would like to see undergirding economic valuation of alternative prescriptions for forest management and capital accounting.

Quite apart from the issue of accounting-quality data (data good at least to a first approximation) is the problem of the way linear programming models are used and FORPLAN is structured in the current planning effort. One important desideratum is to have the FORPLAN models structured geographically around land units so as to capture site interactions and the role of location-specific features such as roads. An appropriately sized drainage would be the most likely naturally defined area for this purpose. Most of the site interaction will occur within, rather than between, drainages. Within a drainage there is certain to be hydrologic coherence, and there is likely to be a coherent transportation net in a geographic area defined by drainage boundaries.

With the advent of improved hydrologic models and models of wildlife responses to changes in the forest condition, greater flexibility in production relations seems possible than is assumed to underlie various current FORPLAN model runs. It would appear that nontimber outputs could often be achieved with a more limited reliance on timber harvests to alter the forest conditions. The inclusion of such alternative management options in FORPLAN is needed in order to have greater confidence in the economic justification for the selected timber harvest programs. In the current use of FORPLAN, the mix of alternatives considered for each area appears to be too restrictive. For each geographical subdivision of the forest, we need a set of plans that reflects production possibilities for each of several budget amounts; for each budget level, a range of output emphases should be considered.

Admittedly, the increase in detailed analysis that these suggestions imply will have implications for the cost of forest planning, and there is already considerable concern over the resources that have been required by the cruder, more aggregated analyses associated with the current round of forest plans. This is a legitimate concern, and we would not press for such a scaled-up analysis unless the Congress desires a more adequate economic accounting of national forest management. But if that is desired, Congress must appropriate the means to implement the more discriminating analyses.

With the recreation resources of the national forests becoming so dominant in satisfying the nation's demands for forest-related recreation and also becoming such a large part of the value of national forest resources, we believe that more attention should be paid to geographically specific characteristics of subdivisions of the forests in estimating the demand for, and value of, recreational features of the forest. Currently in forest planning rather ad hoc values are employed as rough surrogates for the value per recreation visitor day. This practice reflects insufficient research on demand for forest recreation, and should be remedied.

We now come to a major conundrum for multiple-use management. Other things being equal, the Congress is interested in the management of national forests in an economic manner. This motive is apparent in the directive given the Forest Service by Congressman Yates of the Appropriations Subcommittee to prepare an accounting system that would reveal the cost of producing timber, with the costs of multiple-use management on behalf of other forest products netted out. Such concern about economic management, it may be presumed, would motivate the subcommittee in other areas of its responsibility as well.

In chapter 9 we addressed the problem created for efficient forest management by the budget line-item format used by OMB and the Congress in preparing the Forest Service budget and appropriations. Given jointness in production and interdependencies among national forest programs, appropriations that respond to a line-item budget format—and the personnel and program targets specified—are almost certain to result in a program that is inconsistent with the forest plans based on any level of appropriations assumed for forest budget proposals. It is fair to ask whether one would be justified in improving the detail and content of the forest plans in the interest of responding (in part) to congressional accountability, if annual congressional appropriations and targets conformed poorly if at all to a given forest's production plan for any level of appropriated budget. We do not mean to suggest that it is not the congressional prerogative to set the size of the annual budget, but we do suggest that the process by which the appropriated budgets and targets are specified should be consistent with the actual production relations on the forests that they are meant to support. Regardless of improvements in forest planning and economic accountability, unless consistence is achieved the improvement in efficiency at the forest level may not survive the budget and appropriations process at the Washington level.

It seems to us, then, that before any further effort and expense are incurred to improve the economics of multiple-use management at the forest level, thought needs to be given to the gap in effective communication about budgets and program targets between those supplying the means and those implementing the management programs. There is little point in requiring national forest management to achieve efficiency goals when the means provided to accomplish those goals preclude their realization.

This volume has concentrated on the economics of multiple-use forestland management. Multiple use is generally associated with the provision of public goods and other nonmarketed services. When nonpriced resource services are supplied, their external funding is a necessary function. Now, when agencies of the public (whether of the executive or legislative branches) become involved, the efficiency criteria from neoclassical resource allocation theory are bound to be supplemented by additional criteria arising out of both administrative and political considerations. Indeed, in chapter 9 we discovered that the budgeting, appropriations, and accounting procedures established for administrative and political convenience have created some difficult problems for the economic management of public forestlands. Consequently it is probably true that on almost any Tuesday afternoon, somewhere in the National Forest System, one could find a land manager engaged in doing something an economist would regard as foolish. The answer to this problem, we submit, is not to call for a transfer of public lands to private ownership. That issue has been thoroughly reviewed by the Public Land Law Review Commission (in 1970), and legislation pursuant to the commission's recommendations was enacted in 1976 for both the national forests and the public domain lands. Moreover, there is no evidence that the land would be managed more efficiently in private ownership.[11]

It is important to realize that much of the national forestland is not particularly productive from a silvicultural point of view, and that is one of the reasons it has remained in public ownership. It is equally important to realize that this land, largely because of its mountainous character, is highly valued by the public for its forest and stream recreational resources. This great value is not apparent in the financial analysis of the public forests, since to date neither the land management agencies nor the Congress has been disposed to rely extensively on user charges for recreational services. The absence of user fees is largely a result of administrative infeasibility, as most forest recreation is dispersed. Fee collection would be equally problematic under private ownership; thus there would be no private motivation to provide adequate services from nonmarketed but highly valued resources of the national forests.[12] This suggests that there are bound to be economic inefficiencies, of perhaps different kinds, regardless of the ownership of the public lands. One thing is certain: there would be a different mix of uses on public forestland than on private forestland. It seems, however, that the uses provided under public ownership are the most popular—and these are precisely the resource services that would not be provided adequately under private ownership.[13]

Given the flawed management of the public lands when viewed through the also-flawed prism of neoclassical resource allocation theory, we feel that the answer is not to exchange one economically flawed land ownership system for another. Rather, we feel that it is important to realize that there is potential for improvement of management under any system of land ownership, and that this potential merits serious and continuing attention. With that conviction motivating our efforts, we have sought to bring to the economic analysis of multiple-use management of public forestlands an updated theoretical framework and an illustration of its applicability to the management of our public forestlands.

Notes

1. The first to have recognized the emergence of outdoor recreation as a major public policy concern was Marion Clawson (1959).

2. See M. Clawson, "The National Forests," *Science* vol. 191, no. 4227 (February 1976). Here it is implied that National Forest System budgets should be allocated to individual regions and forests on the basis of cash receipts. With timber sales accounting for upwards of 90 percent of the total cash receipts in the years examined (FY 1970–1974), the prescription is equivalent to returning to a single-use timber management regime.

3. Clawson, "The National Forests"; J. V. Krutilla and S. Brubaker, "Alaska National Interest Land Withdrawals and Their Opportunity Costs," in Committee Print 4, *Background Information for Alaska Lands Designations,* Subcommittee on General Oversight and Alaska Lands of the House Committee on Interior and Insular Affairs, 95 Cong. 1 sess. (1977); W. F. Hyde and J. V. Krutilla, "The Question of Development or Restricted Use of Alaska's Interior Forests," *Annals of Regional Science* vol. 13, no. 1 (March 1979).

4. See, for example, S. G. Boyce, *Management of Eastern Hardwood Forests for Multiple Benefits,* USDA Forest Service Research Paper SE-168 (Asheville, N.C., Southeastern Forest Experiment Station, July 1977); R. S. Holthausen, "Use of Multistand Projection Models for Coordinated Resource Management," in J. Verner, M. L. Morrison, and C. J. Ralph, eds., *Modeling Habitat Relationships of Terrestrial Vertebrates* (Madison, University of Wisconsin Press, 1986); J. W. Thomas, *Wildlife Habitats in Managed Forests: The Blue Mountains of Oregon and Washington,* Agricultural Handbook no. 553 (Washington, D.C., USDA Forest Service, 1979).

5. Robert H. Nelson, "The Public Lands," in Paul R. Portney, ed., *Current Issues in National Resource Policy* (Washington, D.C., Resources for the Future, 1982).

6. J. V. Krutilla and A. C. Fisher, *The Economics of Natural Environments: Studies in the Valuation of Commodity and Amenity Resources* (rev. ed., Washington, D.C., Resources for the Future, 1985), chaps. 3 and 4.

7. W. F. Hyde, *Timber Supply, Land Allocation, and Economic Efficiency* (Baltimore, The Johns Hopkins University Press for Resources for the Future, 1980).

8. M. D. Bowes and J. V. Krutilla, "Economic Foundations of Multiple Use Management of Public Forest Lands," Discussion Paper D-104, Resources for the Future (Washington, D.C., 1983).

9. Boyce, *Management of Eastern Hardwood Forests;* Holthausen, "Use of Multistand Projection Models"; Thomas, *Wildlife Habitats;* R. S. Holthausen and N. L. Dobbs, "Computer Assisted Tools for Habitat Capability Evaluations," paper presented at the Society of American Foresters Conference, Fort Collins, Colo., July 30, 1985.

10. For a state-of-the-art update, see H. Salwasser, "Overall Summary: A Manager's Perspective," in J. Verner, M. L. Morrison, and C. J. Ralph, eds., *Modeling Habitat Relationships of Terrestrial Vertebrates* (Madison, University of Wisconsin Press, 1986).

11. Richard R. Nelson, "Assessing Private Enterprise: An Exegesis of Tangled Doctrine," *Bell Journal of Economics* vol. 21, no. 1 (Spring 1981) pp. 93–111, favors the view that economists believe too uncritically that private ownership is ipso facto the most efficient property rights system. Evidence of inefficiencies in the private sector, including public bailouts, are sufficiently frequent to counsel a more careful view of the total relevant evidence. Since property ownership and economics involve imperfect institutions, and less than perfect agents, it would be surprising if one imperfect system would have a preponderance of advantage over another flawed system in every circumstance. It is sobering to reflect, in this connection, that subsidies exceeding $20 billion a year have gone to owners of private farm and forest lands in recent years.

12. R. H. Nelson, "The Public Lands."

13. Ibid.

Index

Accounting methods, 289, 293–294
    capital accounting, 16, 285–286, 296,
      302–312
    cash-flow accounting, 294–295
    joint cost allocation, 286–288
    see also FORPLAN model
ADVENT software, 276, 281, 283
Amenity services, *see* Nonmarket services
American Metals Climax (AMAX),
    Molybdenum Division, 266
American Smelting and Refining Co.
    (ASARCO), 252
Asset value and depreciation, 306–307, 309

Barlow, Tom, 293
Baumol, W. J., 59
Below-cost timber sales
    accounting and, 293–295, 300, 302–312
    community stability and, 300
    evaluation by FORPLAN model, 325–328
    justification for, 16, 296–297, 300,
      328–329
    in Mt. Baker-Snoqualmie National
      Forest, 320–325
    multiple-use framework for, 297–301
    separable costs, 85–86, 295, 302–305,
      307
    in Shoshone National Forest, 314–320

Benefit valuation
    isovalue curves, 39–40
    marginal benefits, 38–39
    recreational amenities application, 44–48
    timber harvesting application, 41–44
Biological management, 1, 7
Bitterroot National Forest timber harvesting
    case, 35
Black Hills National Forest
    deer hunter survey, 200–201
    description of, 199
    hunting quality valuation, 199–200,
      202–209
    Norbeck timber sale, 206–209
    scenic views in, 299
    site characteristics, 201–202
Boundary Waters Canoe Area, 24
Budget of the Forest Service
    ADVENT software, 276, 281, 283
    appropriation and function codes, 276,
      278, 283–284, 289
    cost components, 275
    development process, 274–275
    interdependencies in, 281–282, 283–284,
      285
    line-item format of, 277–278, 284, 289,
      347
    regional office activities, 276, 281
    Washington activities, 276–280
    see also Accounting methods

351

Bureau of Land Management (BLM), 18
Burt, O. D., 177

Calish, S., 107, 108-109
Campgrounds, 22, 244
Capital accounting, 16, 285-286, 296,
    302-312
Cash-flow accounting, 294-295
Cicchetti, C. J., 177
Clark-McNary Act of 1924, 5
Clawson, M., 177
Colorado River Basin water-flow
    management, 157, 158-163
Congress
    Forest Service accounting and, 16,
        285-286, 293-294
    Forest Service budget and, 276-280,
        281, 347
    see also House Appropriations
        Committee; see also names of
        specific legislation
Congressional Research Service, 293
Council of Forest School Executives, 3
Cost function of multiple-use production
    characteristics of, 52
    components of, 51-52
    incremental and separable costs, 54-55
    isocost curves, 52-53
    marginal costs, 53-54
    usefulness of, 51
Costs
    common cost, 295
    definition under capital accounting,
        305-308
    direct costs, 275, 303
    incremental costs, 54
    joint production costs, 287
    joint support costs, 275
    resource coordination costs, 275
    site versus forestwide costs, 70-71
    see also Accounting methods, Marginal
        costs; Separable costs
Creative Act of 1981, 2

Department of Agriculture, 276, 277, 280,
    281
Department of the Interior, 19
Dominant-use management, 32-34
Duval Corp., 266

Eckstein, Otto, 295
Economic efficiency, 3-5
    Forest Service budget process and, 283
    income distribution and, 40-41
    incremental cost tests, 304-305
    jointness in production and, 58-59
    marginal cost tests, 84, 85
    pricing and, 26
    separable cost tests, 84-85, 303, 304,
        305, 310-311
Endangered Species Act of 1973, 21
Energy minerals receipts, 14

Faustmann, Martin, 4, 337, 338
Faustmann model
    compared to Hartman model, 106-107
    description of, 95-96
    illustration of, 97-98
    maximum sustained yield under, 101-104
    multiple-use version of, 115, 117, 119,
        120-132
    optimal harvest rule interpretation, 96-97
    separable costs computation and, 311
    solution characteristics, 99-101
Fees, see User fees
Fight, R. D., 107, 108-109
Fish and wildlife habitat management, 21-22
Fisher, A. C., 177, 267
Forage, see Grazing
Forest and Rangeland Renewable Resources
    Planning Act of 1974 (RPA), 6-7
    budget process and, 274, 284
    habitat management under, 21
    multiple-use management under, 35,
        92-93
Foresters
    accountability of, 33-34
    functions of, 2, 3,
    stewardship role of, 5
    training of, 3
Forestry, defined, 2, 3-4
Forest Service
    accounting methods of, 284-288, 289,
        293-294
    divisions of, 274
    economic criteria in supply decisions, 13
    management philosophy of, 1-2, 3-4,
        6-7, 92-93
    receipts for market commodities, 14-15

role of, 5, 19
see also Budget of the Forest Service
FORPLAN model, 288, 296
  allocation zones, 313–314, 332
  application of, 312–314, 345
  constraints, 334
  early use of, 118–119
  establishment of, 35
  Hartman model and, 117
  improvement of, 327–328
  limitations of, 305, 325–327
  Mt. Baker-Snoqualmie National Forest
    application, 320–325, 326
  objective function, 335–336
  outputs and costs, 332–334, 335
  separable costs computation, 308,
    311–312
  Shoshone National Forest application,
    314–320, 326
  structure of, 332–336, 345, 346
  timber analysis areas, 332

Gaffney, Mason, 4
General Accounting Office (GAO),
  below-cost timber sales and, 293, 294
Gila National Forest, 23
Gisborne, H. T., 3
Grazing
  acreage of, 18
  permits for, 18–19
  receipts for, 14, 19, 27
  in White Clouds Peaks, 252, 260–265
  on wilderness areas, 23

Habitat management, 21–22
Hartman model
  compared to Faustmann model, 106–107
  description of, 104–105
  FORPLAN model and, 117, 119
  optimal rotation age under, 105
  solution characteristics, 105–106
Hedonic valuation of recreational quality,
  209–210
  Black Hills National Forest study,
    199–209
  estimation principles, 190–192, 198–199
  gross benefit approach, 192–199
  net benefit approach, 187–192
  utility maximization, 181–183

value of site attributes changes,
  188–190, 194–198
House Appropriations Committee, 284, 287,
  293, 296, 347
Howe, C. W., 163
Humphrey, Hubert H., 6
Hunting activities, 5, 6
  Black Hills hedonic valuation, 199–209
Hurd, R. M., 202

Interest rates and maximum sustained yield,
  103
Isocost curves
  forest site illustration of, 71–74
  product mix and, 79–80
  site productivity and, 62–63, 67–68
  slope and curvature of, 61–62
  specialization and, 65–66, 68–69
Isovalue curves, 39–40

John Muir Wilderness Area, 24
Johnson, K. N., 117
Joint cost allocation, 286–288
Jointness in production
  budget process and, 275, 283–284
  capital stock and product mix, 13–14
  characterization of, 49–50, 55–59
  cost allocation and, 295, 301, 303
  costs, 54–55, 287
  complications of, 13, 251, 285
  economies and diseconomies of, 58–59
Jones, J. G., 299

Kennecott Copper Corp., 266
Krutilla, J. V., 267

Land and Water Conservation Fund Act of
  1965, 19–20
Law Enforcement Management and
  Reporting System (LEMARS)
  description, 221–223
  sample bias, 223–224
  sample comparisons, 224–225
Leopold, Aldo, 23

McClure, James A., 252, 255
Magma Copper Co., 266

Management approaches
  biological management, 1, 7
  dominant-use management, 32–34
  maximum sustained yield, 92, 101–104
  technocratic management, 6–7
  see also Multiple-use management
Management Information Handbook (MIH)
  budget process and, 274–280, 283, 303
  coding simplification, 289
Marginal costs
  efficiency tests, 84, 85
  product mix and, 81–82, 88
  timber harvest illustration, 74–76
Maximum sustained yield policy, 92
  Faustmann model and, 101
  illustrations of, 102–104
  interest rates and, 103
Mendelson, R., 179, 192
MIH, *see Management Information Handbook*
Mining
  administration of, 19
  funding for, 19
  receipts from, 14, 19, 27
  in White Clouds Peaks, 252–253, 265–270
  on wilderness areas, 23
Mining Law of 1872, 19
Molybdenum Corporation of America (MOLYCORP), 266
Molybdenum mining and production, 265–267
Monongahela National Forest timber harvesting case, 7, 35
Mt. Baker-Snoqualmie National Forest
  below-cost timber sales, 324–325
  description of, 320
  FORPLAN model components, 321–322
  FORPLAN runs and results, 322–323
  Nooksack River area, 320–321
  separable costs of timber, 323–324
Multiple-output production
  benefits from, 37–48
  complementary products, 57–58, 87
  independent outputs, 58
  isocost curves, 71–74
  marginal costs, 74–76
  observed production data and, 50–51
  production function defined, 49
  ray average cost, 59–60
  returns to scale, 60–61

separable costs, 54–55, 77–78
  site productivity and, 62–63, 67–68
  specialization in, 64–71
  studies of, 48–49
  substitutes, 57–59, 69–70, 87
  see also Jointness in production
Multiple-Use and Sustained-Yield Act of 1960 (MUSY)
  definitions under, 32
  fish and wildlife habitats and, 21
  forestry management under, 3, 92–93
  recreation authority under, 22
  wilderness areas under, 24
Multiple-use management
  amenity values, 109–111, 121–122, 134–135
  analytic implications of, 337–341
  benefit function, 37–40
  cost function, 51–78,
  defined, 32
  difficulties of, 88, 337–348
  dominant-use management versus, 32–34
  linear programming models for, 117–119
  operational significance of, 338–341
  planning for, 32–37
  production function, 48–51
  product mix under, 64–71, 78–86
  related-stand timber model, 120–140, 341–344
  single-stand timber models, 95–120, 341–344
  timber-water illustration, 149–173
MUSY, *see* Multiple-Use and Sustained-Yield Act of 1960

National Environmental Policy Act of 1969 (NEPA), 34–35
National Forest Management Act of 1976 (NFMA)
  accounting requirements under, 284
  budget process and, 284
  multiple-use management under, 35–36, 92–93
  planning under, 7–8, 344
National Forest System
  data sources, 220–225
  early management of, 5, 92
  production capabilities of, 7
  recreation versus timber value, 5–6, 337
  unit plans for, 34–35

withdrawals for, 3
National Park System, 5-6
National Wilderness Preservation System (NWPS), 23
Natural Resources Defense Council (NRDC), 293, 294, 297
NEPA, *see* National Environmental Policy Act of 1969
NFMA, *see* National Forest Management Act of 1976
Nonmarket services
    demand for, 20
    Forest Service budget and, 280-281
    pricing of, 24-27
    provision of, 20-24
    value of, 26-27
    see also specific nonmarket services
NWPS, *see* National Wilderness Preservation System

Office of Management and Budget (OMB), 276-280, 281, 347
Organic Act of 1897
    forest management under, 2-3, 5, 91-92, 177
    inadequacy of, 7
    water-flow management under, 3, 5, 20, 155, 177
Outdoor Recreation Resources Review Commission, 22
Ozarks, forest withdrawals, 3, 5

Pacific Coast forests, 6, 293
Pase, C. P., 202
Pearse, Peter, 4
Pinchot, Gifford, 92
Pisgah National Forest, scenic views, 299
Planning
    comprehensive approach to, 34-35
    information use in, 345-346
    linear programming methods for, 93, 117-119
    production possibilities and, 36
    scale and, 36-37
    under National Forest Management Act, 35-36, 344
    see also FORPLAN models

Pricing
    competitive, 15
    institutional arrangement limit, 25
    of nonmarket services, 24-27
    on site entry, 25-26
    rationale for market pricing, 26-27
    zero pricing, 24, 26
Primitive areas, *see* Wilderness areas
Product mix
    budget-constrained choice, 78-83
    cost-minimizing solution, 65-66
    relative marginal values, 87
    site productivity and, 63
    solving for optimal mix, 86
    specialization in, 64-71
    unconstrained optimal choice, 83-86
Public Land Law Review Commission, 348
Public Rangelands Improvement Act of 1978, 18

Range, *see* Grazing
Recreation amenities and services
    benefits valuation, 44-48
    construction funding, 22
    consumer-choice theory, 178-179
    demand for, 6-7, 215-228
    fees for, 19-20
    hedonic valuation methods, 181-183, 187-199
    marginal value of, 45-48, 178
    receipts from, 15
    relation to timber management, 244-245
    role of Forest Service, 22
    site selection, 25-26
    special use permits, 20
    travel-cost valuation methods, 177-178, 180-181, 183-187, 215-219
    value of, 22, 27-28
    White Mountain net value from, 232-233
    wilderness value in White Clouds Peaks, 255-260
Recreation Information Management System (RIM), 220
Recreation Opportunities Spectrum (ROS), 220
Renewable Resources Planning Act, *see* Forest and Rangeland Renewable Resources Planning Act of 1974

Roads
  impact on future flows, 305–306
  for timber access, 153–154, 298–299
  in White Clouds Peaks, 252
  in White Mountain National Forest,
    236–237
Rocky Mountains
  Douglas fir sites, 150
  forest withdrawals, 5
  sawtimber inventories, 6
Roosevelt, Franklin D., 7
Roosevelt, Theodore, 7
Rosen, S., 192
RPA, see Forest and Rangeland Renewable
  Resources Planning Act of 1974

Samuelson, P. A., 102
Scheurman, H., 117
Schuster, E. G., 299
Separable costs
  below-cost timber sales and, 85–86, 295,
    302–305, 307
  computation of, 305, 308–312
  defined, 77, 302–303, 307
  determination of, 286–287, 288, 295,
    307–308
  direct costs versus, 303
  efficiency tests and, 84–85, 303, 304,
    305, 310–311
  multiple-output production and, 54–55,
    77–78
  in Shoshone National Forest model,
    318–319
  support costs, 275
  use of, 303
Shoshone National Forest
  below-cost timber sales, 319–320
  description of, 314
  Dunoir/Doby Cliff allocation zone, 315
  FORPLAN model components, 315–316
  FORPLAN runs and results, 316–318
  separable costs of timber, 318–319
  separable versus direct costs, 319
Site productivity
  measures of, 62–63
  specialization and, 67–68
Site selection
  forest benefits and, 184
  hedonic view, 181–182
  site attributes and, 184–187

travel-cost view, 180–181
  utility theory framework, 179–183
Ski areas, special use permits, 20
Smith, V. K., 177
Snoqualmie National Forest, see Mt.
  Baker-Snoqualmie National Forest
Society of American Foresters, 3–4, 296
Soil conservation, 3, 4, 5
Southeastern states timber sales, 293
Southwest Power Administration, 161
Specialization
  convexity of production possibilities and,
    115–116
  cost-minimizing solution, 64–66
  defined, 64
  isocost curves and, 68–69
  site productivity and, 67–68
  site versus forestwide costs, 70–71
  substitute products and, 60–70
Steady-state solutions
  approaches to, 131–132
  cyclical steady states, 130–131
  even-flow steady states, 129–130
  multiple steady states, 131
  multiple-use values and, 131
  optimality of, 128–129

Technical efficiency doctrine, 4
Teeguarden, D. E., 107, 108–109
Timber
  competitive sales, 15–16
  even-flow harvesting policy, 16–17, 92,
    140, 299–300
  harvest age, 17
  harvest benefits computation, 41–44
  harvest costs, 16, 123, 135, 151–152
  harvest scheduling, 128, 135–136, 150–151
  harvest value, 122–123, 135, 152–153
  harvest volume, 6, 298
  isocost curve illustration, 71–74
  management under Organic Act, 3
  marginal costs illustration, 74–76
  marginal value of, 42–44
  maximizing net present value, 124
  pricing of, 151
  production equations, 123–124
  receipts for, 14, 15
  sawtimber inventories, 6
  White Mountain resources and
    management, 233–243

see also Below-cost timber sales; Timber harvesting solutions
Timber harvesting solutions
  alternative variations, 108–113
  Faustmann single-stand model, 95–104
  hardwood harvesting, 240–242
  Hartman single-stand model, 104–108
  insensitivity to multiple-use values, 111–113
  multiple-use related-stand model, 120–140
  multiple-use single-stand model, 108–120
  optimal harvest choice, 125–127, 136–140
  optimal solutions based on amenity values, 109–111
  paper birch harvesting, 237–240
  present-value production possibility curve, 113–117
  recreation management solution, 139–140
  recreation/timber solution, 139
  steady-state solutions, 128–132
  timber and low-amenity value solutions, 138–139
  timber management solution, 138
  timber/water-flow solution, 139
  uneconomic timber stands and, 113
  water-flow management solution, 139
  White Mountain application, 235–236
Trails, mileage of, 22
Travel-cost valuation of recreational quality, 209–210, 215
  benefit computation, 216–217, 218
  consumer surplus, 229–232
  demand estimation, 215–216, 218–219, 225–229
  demand function, 183–184
  focus on total site value, 177
  model application problems, 219
  model specification, 217–219
  utility maximization, 180–181, 182–183

U.S. Forest Service, *see* Forest Service
User fees
  additional funding requirements, 273
  collection of, 24–25, 26
  for commercial enterprises, 14–15

expectations regarding, 14, 27
  legislative authority for, 19–20

Water
  economic dimensions of, 158–167
  Forest Service management role, 20–21
  hydroelectric power and, 160–161
  management under Organic Act, 3, 5, 20, 155, 177
  marginal value of, 160, 162, 163–167
  market for water rights, 161–162
  valuation of, 161
  vegetation and water augmentation, 155–157
Weeks Act of 1911, 3, 20
White Clouds Peaks
  description of, 252, 253
  grazing activity, 260–265
  land use conflicts in, 252–255
  mining activity, 265–270
  wilderness recreation value, 255–260
White Mountain National Forest
  consumer surplus, 229–232
  data sources, 221–222
  demand estimation, 223–229
  description of, 213–214
  land classes, 234–235
  recreation value, 232–233, 244
  roads in, 236–237, 244, 245
  scenic views, 244–245
  timber resources and management, 233–242
White River water-flow management, 158–160
Wilderness Act of 1964, 23
Wildlife management
  habitats, 21–22
  isocost curve illustration, 71–74
  in White Clouds Peaks, 254
Wilderness areas
  acreage of, 23
  establishment of, 23
  management of, 23–24
  recreation use of, 23
  recreation value in White Clouds Peaks, 255–260
Wilman, Elizabeth A., 179, 192, 199, 201, 202, 203, 210

Yates, Sidney, 287, 293–294, 296, 347